Digital Electronics
for Musicians

Alexandros Drymonitis

■ ■ ■

Apress®

Digital Electronics for Musicians

ISBN-13 (pbk): 978-1-4842-1584-5

ISBN-13 (electronic): 978-1-4842-1583-8

Managing Director: Welmoed Spahr
Lead Editor: Michelle Lowman
Technical Reviewer: Johan Eriksson
Editorial Board: Steve Anglin, Pramila Balan, Louise Corrigan, James T. DeWolf, Jonathan Gennick, Robert Hutchinson, Celestin Suresh John, Michelle Lowman, James Markham, Susan McDermott, Matthew Moodie, Jeffrey Pepper, Douglas Pundick, Ben Renow-Clarke, Gwenan Spearing
Coordinating Editor: Mark Powers
Copy Editor: Kimberly Burton
Compositor: SPi Global
Indexer: SPi Global
Artist: SPi Global

Distributed to the book trade worldwide by Springer Science+Business Media New York, 233 Spring Street, 6th Floor, New York, NY 10013. Phone 1-800-SPRINGER, fax (201) 348-4505, e-mail orders-ny@springer-sbm.com, or visit www.springeronline.com. Apress Media, LLC is a California LLC and the sole member (owner) is Springer Science + Business Media Finance Inc (SSBM Finance Inc). SSBM Finance Inc is a Delaware corporation.

For information on translations, please e-mail rights@apress.com, or visit www.apress.com.

Apress and friends of ED books may be purchased in bulk for academic, corporate, or promotional use. eBook versions and licenses are also available for most titles. For more information, reference our Special Bulk Sales–eBook Licensing web page at www.apress.com/bulk-sales.

Any source code or other supplementary materials referenced by the author in this text is available to readers at www.apress.com/9781484215845. For detailed information about how to locate your book's source code, go to www.apress.com/source-code/. Readers can also access source code at SpringerLink in the Supplementary Material section for each chapter.

Contents at a Glance

Contents

About the Author

Alexandros Drymonitis is a musician from Athens, Greece. He studied at the Conservatory of Amsterdam, where he got his first exposure to music technology. Ever since, he has been creating electronic music using open source software and hardware such as Pure Data and Arduino, as well as giving workshops on electronic music programming and digital synthesizer building. He is also very keen on community building, and is a founding member of the Patching Circle Athens group, a group of users of visual programming languages.

About the Technical Reviewer

Johan Eriksson is a composer and electronic musician from the north of Sweden. He has a first class degree in composition from the Birmingham Conservatoire in the UK and has had his work commissioned and performed across the UK and Sweden. Johan has been releasing records as "Monolog X" frequently since 2007. Modular synthesis is very dear to him, especially the Pure Data language. In early 2015, he released XODULAR, which is a virtual modular synthesizer environment in Pure Data that was given a very warm welcome by the Pure Data community and introduced new people to the language.

Acknowledgments

The communities of Pure Data and Arduino have been of great assistance prior and during the writing of this book. Also, I would like to thank Michelle Lowman from Apress for asking me to write this book, as well as Miller Puckette for creating Pure Data.

It wouldn't have been possible to reach a point where I would be able to write a book on these subjects without the long support of my parents, and I wouldn't have been able to write this book without the support and patience of my lovely wife.

Introduction

This book aims at giving insight on a few of the most widely used tools in the fields of creative coding and DIY digital electronic musical interfaces. It is a result of personal exploration in these fields and an attempt to gather information about the combination of the very popular prototyping platform, the Arduino, with the also very popular visual programming language for multimedia, Pure Data (a.k.a. Pd).

The main focus of the book is interactivity with the physical world, and how to make this musical. It is split among several projects where each project brings a fresh idea on how to combine musical instruments with computers, whereas the use of programming builds up gradually. Also, this book uses only open source software and hardware, because of the great advantages one can have from an open source community, but also in order to bring the cost of every project to its minimum.

At the time of writing (December 2015) Pd is at a turning point. Being split in two major version up to now, Pd-vanilla and Pd-extended, the latter version is used throughout the book, since it includes various external packages, some of which are constantly used in this book. Pd-extended is not maintained any longer, which leaves Pd-vanilla as the only actively maintained major Pd flavor. This version (which is the original version maintained by the maker of Pd itself, Miller Puckette) consists of the very core of Pd, lacking the external packages Pd-extended has. A new plug-in has been introduced though in vanilla which will be part of the next release, Pd-0.47, to be released during December (but maybe a bit later). This is the deken plug-in which simplifies the addition of certain external packages to a great extent.

I strongly suggest the reader uses Pd-vanilla once the 0.47 version is published and to install a few external packages using this plug-in. You can download it from Miller Puckette's personal website. If you do so, you'll need to go to the **Help** menu and choose **Find externals**. In the window that will open, search for the following packages: **comport**, **zexy**, **ggee** and **iemmatrix**. If you do use Pd-vanilla, the following changes should be applied to all projects of this book. All Pd objects (actually abstractions, but what this is has not been explained yet) that end with the word "extended" should be replaced by the same object without this word. For example, "serial_print_extended" should be replaced by "serial_print". All "arraysize" objects should be replaced by "array size" (there's a white space between the two words). The "import" object is not used at all. In chapter three you'll read how you can configure Pd to find these external packages under Linux, but the process is very similar for the other operating systems. All this will make sense as you read through the book.

Another issue at the time of writing is that the **comport** object used in Pd to communicate with the Arduino seems to malfunction in Windows 10. Hopefully this bug will be fixed shortly. If not, I suggest you sign up for Pd's mailing list or forum (their websites are mentioned in chapter 1) and search their archives for solutions, or even ask people there. Audio issues have also been reported under OS X El Capitan, but that applies to other audio software too. In general, a brand new version of an operating system is very likely to have various issues, so be aware if your preference is to upgrade as soon as an OS update is released.

If you want to contact me for any reason regarding this book, drop me a line at alexdrymonitis@gmail.com.

Now you can start your journey in the world of creative coding and DIY electronics.

CHAPTER 1

■ ■ ■

Introduction to Pure Data

Pure Data, a.k.a. *Pd*, is a visual programming language and environment for audio and visuals. It is open source and it was made by Miller Puckette during the 1990s. *Visual programming* means that instead of writing code (a series of keywords and symbols that have a specific meaning in a programming language), you use a graphical interface to create programs, where in most cases, a "box" represents a certain function, and you connect these "boxes" with lines, which represent patch cords on analog audio devices. This kind of programming is also called *data flow programming* because of the way the parts of a program are connected, which indicates how its data flows from one part of the program to another.

Visual programming can have various advantages compared to textual programming. One advantage is that a visual programming language can be very flexible and quick for prototyping, where in many textual programming cases, you need to write a good amount of lines of code before you can achieve even something simple. Another advantage is that you can say that visual programming is more intuitive than textual programming. When non-programmers confront visual code, it's very likely that they will get a better idea as to what this code does than when confronting textual code. On the other hand, there are also disadvantages and limitations imposed by visual programming. These are technical and concern things like DSP chains, recursion, and others, but we won't bother with these issues in this book, as we'll never reach these limits. Nevertheless, Pd is a very powerful and flexible programming language used by professionals and hobbyists alike around the world.

Throughout this book, we'll use Pd for all of our audio and sequencing programming, always in combination with the Arduino. The Arduino is a prototyping platform used for physical computing (among other things), which enables us to connect the physical world with the world of computers. A thorough introduction to Arduino is given in Chapter 2. This chapter is an introduction to Pd, where we'll go through its basics, its philosophy, as well as some general electronic music techniques. If you are already using Pd and know its basics, you can skip this chapter and go straight to the next one. Still, if you're using Pd but have a fuzzy understanding on some of its concepts, you might want to read this chapter. Mind that the introduction to Pd made in this chapter is centralized around the chapters that follow, and even though some generic concepts will be covered, it is focused on the techniques that will be used in this book's projects.

In order to follow this chapter and the rest of this book, you'll need to install Pd on your computer. Luckily, Pd runs on all three major operating systems: OS X, Linux, and Windows. You can download it for free from its web site at `https://puredata.info/`. You will find two version of Pd: vanilla and extended. Pd-vanilla (simply Pure Data) is the "vanilla" version of Pd, as its nickname states. It's the version made and maintained by Miller Puckette, and it consists of the core of Pd. Most of the things we'll be doing in this book can be made with vanilla, but Pd-extended adds some missing features to Pd that we will sometimes use. For example, the communication between Pd and Arduino is achieved with Pd-extended and not vanilla. Of course, you can add these features to vanilla, but it's beyond the scope of this book to explain how to do this, so we'll be using Pd-extended in all of our projects. Find the version for your OS and install it on your computer before you go on reading.

1

By the end of this chapter, you'll be able to

- understand how a Pd program works

- create small and simple programs in Pd

- find help in the Pd environment

- create oscillators in Pd

- make use of existing abstractions in Pd and create your own

- realize standard electronic music techniques in Pd

Pd Basics: How It Works

Pd consists of several elements that work together to create programs. The most basic elements are the *object* and the *message*. An object is a function that receives input and gives output. Figure 1-1 shows the *osc~* Pd object.

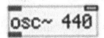

Figure 1-1. A Pd object

This specific object is a sine wave oscillator with a 440-hertz (Hz) frequency. There's no need to understand what this object does; we'll go through that in a bit. There are a few things we need to note. First of all, there is specific text inside the object box, in this case "osc~ 440". "osc" stands for *oscillator*, and the ~ (called the *tilde*) means that this object is a signal object. In Pd, there are two types of objects: signal and control. A *signal object* is a function that deals with signals (a digital form of an electric signal). A signal object will run its function for as long as the audio is on (the audio is also called the DSP, which stands for *digital signal processing*, or the DAC, *digital-to-analog converter*). A *control object* is independent of audio and runs its function only when it is told to. We'll get a better picture of the difference between the two as we go. The last part of the text, "440", is called an *argument*. This is the data that a function receives, and we provide it as an argument when we want to initialize an object with it. It is not necessary to provide an argument; when there's no argument in an object, the object is initialized with the value(s) of zero (0).

The second main element in Pd is the message, which is shown in Figure 1-2.

Figure 1-2. A Pd message

It is a little bit different from the object, because on its right side, it is indented, and it looks a bit like a flag. The message delivers data. There's no function here, only the data we write in the message (sometimes referred to as a *message box*). One thing the object and the message have in common is the inlets and the outlets. These are the little rectangles on the top and the bottom, respectively, of the object and the message. All messages have the same form, no matter what we type in them. They all have one inlet to receive data and one outlet to provide the data typed in them. The objects differ, in the sense that each object has as many inlets as it needs to receive data for its function, and as many outlets as it needs to give the output(s) of the function. With the osc~ object, we can see that it has two inlets and one outlet. The left inlet and the outlet are different than the right inlet. Their rectangle is filled, whereas the right inlet has its rectangle blank, like the message does. The filled inlets/outlets are signal inlets/outlets and the blank ones are control inlets/outlets. Their differences are the same as the signal and control objects. Note that a signal object can have control inlets/outlets, but a control object cannot have signal inlets/outlets.

Objects and messages in Pd are connected with lines, which we also simply call *connections*. A message connected to the osc~ object is shown in Figure 1-3.

Figure 1-3. *A message connected to an object*

A connection comes out the outlet of the message and goes to the inlet of the object. This way we connect parts of our programs in Pd.

Our First Patch

Now let's try to make the little program (programs in Pd are called *patches*, which is what I will call them from now on). Launch Pd like you launch any other application. When you launch it, you get a window that has some comments in it. Don't bother with it; it is just some information for some features it includes. This is the Pd window, also called the *Pd console*, and you can see it in Figure 1-6. It is very important to always have this window open and visible, because we get important information there, like various messages printed from objects, error messages, and so forth.

Go to **File ➤ New** to create a new window. You will get another window that is totally empty (don't make it full-screen because you won't be able to see the Pd console any more). Note that the mouse cursor is a little hand instead of an arrow. This means that you are in Edit Mode, so you can edit your patch. In this window, we will put our objects and messages. In this window's menu, go to **Put ➤ Object** (in OS X there's a "global" menu for the application; it's not on every window). This will create a small dotted rectangle that follows the mouse. If you click once, it will stop following the mouse. Inside the rectangle, there's a blinking cursor. This means that you can type in there. For this patch, you will type **osc~**.

After you type this, click anywhere in the window, outside the object, and you'll see your first Pd object, which should look like the one shown in Figure 1-1. (Note the shortcut for creating objects; it's Ctrl+1 for Linux and Windows, and Cmd+1 for OS X. We'll be using the shortcut for creating objects for now on). Now go to **Put ➤ Message** (the second choice in the menu, with the Ctrl/Cmd+2 shortcut). This will create a message. Place it somewhere in the patch, preferably above the object. Once you've already placed a message or an object in a patch, to move it, you need to select it by dragging. You can tell that is has been selected because its frame and text is blue, as shown in Figure 1-4.

Figure 1-4. *A selected message*

3

If you click straight into the message or object, it will become editable, and it will be blue like in Figure 1-4, but there will also be a blue rectangle inside it, like in Figure 1-5. When an object or message looks like the one in Figure 1-5, you cannot move it around but only edit it.

Figure 1-5. *An editable message*

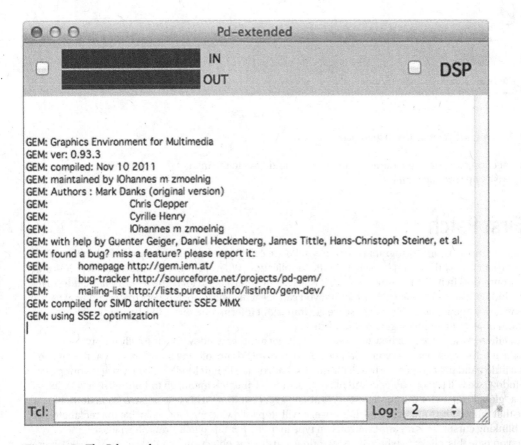

Figure 1-6. *The Pd console*

Type **440** and click outside it. To connect the message to the object, hover the mouse above the outlet of the message (on its lower side). The cursor should change from a hand to a circle and the outlet should become blue (on Linux, the cursor changes from a hand to an arrow, with a small circular symbol next to it). Click and drag. You will see a line coming out the outlet of the message. When holding the mouse click down, if you hover over the left inlet of the object, the cursor will again become a circle and the inlet will become blue. Let go of the mouse click, and the line will stay between the message and the object. You have now made your first connection.

What we have until now is a patch, but it doesn't really do anything. We need to put at least one more object for it to work. Again, put an object in your patch (preferably with the shortcut instead of using the menu) and place it below the [osc~] object (using square brackets indicates a Pd object when Pd patches are explained in text). This time, type **dac~**. This object has two inlets and no outlets. This is actually your computers' left and right speakers. Connect the outlet of [osc~] to both inlets of [dac~], the same way you connected the message to [osc~]. Your patch should look like the one in Figure 1-7.

Figure 1-7. *A simple Pd patch*

You might notice that the connections that come out from [osc~] are thicker than the one coming out from the message. These are signal connections, whereas the thin one is a control connection. The difference is the same as with signal/control objects.

Now we have a fully functional patch. In order to make it work, we need to take another two steps. First, we need to get out of Edit Mode. Go to the Edit menu and you'll see Edit Mode; it is ticked, which means that you are in this mode. If you click it, the cursor will turn from a little hand to an arrow. This means that you are no longer in Edit Mode and cannot edit the patch, but you can interact with it. If you go to the Edit menu again, you'll see that Edit Mode is not ticked anymore. From now on, we'll use the shortcut for Edit Mode, which is Ctrl/Cmd+E. The last thing you need to do to activate the patch is to turn on the DSP. Go to **Media ➤ DSP On** (Ctrl+/or Cmd+/). On the Pd console, there is a DSP tick box. To turn on the DSP, select the tick box; DSP becomes green, as shown in Figure 1-8. When the DSP is on, all signal objects are activated. Make sure to put your computer's volume at a low level before you turn the DSP on, as it might sound rather loud.

Figure 1-8. *DSP on indication on Pd's console*

Even though you turned on the DSP, you still hear nothing. Hover the mouse over the message (you'll see the cursor arrow change direction, which means that you can interact with the element you hover your mouse over) and click. Voila! You have your first functional Pd patch! This is a very simple program that plays a sine wave at a 440 Hz frequency.

Before getting overly excited, it's good practice to save your first patch. Before you do that, you might want to turn the DSP off by going to **Media ➤ DSP Off** (Ctrl/Cmd+.) Now the DSP tick box should be unticked. Saving a patch is done the same way that you save a text file. Go to **File ➤ Save As...** (Shift+Ctrl/Cmd+S) and a dialog window will open. Here you can set a name for your patch (that could be my_first_patch) and a place to save it. If you haven't done so yet, create a folder somewhere in your computer (a good place is Documents/ pd_patches, for example, definitely not the Program Files folder) and click **Save**. It's good practice to avoid using spaces both in Pd patch names and folders used by Pd, as it's pretty difficult to handle them. It's better to use an underscore (_) instead. Also, notice the file extension created by Pd, which is .pd (not too much of a surprise…). These are the files that Pd reads.

Now that we've saved our first patch, let's work on it a bit more. Go back to Edit Mode (Ctrl/Cmd+E) to edit your patch. The cursor should again turn to a little hand. Now we'll replace the message with another element of Pd, the number atom. First, we'll need to delete the message. To do this, drag your mouse and select it, the same way you select it to move it around. Hit **Backspace** and the message (along with its connections) will disappear. Go to **Put ➤ Number** (Ctrl/Cmd+3) and the number atom will be added to your patch, which is shown in Figure 1-9.

Figure 1-9. *A number atom*

Connect its outlet to the left inlet of [osc~] (it actually replaces the message) and get out of Edit Mode. Turn the DSP on and again you'll hear the same tone as before. This is because [osc~] has saved the last value it received in its inlet, which was 440. Click the number atom and type a number (preferably something different than 440) and hit **Return** (a.k.a. Enter). You have now provided a new frequency to [osc~] and the pitch of the tone you hear has changed to that. Another thing you can do with number atoms is drag their values. Click the number and drag your mouse. Dragging upward will increase the values and dragging downward will decrease them. You should hear the result instantly. When done playing, turn off the DSP and save this patch with a different name from the previous one.

The Control Domain

Our next step will be dealing with the control domain. As mentioned, the control objects are independent of the DSP and run their functions only when they are instructed to do so, regardless of the DSP being on or off. Let's create a simple patch in the control domain. Let's open a new window and put an object in it. Inside the object type +. This is a simple addition object. It has two inlets and one outlet, because it adds two numbers and gives the result of the addition. Now put three number atoms and connect two of them to each inlet of [+] and the outlet of [+] to the inlet of the third number. Make sure that your patch is like the one in Figure 1-10.

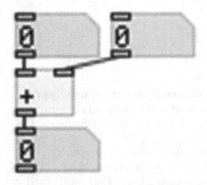

Figure 1-10. *A control domain patch*

Go out of the Edit Mode (from now on we'll refer to this action as "lock your patch") and click the top-right number. Type a number and hit **Return**. Doing this gives no output. Providing a value to the top-left number, will give the result of the addition of the two values, which is displayed on the bottom number atom. What we see here are the so-called cold and hot inlets in action. In Pd, all control objects have cold and hot inlets. No matter how many inlets they have (unless they have only one), all of them but the far left are cold. This means that providing input to these inlets will not give any output, but will only store the data in the object. The far left inlet of all control objects is hot, which means that providing input to that inlet will both store the data and give output. This is a very important rule in Pd, as its philosophy is a right-to-left execution order. It might take a while to get used to this, and not bearing it in mind might give strange results sometimes; but as soon as you get the grasp of it, you'll see that it is a very reasonable approach to visual programming.

Execution Order

Before moving on to some other very important aspects of Pd, I should talk a bit more about the order of execution, since you saw a small example earlier. In a new patch, put a number atom and put an object below it. Type * inside the object. This is a multiplication object, which works in a way similar way to the addition object you saw. Connect the number to both inlets of [*], but first connect it to the left inlet of [*] and then to the right one. Put another number and connect the outlet of [*] to the inlet of the new number. You should have the patch shown in Figure 1-11.

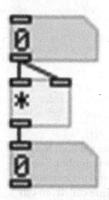

Figure 1-11. *A fan out connection*

As you can imagine, this patch gives the square of a given number by multiplying it to itself. Lock your patch and type the number **2** in the number atom. What you would expect to receive is 4, right? But instead, you got 0. Now type **3**. Again, you would expect to get 9, but you got 6. Now type **4**. Instead of 16, you got 12.

Even though I said that Pd executes everything from right to left, another rule is that execution will follow the order of connection. That means that if you connect the top number atom to the left inlet of [*] first, and then to the right one, whatever value you provide through the number atom will first go to the left inlet of [*] and then to the right. But I've already mentioned that all left inlets in all control objects are hot, and all the rest are cold. So what happens here is that when we gave the number 2 to [*], it went first to the left inlet and we immediately received output. That output was the number provided in the left inlet, and whatever other value was stored in [*]. But we hadn't yet stored any value, so [*] contained 0,[1] and gave the multiplication 2 * 0 = 0. Immediately after this happened, the number 2 went to the right inlet of [*] and was stored, but gave no output, as the right inlet is cold. The next time we gave input to [*], we sent the number 3.

---

[1]In contradiction to many programming languages, Pd has 0 when no argument is provided, instead of NULL.

Again, we got the same behavior. [*] first gave the multiplication of 3 by the number already stored, which was 2 from the previous input, so we got 3 * 2 = 6; and then the number 3 was stored in [*] without giving output. The same thing happened with as many numbers we provided [*] with.

If we had connected the number atom to the right inlet of [*] first and then to the left one, things would have worked as expected. But connecting one element to many can be confusing and lead to bugs, which can be very hard to trace. In order to avoid that, we must force the execution order in an explicit way. To achieve this, we use an object called *trigger*. Disconnect the top number atom from [*] and put a new object between them. In it, type **t f f**. "t" is an abbreviation for trigger and "f" is an abbreviation for float. A float in programming is a decimal number, and in Pd, all numbers are considered decimals, even if they are integers. [t f f] has one inlet (which is hot) and two outlets. Connect the top number atom to the inlet of [t f f] and the outlets of [t f f] to the corresponding inlets of [*]. You should have a patch like the one shown in Figure 1-12.

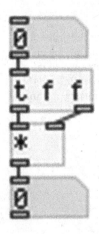

Figure 1-12. *Using [trigger] instead of fan out*

[t f f] follows Pd's right-to-left execution order, no matter which of its inlets gets connected first. Now whichever number you type in the top number atom, you should get its square in the lower number. This technique is much safer than the previous one and it is much easier for someone to read and understand. By far, it's preferred over connecting one outlet to many inlets without using trigger, a technique called *fan out*.

Bang!

It's time to talk about another very important aspect of Pd, the "bang." The bang is a form of execution order. In simple terms, it means "do it!". Imagine it as pressing a button on a machine that does something—the elevator, for example. When you press the one and only button to call the elevator, the elevator will come to your floor. In Pd language, this button press is called a *bang*. In order to understand this thoroughly, we'll build a simple patch that counts up, starting from zero. Open a new window and put two objects. In one of them, type **f**, and in the other **+ 1** (always use a space between object name and argument). "f" stands for *float*, as in the case of [trigger], and [+ 1] is the same as [+] we have already used, only this time it has an argument, so we don't need to provide a value in its right inlet. Whatever value comes in its left inlet will be added to 1 and we'll get the result from its outlet. Connect these two objects in the way shown in Figure 1-13.

Figure 1-13. *A simple counter mechanism*

Take good care of the connections. [f] connects to the left inlet of [+ 1], but [+ 1] connects to the right inlet of [f]. If you connect [+ 1] to the left inlet of [f], then you'll have connected each object to the hot inlet of the other. In this case, as soon as you try to do anything with this, you'll get what is called a *stack overflow*, as this will cause an infinite loop, since there will be no mechanism to stop it.

Above these two objects put a message and type **bang** in it. Connect its outlet to the inlet left of [f]. Lastly, put a number atom below the two objects and connect the outlet of [f] to it. You should have the patch in Figure 1-14.

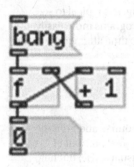

Figure 1-14. *A simple Pd counter*

Note that, even though in the previous section I mentioned the importance of using [trigger], here we're connecting [f] to two objects (one object and one number atom) without using [trigger]. This specific kind of patch is one of the very rare cases where execution order doesn't really matter, so we can omit using [trigger]. Still, in most cases it is really not a good idea not to use it.

What we have now created is a very simple counter that counts up from zero (it starts from zero because we haven't provided any argument to [f], and as already stated, no argument defaults to zero). [f] will go to [+ 1], which will give 0 + 1 = 1. This result will go to the right inlet of [f], meaning that the value will only be stored and we'll get no output. The value that comes out of [f] will also go to the number atom, where it will be displayed. The next time [f] will throw its value, it will be 1, which will go to [+ 1], which will give 1 + 1 = 2, which will be stored in [f], and 1 will be displayed in the number atom, and so on.

For [f] to output its value, it must receive some kind of trigger. This is where the "bang" comes in. Lock your patch and click the message. The first time you'll see nothing because [f] will output zero, but the number atom displays zero by default. Hit the message again and you'll see the value 1 in the number atom. The next time you hit the "bang" message, you'll see 2, and so on.

This is how we create a simple counter, which a very valuable tool in programming—for example, when building a sequencer, which we'll do in Chapter 4. We've also seen bang in action, a very important aspect of Pd.

Comments

In programming, one common element between different languages is the *comment*. A comment is just this, a comment. It's there to provide information about some features of a program or a part of it. Pd is no different when it comes to comments. In a new patch, go to **Put ➤ Comment** (Ctrl/Cmd+5) and the word "comment" will appear, following your mouse. As with all other elements, click so that it stops following the mouse. By clicking, you'll also see a blue rectangle around the comment. This means that you can edit it. Go ahead and type anything you want. Figure 1-15 shows a Pd comment.

```
this is just a comment, it does nothing else than display
the text typed in it.
```

Figure 1-15. *A Pd comment*

From the early stages in programming learning up to professional programming, it is typical to see comments, which are extremely helpful. Comments help others understand your programs more easily, but also help you to understand your own programs when you come back to them some time after you've made them. With comments, we covered the most basic and useful elements of Pd.

Getting Help

Pd is a programming language that is very well documented. Even though it's open source, and nobody gets paid for creating it, maintaining it, developing it, or documenting it, it still has a great amount of documentation. When we say documentation, we don't really mean tutorials in its usual sense, but help files, which themselves are some kind of short tutorials. Every element, like the object, the message, and so forth in Pd has a help file, which we call a *help patch*. To get to it, all you need to do is right-click the element. You get a menu with three choices: Properties, Open, and Help. The first two are very likely to be grayed out, so you can't choose them, but Help is always available. Clicking it will open a new patch, which is the help patch of the element you chose. All elements but the object have one help patch, as they do something very specific (the message, for example, delivers a message, and that's all it does). But the object is a different case, as there are many of them in Pd. So, depending on the object you choose (which is defined by the text in it), you'll get the help patch for that specific object. For example, [osc~] has a different help patch than [dac~].

In a patch, put an object ([osc~] for instance), a message (no need to type anything in it), a number atom, and a comment (also no need to type anything), and right-click each of them and open their help patches. In there, there's text (actually comments) describing what the element does, and providing examples and other information. Don't bother to read everything for now, you're just checking it to get the hang of finding and using help patches. You need to know that you can copy and use parts or the whole patch into your own patch. Go ahead and play a bit with the examples provided, and if you want, click the **Usage Guide** link on the bottom left. This will open a help patch for help patches, describing how a help patch is structured to get a better understanding of them. Mind that objects starting with pd (for example, [pd Related_objects]) are clickable and doing so (in a locked patch) will open a window. This is called a *subpatch*, which we'll cover further on.

Lastly, right-clicking a blank part of a patch, will open the same menu, but this time **Properties** is selectable. You won't select it now, but instead click **Help**. This will open a help patch with all the vanilla objects (you might get some red messages in Pd's console, called *error messages*, but it's not a problem). If you want, you can go through these, but don't bother too much, as it might become a bit overwhelming. By using Pd more and more, you get to know the available objects or how and where to look for what one needs.

GUIs

The next step is the GUI. GUI stands for *graphical user interface*. In computers, GUIs are very common. Actually, Pd itself runs as GUI and your computer system too (most likely). All the windows you open from various programs are considered GUIs. This approach is usually much preferred over its counterpart, the CLI (command-line interface), where the user sees only text and interacts with it in a terminal window, for example.

Even though Pd itself runs as GUI (since it is visual and not textual) there are some elements of it that are considered to be its GUIs (the elements covered so far, but the number atom are not GUIs). If you click the **Put** menu, the second group of elements contains GUIs: the Bang, the Toggle, the Number2, and so forth. The ones we'll use most are the bang, the toggle, the sliders, and the radios, which you can see in Figure 1-16.

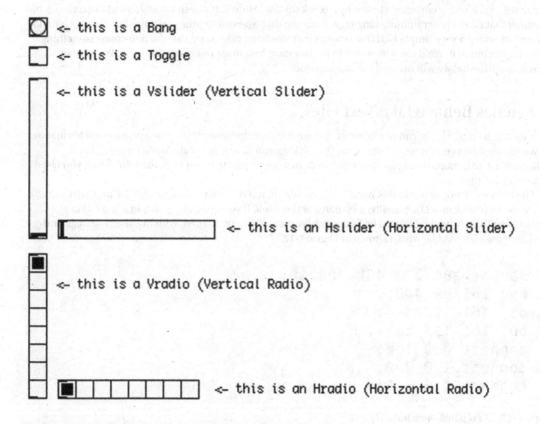

Figure 1-16. *The bang, the toggle, the sliders, and the radios*

Open a new patch and put a bang from the Put menu (Shift+Ctrl/Cmd+B). This is the graphical representation of the [bang] (a Pd message in textual form starts with an opening square bracket and ends with an opening parenthesis, imitating the way it looks in Pd). The Bang is clickable and it outputs a bang. Right-click it and open its help patch (here the Properties are not grayed out and are selectable, but we won't delve into this for now). On the top-right of the help patch, there's an object [x_all_guis ...]. Click it and a new window will open with all the GUIs in Pd. From there you can right-click each and check its help patch to see what it does. Focus on the GUIs that we'll typically use, which I've already mentioned. Let's talk a bit about these.

We've already covered the Bang, so let's now talk about the Toggle. The Toggle functions like a toggle switch; it's either on or off. It's a square, and when you click it (in a locked patch), an X appears in it. That's when it's on. When there's no X in it, it's off. By "on" and "off" here, I mean 1 and 0. What the Toggle actually does is output a 1 and a 0 alternately when clicking it, and we can tell what it outputs by the X that appears in it.

The Slider (Vslider stands for *vertical slider* and Hslider for *horizontal slider*) is a GUI imitating the slider in hardware; for example, in a mixing desk. Clicking and dragging the small line in it outputs values from 0 to 127 by default, following the range of MIDI; but this can be changed in its properties. You can get these values from its outlet.

The Radio (again Vradio and Hradio stand for vertical and horizontal) works a bit like a menu with choices (like the one that appears when you right-click a Pd element). Only instead of text, it consists of little white squares next to each other, and clicking them outputs a number starting from 0 (clicking the top of the Vradio will output 0, clicking the one below will output 1, and so forth). The Hradio counts from left to right. You can tell which one is currently clicked by a black square inside it. It doesn't really sound like a menu, but remember that Pd is a programming language, meaning that we need to program anything we want it to do. This way, provided a very simple GUI that outputs incrementing values, we can use it to create something more complex out of it. We'll see it in action in the interface building projects in this book. We've now covered the GUIs that we will use and we can move on.

Pd Patches Behave Like Text Files

When we edit a Pd patch, we can use features that are common between text editing programs. This means that we can choose a certain part of the patch (by clicking and dragging in Edit Mode), copy it, cut it, duplicate it, paste it somewhere else in the patch, or in another patch. If you click the **Edit** menu, you'll see all the available options.

The ones we'll mostly use in this book are Copy (Ctrl/Cmd+C), Paste (Ctrl/Cmd+V), Cut (Ctrl/Cmd+X), and Duplicate (Ctrl/Cmd+D). Actually, a Pd patch is text itself. If you open any patch in a text editing program, you'll see its textual form. The first patch we created in this chapter, with the [440] message and [osc~] and [dac~], looks like what's shown in Figure 1-17.

```
#N canvas 384 273 450 300 10;
#X msg 161 69 440;
#X obj 161 91 osc~;
#X obj 161 113 dac~;
#X connect 0 0 1 0;
#X connect 1 0 2 0;
#X connect 1 0 2 1;
```

Figure 1-17. A Pd patch in its textual form

Even though this is pretty simple, you don't really need to understand it thoroughly, as we're not going to edit patches in their textual form in the course of this book. Still, it's good to know what a Pd patch really is.

Making Oscillators in Pd

Now that we've covered the real basics of Pd, and you know how to create patches, let's look at some sound generators that we will use in later chapters. First, I need to explain what an *oscillator* is. In analog electronics, it is a mechanism that creates certain patterns in an electrical current, which is fed to the sound system, causing the woofers of the speakers to vibrate in that pattern, thus creating sound. In digital electronics, this mechanism consists of very simple (or sometimes more complex) math operations (here we're talking about multiplication, addition, subtraction, and division, nothing else) that create a stream of numbers that is fed to the sound card of a computer, turning this number stream to an electrical current. From that point on, up to the point the electrical current reaches the sound system's speakers, everything is the same between analog and digital electronics. The patterns that the oscillators create are called *waveforms*, because sound is perceived in auditory waves.

There are four standard waveforms in electronic music, which we will create in this section. These are the sine wave, the triangle, the sawtooth, and the square wave. They are called like this because of the shapes they create, which you can see in Figure 1-18.

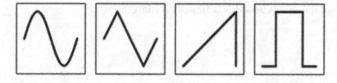

Figure 1-18. *The four standard oscillator waveforms: the sine wave, the triangle, the sawtooth, and the square wave*

Some other audio programming environments have classes for all these waveforms, but Pd has an object only for the first one, the sine wave. Some consider this to be a bad thing, but others consider it to be good. Not having these oscillators available means that you need to build them yourself, and this way you learn how they really function, which is very important when dealing with electronic music. The object for the sine wave oscillator is [osc~], which you've already seen, so we're not going to talk about it here.

Before moving on to the next waveform, we need to talk about the range of digital audio. Audio in digital electronics is expressed with values from –1 to 1 (remember, a *digital signal* is a stream of numbers representing the amplitude of an electric signal). In the waveforms in Figure 1-18, –1 is expressed by the lowest point in the vertical axis, and 1 by the highest (waveforms are represented in time, which lies at the horizontal axis). Paralleling this range to the movement of the speaker's woofer, –1 is the woofer all the way in, 1 is the woofer all the way out, and 0 is the woofer in the middle position (this is also the position when it's receiving no signal). Figure 1-19 represents these positions of the woofer by looking at it from above.

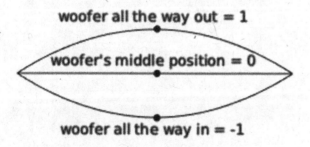

Figure 1-19. *The speaker's woofer's positions and their corresponding digital values*

Making a Triangle Wave Oscillator

Now to the next wave form, which is the triangle. We will create this with simple objects. The driving force for every oscillator (and other things we'll build along the way) is the [phasor~]. [phasor~] is a rising ramp that looks like the sawtooth wave form, only it outputs values from 0 to 1. To create a triangle out of this, we need a copy of [phasor~]; but instead of rising, we need it to be falling from 1 to 0. To achieve this, we must multiply [phasor~]'s output by –1 and add 1. This is very simple math if you think about it. If you multiply [phasor~]'s initial value, which is 0, by –1, you'll get 0, and if you add 1 you'll get 1. If you multiply [phasor~]'s last value, which is 1, by –1, you'll get -1, and if you add 1, you'll get 0. All the values in between will form the ramp from 1 to 0. Mind, though, that we are now in the signal domain and all the objects we'll use are signal objects. So for multiplying, we'll use [*~] and for adding we'll use [+~] .

Once we have the two opposite ramps, we'll send them both to [min~]. This object takes two signals and outputs the minimum value of the two. The two ramps we have intersect at the value 0.5 during their period (a *period* in a wave form is one time its complete shape, like the preceding wave form images). For the first half of the period, the rising [phasor~] is always less than the falling one (the rising one starts from 0 and the falling from 1), so [min~] will output this. For the second half of the period, the falling [phasor~] will be less than the rising one, so [min~] will output that. What [min~] actually gives us is a triangle that starts from 0, goes up to 0.5, and comes back to 0. Figure 1-20 illustrates how this is actually achieved.

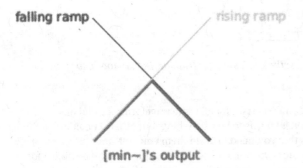

Figure 1-20. *Getting a triangle out of two opposite ramps*

As I've already mentioned, the range of oscillators is from –1 to 1. This is 2 in total. So, multiplying the output of [min~] by 4, will give us a triangle that goes from 0 to 2. Subtracting 1, will bring it to the desired range. These last two actions—the multiplication and the subtraction—are called *scaling* and *offset*, respectively. So, our triangle oscillator patch should look the patch in Figure 1-21.

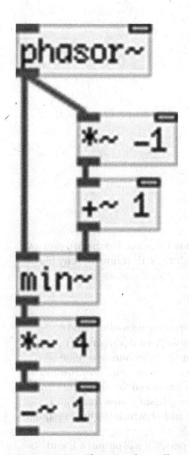

Figure 1-21. *The triangle oscillator Pd patch*

Connect a number atom to [phasor~] to give it a frequency (lock your patch before trying to type a number into the number atom) and place a [dac~] and connect [-~ 1] to it. Turn on the DSP and listen to this wave form. Compare its sound to the sound of [osc~]. The triangle wave form is brighter than the sound of the sine wave, which is because it has more harmonics than the sine wave. Actually, the sine wave has no harmonics at all, and even though it is everywhere in nature, you can only reproduce it with such means, as you can't really isolate it in nature.

Note that execution order doesn't apply to the signal domain, because signal objects calculate their samples in blocks, and they have received their signals from all their inlets before they go on and calculate the next audio block.

Making a Sawtooth Oscillator

The next waveform we'll build is the sawtooth. This one is very easy, since we'll use [phasor~], which is itself a sawtooth that goes from 0 to 1, instead of -1 to 1. All we need to do here is correct its range, meaning apply scaling and offset. Since [phasor~] has a value span of 1, and oscillators have a value span of 2, we have to multiply its output by 2; so now we get a ramp from 0 to 2. Subtracting 1 gives us a ramp from -1 to 1, which is what we essentially want. The patch is illustrated in Figure 1-22.

Figure 1-22. *The sawtooth oscillator Pd patch*

Supply [phasor~] with a frequency and connect [-~ 1] to [dac~] to hear it. Compared to the two previous oscillators, this one has even more harmonics, which you can tell by its brightness; its sound is pretty harsh.

Making a Square Wave Oscillator

Finally, let's build the last wave form, the square wave. This oscillator is alternating between -1 and 1, without any values in between. Again, we'll use [phasor~] to build it. Now we'll send [phasor~] to a comparison object, [<~], which compares if a signal has a smaller value than another one, or a value smaller than its argument (if one is provided). If the value is smaller, [<~] will output 1, else it will output 0. Connecting [phasor~] to [<~ 0.5] (don't forget the space between the object name and the argument), will give 1 for the first half of [phasor~]'s period, and 0 for the other half, because [phasor~] goes from 0 to 1, linearly. Multiplying this by 2 and subtracting one will give an alternating 1 and -1, which is what the square wave oscillator is.

This oscillator has one more control feature, which is how much of its period it will output a 1, and how much it will output a 0 (for example, it can output a 1 for 75% of its period and a 0 for the rest 25%, or vice versa, or any such combination). This is called the *duty cycle*, which is easy to make in Pd. All you need to do is control [<~] with a value that ranges from 0 to 1 (actually from something over 0, like 0.01, to something less than 1, like 0.99). If you connect a number atom to the right inlet of [<~ 0.5] you'll override the argument with whatever number you provide (mind that the right inlet of [<~ 0.5] is a control inlet, and that is because you have provided an argument. If you create the object without an argument, both its inlets will be signal inlets). Your patch should look Figure 1-23.

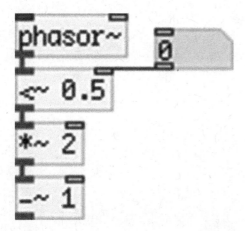

Figure 1-23. *The square wave oscillator Pd patch*

Try some different values by typing into the number atom (in a locked patch), always staying between 0.01 and 0.99. You can also hold down the Shift key and click and drag the number atom. This way, it will scroll its values with two decimal places.

Mind that it is possible to create the same oscillator with [>~] instead of [<~]; the only difference is that it will first output −1 and then 1, but that's a difference that is not recognizable by the human ear. Comparing this oscillator to the others, we see that this one also has a lot of harmonics, as its sound is very bright.

We have now created the four basic oscillators of electronic music. Their raw continuous sound might not be very musical or inspiring, but the way we'll use them in some of this book's projects will be quite different and will provide more musical results.

Using Tables in Pd

The next feature we're going to look at is tables. You'll learn how to use them in Pd. You learned about tables in school math; a table stores values based on an index. In Pd, this is either called a *table*, and you can create it with [table], or array, which we can put from the Put menu. Open a new window and go to **Put ➤ Array** (there's no shortcut for this one). A properties window will open, where you can set its name, its size, whether you want to save it contents, the way to draw its contents, and whether to put the array in a new or in the last graph. From these options, you'll only deal with the first three. For now, you can keep the default name, which is **array1**. You'll also keep the size for now, which is 100, but you'll untick the **Save contents** field, because we don't care to save whatever you'll store in it.

Click **OK** and you'll see a graph in your patch. If you move it, you'll also see its name projected on top of it, looking like the one shown in Figure 1-24.

array1

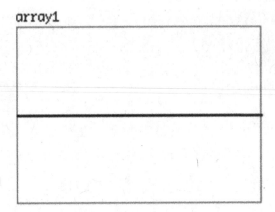

Figure 1-24. *A Pd array*

Inside the array's window, there's a straight line, right in the middle, spanning from left to right. These are the values stored in the array, all of which are 0 for now. The values in an array are graphed in the Y axis and the indexes in the X axis. Indexes start counting from 0 and go from left to right. In our case, the last index is 99, as we have an array of size 100 and the first index is 0.

There are a few ways to store values in an array. The simplest one is to draw values by hand. Lock you patch and hover the mouse over the line in the middle of the array. Click and drag (both up and down, as well as right and left) and you'll see that the line follows the mouse. This way is not very useful because you generally want to store certain patterns in arrays that are actually impossible to draw by hand.

Another way to store values is by using [tabwrite], where "tab" stands for *table*. This object has two inlets and no outlet. The right inlet sets the index and the left sets the value. It also takes an argument, which is the name of the array to write to. Put a [tabwrite array1] to your patch and connect a number atom to each inlet. Lock your patch and store a value to an index; for example, store 0.75 to index 55 (indexes are always integers). Mind to first provide the index to the right inlet, and then the value—again, the right to left execution order and hot and cold inlets apply. You should immediately see the dot at the 55th place jump to the value 0.75 (a bit lower than the upper part of the frame). To double-check it, put a [tabread array1], which reads values from an array. This one has one inlet and one outlet. In the inlet. you provide an index and it spits the value at that index out its outlet. Give it the value 55 and it should give 0.75.

All of this isn't likely very intuitive and the point might seem a bit hidden. Let's look at another way to use arrays. Right-click Array and click Properties. Now you get two windows, but we only care about the first one. Change the size of the array to 2048, click **OK**, and close both of these windows (whatever values we've already stored will shrink to the right, as there are now in the very first indexes). Copy one of the oscillator patches (the triangle, for example) built in the previous section, but instead of [dac~] at the bottom, put [tabwrite~ array1] (mind the tilde that makes it different from [tabwrite array1]). Connect a number atom to [phasor~] and a bang (Shift+Ctrl/Cmd+B) to [tabwrite~ array1] (this object has one inlet only and it takes both signals and bangs). Figure 1-25 shows what your patch should look like.

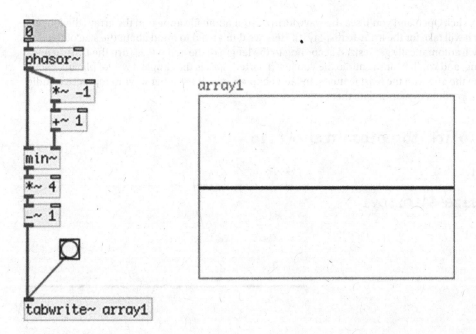

Figure 1-25. The triangle oscillator connected to [tabwrite~] in order to be stored in an Array

Provide a frequency via the number atom (don't forget to lock you patch), turn the DSP on and click the Bang ([tabwrite~] will store any signal connected to its inlet, whenever it receives a bang). You should see the wave form of the oscillator stored in the Array, similar to Figure 1-26.

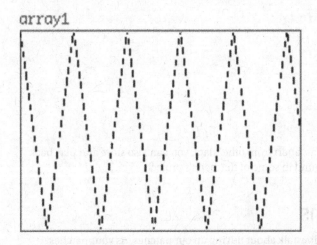

Figure 1-26. The triangle oscillator wave form stored in an array

You can display all four oscillators we've already made to see their shape in action.

Another very useful feature of tables in Pd is that we can upload existing audio files to them. Create the patch in Figure 1-27. [tabplay~] is designed to play audio stored in a table. Clicking the top bang will open a dialog window, where you can navigate to a folder where you have an audio file (a .wav or .aiff file, no .mp3). Once you select an audio file (not a very long one, up to 5 minutes, more or less; usually tables have files that are a few seconds), click Open and you'll see the wave form of your audio file appear in the array (the longer the file, the longer it will take for the array to display it). Here we don't need to mind about the size of the array, because it will automatically get resized according to the length of the audio file. Turn the DSP on and click the lower bang, and you'll hear the audio file you just inserted. This is the simplest way of playing back audio files, but also the one with the least features. In later chapters, you'll see other ways to reproduce audio files that give more freedom to manipulate them.

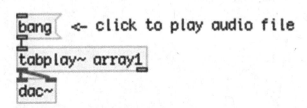

Figure 1-27. *An audio file playback patch*

So you can see that tables can be very useful as, apart from other data, you can also store and play back audio. We'll use arrays to store and manipulate sound in some of this book's projects.

Subpatches and Abstractions

Since you've done a little bit of patching, we can now talk about tidying up our patches. As your patches grow more and more complex, you'll see that having all objects, messages, number atoms, and so forth, visible in your patch will be more and more difficult. It will also be difficult to keep track of what each part of your patch does. This is where the subpatch comes in. A subpatch is an object that contains a patch. Open a new window and put an object. Inside the object type **pd percentage**. (We'll make a subpatch that gives a percentage value, although the name of the subpatch could be anything. Naming it "percentage" makes it clear as to what the subpatch does.) A new window will open, titled "percentage". This window is a Pd patch, more or less as you already know it. The only difference between this and a normal Pd patch is that the subpatch cannot be saved independently from the patch that contains it, which is called the *parent patch*. A subpatch is part of its parent patch, and will be saved only if you save the parent patch.

Using subpatches is very useful for tidying up our patches, and helps us create programs in a self-explanatory way. We can put any Pd element in a subpatch, but in order to have access to it, we need to use [inlet] and [outlet]. In the "percentage" subpatch, put an object and type **inlet**. If you look at the parent patch, [pd percentage] now has one inlet. If you put more [inlet]'s in the subpatch, you'll see them in the parent patch object. The same goes for the [outlet]. The order of their appearance in the parent patch follows the order of their placement inside the subpatch, meaning that the far left [inlet] in "percentage" is the far left inlet in [pd percentage] in the parent patch. Let's see the subpatch in action. Inside the subpatch put the objects, as shown in Figure 1-28.

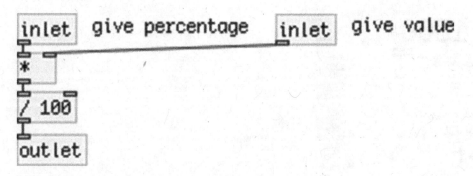

Figure 1-28. *Contents of the "percentage" subpatch*

This subpatch calculates a given percentage of a given value, where the percentage goes into the left inlet and the value into the right one. Lock it and close it.

In the parent patch, you should have a [pd percentage] with two inlets and one outlet. The left inlet of [pd percentage] corresponds to the left inlet of Figure 1-28. Connect a number atom to each inlet and to the outlet. Provide a value to get its percentage to the right inlet, for example, 220 (remember that the left inlet of [*] is hot, so we need to provide input to the right one first) and the percentage, to the left inlet. Figure 1-29 shows the subpatch in action, where we ask for 40% of 220, and we get 88.

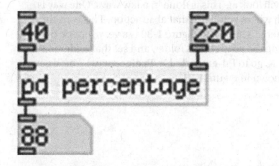

Figure 1-29. *The "percentage" subpatch*

This specific subpatch is quite simple, but we have already enclosed two objects in one. The more complex a function within a patch becomes, the more space we save by placing it in a subpatch, and the more readable our patch is, since we can give a name to the subpatch that corresponds to its function. This way we can even avoid writing comments, as our patch is self-explanatory.

Abstractions are somewhat different than subpatches. They are also Pd patches used as objects, but instead of creating them inside a parent patch (and saving them only through their parent patch), we create them independently of any other patch. *Abstractions* are essentially pieces of code that we very often use, so instead of making that specific piece of code over and over again, we create it once, and use it as is. Take a simple example—a hertz-to-milliseconds converter. This is a very simple patch to create; it is shown in Figure 1-30.

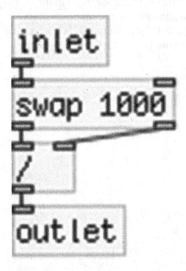

Figure 1-30. Contents of the "Hz2ms" abstraction

In this patch, we provide a value to [swap 1000]. What [swap 1000] does is get a value in its left inlet and output it out its right inlet, and output 1000 out its left inlet; in three words: swaps its values. Check its help patch for more information.

Pd's objects will receive either hertz or milliseconds as time units, so it's very helpful to have an object that converts from one to the other. But Pd doesn't have such an object, and creating this patch (no matter how simple it is) every time you need to make this conversion would be rather painful. What you can do is create this patch once and save it to a place where Pd will look at. This is done in a few ways. One way is to save your abstraction in the same folder with the patch where you'll use that abstraction. This way, the abstraction will be more project -specific rather than generic. The one in Figure 1-30 is a very generic one. What we'll do is create a folder called abstractions, inside the pd_patches folder, and set that folder to Pd's search path. To do this, go to **Edit ➤ Preferences** (on OS X, go to **Pd-extended ➤ Preferences**). You'll get a window where you can set a search path for Pd. This is shown in Figure 1-31.

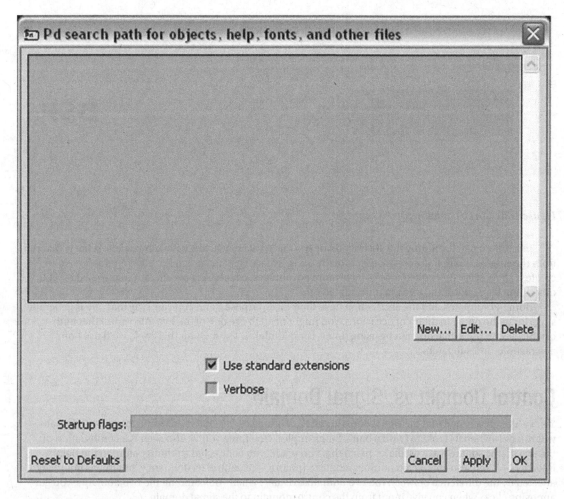

Figure 1-31. *Pd's Preferences window*

Click **New...** and a dialog window will open. Navigate to the newly created abstractions folder and click **Choose**. In Pd's **Preferences**, click **Apply**. You won't see anything happening, but the search path has been stored. Then click **OK** and the window will close. Now save the patch to the abstractions folder with the name Hz2ms, which stands for "hertz to milliseconds."

For Pd to be able to use the newly set search path, you must quit it and restart it. Once you restart Pd, open a new window, put an object, and type **Hz2ms**. If all goes well, you'll see an object with that name. If instead of an object you get a red dotted rectangle and a message in Pd's console, like the one shown in Figure 1-32, check the search path in Pd's Preferences, or make sure that you typed the name of the abstraction correctly.

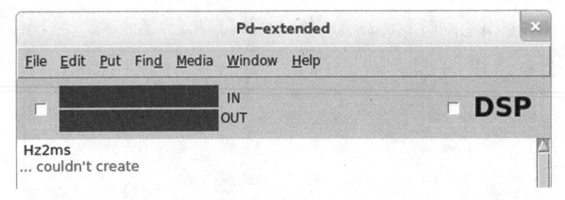

Figure 1-32. An error message in Pd's console

In a locked patch, clicking the abstraction opens the actual patch, like with subpatches. What is different from the subpatch is that the abstraction is ready to use whenever you launch Pd, and you don't need to create it anew every time. Abstractions have more advantages, concerning arguments, name clashes, and others, but we're not going into too much detail now.

Both the subpatch and the abstraction have their own purposes, and you can't say that one is generally superior to the other. In different occasions, you might need to use one of the two. Also mind that both can be used in the signal domain by using [inlet~] and [outlet~]. Throughout this book, we'll use both abstractions and subpatches.

Control Domain vs. Signal Domain

You've already learned that the signal domain runs for as long as the DSP is on, while the control domain will run only when it is told to (with a bang, for example). One thing you've also seen is a combination of the two. In the square wave oscillator patch that you made, you connected a number atom to the right inlet of [<~]. You also connected number atoms to [phasor~] to control its frequency, but that is not really affected by the difference between the two domains. Usually when you combine the two domains, you get annoying clicks when you give input from the control domain to the signal domain.

In the case of the square wave oscillator, these clicks are not really audible, because the square wave form is itself very "clicky." Let's take another example, one where you control the amplitude of a sine wave. Controlling the amplitude of a signal in digital electronics is simple multiplication with values between 0 and 1.

Since a digital signal is a stream of numbers, when you multiply these numbers by 0, you'll get a constant 0 (remember the woofer's positions; this would be silence). Multiplying the number stream by 1 will give you the number stream intact, hence the signal in its full amplitude. All the values in between will give corresponding results. Go ahead and build the simple patch, as shown in Figure 1-33.

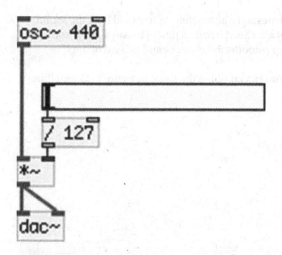

Figure 1-33. *Controlling the amplitude of an oscillator*

You're dividing the output of the Hslider by 127 to get a range from 0 to 1 (sliders have a default range from 0 to 127). Turn the DSP on and use the slider (in a locked patch). The more you move it to the right, the louder you'll hear the sine wave. Pay close attention to these amplitude changes; the faster you move the slider, the more clicks you hear. This is due to the clash between the control and the signal domain. In detail, [*~] refreshes its values in every new DSP cycle (as all signal objects do), which is done in blocks of 64 samples (no need to really grasp this detail though).Therefore, if you change the slider values quickly, [*~] will make sudden jumps from the previous value to the current, which is heard as a click.

To remedy this, you must make the value changes smoother. There's a very useful object in Pd for this: [line~]. [line~] takes two values: the destination output value and an amount of time in milliseconds. [line~] will make a linear ramp in the signal rate, from its current value to the destination value, and this ramp will last as many milliseconds as you provide with the second value. Change the patch in Figure 1-33 to the patch in Figure 1-34.

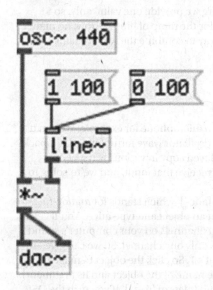

Figure 1-34. *Using [line~] to avoid clicks*

Lock your patch, turn the DSP on, and click the two messages alternately. You should hear the sound of the sine wave come in and go out without any clicks at all. This is because [line~] makes a ramp from 0 to 1 in 100 milliseconds, and the other way around. This ramp smooths the changes and gets rid of the annoying clicks.

But what if you want to have a variable amplitude? You can combine the slider in Figure 1-33 with [line~], as shown in Figure 1-35.

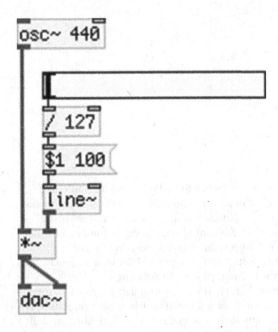

Figure 1-35. *Combining the Hslider with [line~]*

$1 in the message means the first value that comes in its inlet. Here we provide one value only, so $1 will take the value of the slider. 100 is still the amount of milliseconds for the ramp of [line~]. Now, no matter how quickly you move the slider, there are no clicks at all. This is the way to combine the two domains—something that happens very often in Pd.

Audio Input in Pd

Apart from sound generators, in Pd we can also use audio input, from a microphone for example. We can use that input in many different ways. We can store it, like we did with the oscillator wave forms, and play it back in various ways, we can write it in delay lines and use that to play a delayed copy in various ways, we can apply pitch shifting, and many more. For now, we'll talk about how to receive that input, and we're going to play it straight away from the speakers.

The object that receives input from the computer's sound card is [adc~], which stands for *analog-to-digital converter*, which is the opposite of [dac~]. In a new window, put an object and type **adc~**. You'll get an object with no inlets and two outlets. The two outlets are the two input channels on your computer's sound card. But the default input in Pd is the built-in microphone, which has only one channel. So we can give an argument to [adc~], which is the channel we want to use; in this case, it's 1. So click the object to make it editable, and type **adc~ 1** (make sure that you put a space between the name of the object and its argument). Now the object has one outlet. Put a [dac~] and connect [adc~ 1] to both inlets of [dac~]. If you turn the DSP on and your speakers are quite high, you'll get feedback, meaning that the audio that goes out the speakers will immediately go back in through the microphone, creating a loop, and it will most likely create a high and rather loud tone. To avoid that, you can use headphones for this patch. Now if you talk close to your computer's built-in microphone, you'll hear your voice through the headphones. Maybe the output is a bit delayed; that is because digital audio takes some time to make its calculations before it outputs sound. In the following sections, we'll talk about a way to reduce that delay.

Basic Electronic Music Techniques

Now let's cover some basic electronic music techniques with Pd. We've already created the four basic oscillators, but now we're going to create some more interesting sounds by using them.

Additive Synthesis

The first technique we're going to look at is called *additive synthesis*, because we use many oscillators (usually sine waves) added together to create more interesting timbres. Figure 1-36 shows an additive synthesis Pd patch.

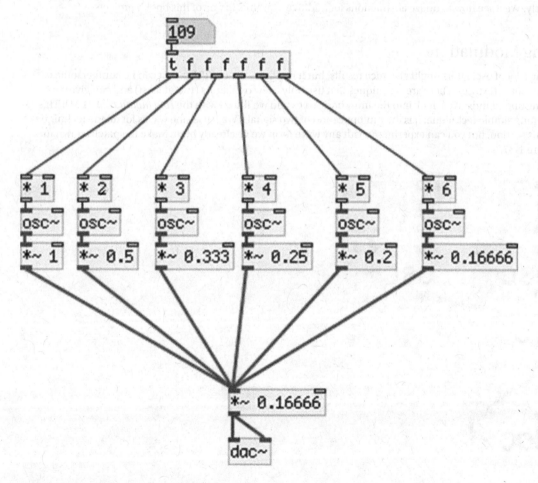

Figure 1-36. *Additive synthesis Pd patch*

In additive synthesis, if we want to create a harmonic sound, we provide a base frequency, and we multiply it by the order of each oscillator. For the first oscillator, we multiply it by 1, for the second by 2, and so forth. Something else we do in the patch shown in Figure 1-36 is set the amplitude of each oscillator to the reciprocal of its order, so the first oscillator will have full amplitude, the second will have one half of its amplitude, the third will have one-third, and so forth. Note that multiplying requires less CPU than dividing, so instead of dividing, we multiply each oscillator's output by the reciprocal of its order. Mind the multiplication at the bottom of the patch. When we output more than one signal as they are being added, therefore we must scale them to make sure that their sum won't go over 1 or below –1. To achieve this, we multiply the total output by the reciprocal of the number of signals we send to the speakers; here it's 1 ÷ 6 = 0.16666.

The more oscillators we use, the more it will sound like a sawtooth oscillator, because this is the algorithm the sawtooth wave form uses. Still, you need many sine wave oscillators to make them sound like a sawtooth. With additive synthesis, we can create more textures by providing frequencies to each oscillator that are not integer multiples of the base frequency. This will create non-harmonic sounds, but you might find very interesting textures this way.

Another aspect to experiment with is the amplitude of each oscillator. By changing these, the timbre of the sound changes drastically. Try some random values between 0 and 1 for each oscillator to hear the result. The way to create additive synthesis shown here is not the most effective one; usually, we prefer to use abstractions, as they reduce patching to a great extent. Still, it shows how additive synthesis works rather clearly. We'll see how to utilize abstractions for additive synthesis in one of this book's projects.

Ring Modulation

Using a lot of oscillators might give nice results, but it requires a lot of CPU, plus it can be cumbersome to create some textures. There are techniques that use only a few oscillators (two at least) and can give very interesting sounds. We'll look into the most basic ones and we'll start with the *ring modulation* (RM). This is a quite simple technique; it is the multiplication of two signals. We'll use sine waves for all the techniques in this section, but you can experiment with any wave form we've already built. Make the patch shown in Figure 1-37.

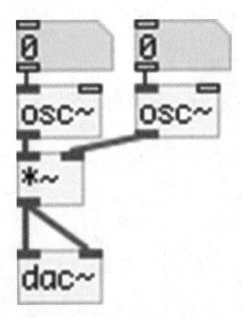

Figure 1-37. *A ring modulation patch*

Lock your patch, turn the DSP on, and start dragging your mouse in the number atoms, or use sliders instead. (Although you might want to multiply their output to a range other than 127; multiply them by 3, for example.) You'll start hearing two tones, which are the addition and the subtraction of the two provided frequencies. For example, if you type **300** in one number atom and **5** in the other, you'll hear the frequencies 305 and 295. Sometimes the subtraction of the two frequencies result in a very low frequency (if, for example, you provide 305 and 300, you'll get 605 and 5), which are not audible by the human ear. Keep on dragging your mouse upward till you start hearing two tones. Experiment a bit till you find some interesting results. You can also combine RM with the additive synthesis patch in Figure 1-36, where one of the two oscillators in Figure 1-37 will be replaced by the additive synthesis patch.

Amplitude Modulation

The next technique is the *amplitude modulation* (AM). It is very similar to the ring modulation. Again we multiply two signals, but we use one signal to modulate the amplitude of the other. As we've already seen, to modulate the amplitude of a signal, we need to multiply it with values between 0 and 1. But oscillators give values from –1 to 1. To bring one of the oscillators to the desired range, we need to apply some scaling and offset. First of all, we need to shrink the oscillator's range to its half. We do this by multiplying it by 0.5 (remember that multiplication in computers require less CPU than division, so instead of dividing by 2, we prefer to multiply by 0.5). Multiplying by 0.5 will make the oscillator go from –0.5 to 0.5. If we add 0.5 (this is the offset), it will go from 0 to 1.

So, the patch in Figure 1-37 changes slightly and becomes the patch in the Figure 1-38.

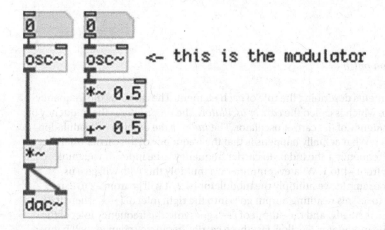

Figure 1-38. An amplitude modulation patch

Give a high enough frequency to the left oscillator (depending on your speakers; if you use laptop speakers, you should give at least 200) and a very low one (like 1) to the right oscillator, which is the modulator. You should hear the tone of the left oscillator come in and out smoothly. The higher you bring the modulator's frequency, the faster these changes will happen. If you bring it too high (above 20), you'll start getting similar results to the ring modulation. This is because humans can hear frequencies as low as 20 hertz. Using lower frequencies are not immediately audible, but they can effectively change the amplitude of another audio generator. In this case, the oscillator is called an LFO (low-frequency oscillator). If we provide a frequency higher than that, then it enters the audible range and we start hearing it immediately (and it's no longer an LFO).

Frequency Modulation

Next in line is the *frequency modulation* (FM). Here we use one oscillator, called the *modulator*, to modulate the frequency of another oscillator. The patch to do this is shown in Figure 1-39.

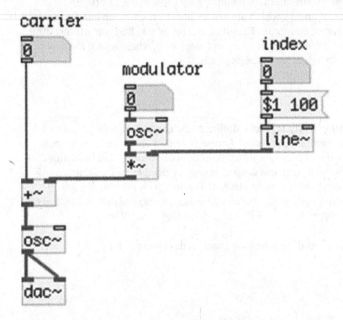

Figure 1-39. *A Frequency modulation patch*

Figure 1-39 includes some comments describing the role of each element. The *carrier* is the frequency of the oscillator that we actually hear, which is called the *carrier oscillator*. The *modulator* is the frequency of the oscillator that modulates the frequency of the carrier oscillator. The *index* is the amount of modulation the modulator will apply to the carrier. What actually happens is that the frequency of the carrier oscillator goes up and down, from the carrier frequency + the index, to carrier frequency – the index. If you think about it, an oscillator outputs values from –1 to 1. Whatever number we multiply this with will give us the multiplier and its negative. If, for example, we multiply the modulator by 2, it will go from –2 to 2; if we multiply it by 5, it will go from –5 to 5. This resulting output goes into the right inlet of [+~], where it is added to the carrier frequency, which is steady, and the output of [+~] goes into the frequency inlet of the carrier oscillator. The frequency of the modulator sets how fast these carrier frequency changes will happen. Figure 1-40 illustrates this.

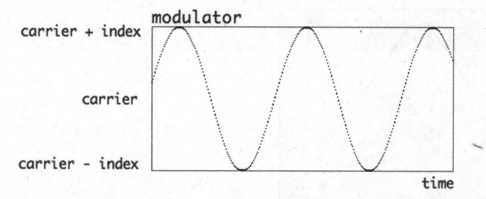

Figure 1-40. *FM illustration*

If the horizontal axis of the graph is 1 second, the carrier frequency is 300, and the index is 50, the carrier oscillator's frequency will go from 350 to 250 about three times a second (as many as the peaks in the graph) in a sine fashion. If the modulator frequency is quite low, what we hear is a vibrato-like sound (especially if we keep the index low as well). The higher the frequency of the modulator, the less we can tell that the frequency of the carrier oscillator is actually oscillating, and what we start to perceive is more tones around the carrier frequency. The higher the index, the broader the spectrum of the resulting sound becomes.

This technique is very common in electronic music, as you can create complex and interesting textures using only two oscillators. Experiment with all three values and try to find sounds that are interesting to you.

Envelopes

The last technique is the *envelope*, which we will use for amplitude. The previous four techniques dealt with timbres by using oscillators in various ways. One thing that none of these techniques included was some sort of amplitude evolution. Although AM did modulate the amplitude of an oscillator, that modulation had almost no variation, but a steady oscillating fading in and out. Some musicians tend to treat sound in a different way, where sounds evolve both in frequency and amplitude, with crescendo, decrescendo, and similar characteristics. An amplitude envelope is the evolution of the amplitude in time. By applying it to an oscillator (or to any of the preceding techniques), we can control its amplitude to a great extent, from simple to very complex ways.

Pd has vanilla objects that can create envelopes, but it can be rather cumbersome and not very intuitive, especially if someone is not very familiar with programming and Pd itself. For this reason, we'll use an external object that is part of Pd-extended. This object is called *envgen* (for envelope generator) and it is part of the "ggee" library. A library is a set of external objects. It is very important to know which library each external that we use belongs to. To create this object, we need to specify its library, and that is done in a few ways. For now, we'll use the simplest one, which is to type the name of the library first, then a forward slash, and then the name of the object. In this case, put an object in a new window and type **ggee/envgen**. Figure 1-41 shows what you should get by typing this.

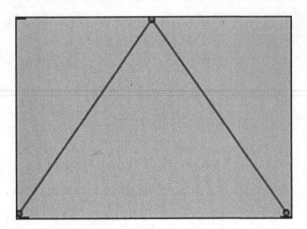

Figure 1-41. *The "envgen" external object from the "ggee" library*

This object is actually a GUI, as you might have already guessed. It is designed to work with [line~], where it sends sequential lists to it. The envelope we'll design is called the ADSR, which stands for Attack-Decay-Sustain-Release. It is the most common amplitude envelope that is used in many synthesizers, commercial or not. The ADSR is a simple imitation of the behavior of the sound of (plucked) acoustic instruments. The Attach part of it goes to full amplitude in a short time. The Decay part goes down to a lower amplitude, where it stays for a while, and this is the Sustain. Lastly, the Release goes down to zero amplitude in a short time. An ADSR envelope is shown in Figure 1-42.

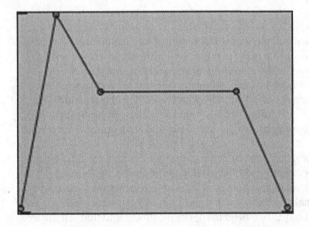

Figure 1-42. *An ADSR envelope with [ggee/envgen]*

To create this, lock you patch and hover your mouse over the peak of the graph in [envgen], where you see a small circle. Drag this to the left a bit. To create more *breakpoints* (that's what the points where the lines break are called), click anywhere inside the GUI. To move an existing point, click it and drag it, like you did with the first point. To delete a point, click it and hit Backspace.

The rising part of the envelope in Figure 1-42 is the Attack; the first falling part is the Decay; the horizontal line is the Sustain' and the last falling part is the Release. Make the patch shown in Figure 1-43. The "duration 2000" message sets the duration of the envelope in milliseconds. The bang activates it. Turn the DSP on, lock your patch, click the "duration" message, and then click the bang message. You should hear the sine wave fade in and out in the fashion of the graph of the envelope, which should last two seconds in total. Go ahead and try different envelopes. Also try them in combination with the other techniques in this section.

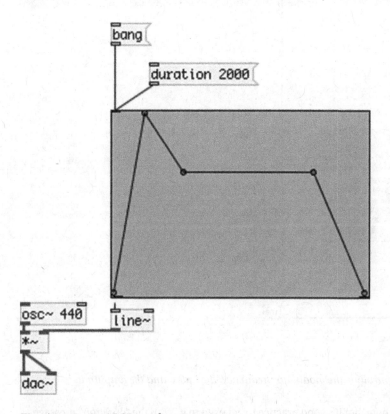

Figure 1-43. *An ADSR envelope in action*

An envelope can be used in any control parameter of a sound, like the frequency, the index in FM, and so forth.

Figure 1-44 shows a patch where the modulator frequency, the index, and the amplitude of FM synthesis are being controlled by an [envgen] each. Mind the [trigger], where instead of f, we type **b**, which stands for "bang." In this patch, [trigger] sends the bang it receives from right to left.

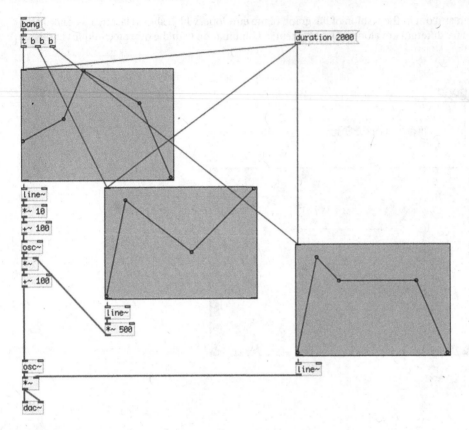

Figure 1-44. *FM with envelopes controlling the modulator frequency, the index, and the amplitude*

It is not very important to use here, but it's good practice to get used to it. With the "duration" message on the other hand, it is even less important where it will go first, as this message causes no execution at all. As with the patch in Figure 1-43, lock your patch, click the "duration" message, turn on the DSP, and click the "bang" message. Experiment with different envelopes and durations.

Note that setting minimum and maximum values for [envgen] can be achieved with arguments. But for now, we don't give any arguments and use the object as is, where it defaults to the range 0-1. Check the help patch to get more information about its use.

We have now covered five basic techniques of electronic music, which we will see later on as build musical interfaces. We've seen these techniques in their simplest form. In the following chapters, we will make more efficient, but also a bit more complex, use of them.

Delay Lines in Pd

A delay line is different than the delay mentioned earlier. A delay line delays sound intentionally. Actually, using delay in many music styles is a much celebrated effect, which has been around for a long time. Pd has built-in objects for that: [delwrite~], [delread~], and [vd~]. [delwrite~] is a bit like [tabwrite~]; it takes two arguments though. The first argument is the name of the delay line (like tables, delay lines need to have names so that we are able to access them), and the second is its length in milliseconds. In the previous patch with [adc~ 1] and [dac~], disconnect [adc~] from [dac~] and connect it to [delwrite~ my_delay 1000]. This is a delay line called "my_delay" and it will store 1 second of audio. Apart from its similarity with [tabwrite~], [delwrite~] will write on the delay line continuously, as long as the DSP is on. It doesn't take a bang, and you can't control the beginning and ending of writing to the delay line.

[delread~] will read audio from a delay line. It also takes two arguments: the first is the name of the delay line and the second the delay time in milliseconds. Put a [delread~ my_delay 500] in your patch and connect it to [dac~]. Figure 1-45 shows what your patch should look like. In this patch, [delwrite~] takes audio from the built-in microphone of your computer and writes 1 second to the delay line, called "my_delay". When that second is over, it goes back to the beginning of the delay line and overwrites whatever was stored there; it does this for as long as the DSP is on. [delread~] reads from that delay line (because its first argument is the same with the first argument of [delwrite~]), but delays its reading by half a second, which is the second argument. If the second argument of [delread~] exceeds the second argument of its corresponding [delwrite~], the delay time is automatically clipped to the length of the delay line (in this case, 1000 milliseconds). If you turn on the DSP and talk into the microphone, you'll hear your voice delayed by half a second.

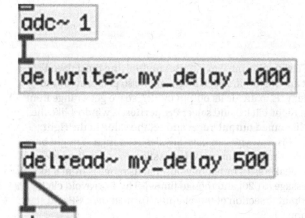

Figure 1-45. *Simple delay line patch*

This might not be a very interesting use of delay, so let's enhance it a bit. One feature of delay lines is feedback. What was not desired in the previous section, where we introduced [adc~], is desired when it comes to delay. If we send the audio read by [delread~] back to its corresponding [delwrite~], along with the audio that comes in from the built-in microphone, all audio stored in the delay line will be stored again, but delayed by 500 milliseconds. If we simply connect these two objects, we won't have any control over it, and the delay line will keep on writing its own audio back to itself along with the audio that comes in through the computer's microphone. To be able to control the feedback, we need to send the output of [delread~] to a [*~], which will control its level, and then to [delwrite~]. Figure 1-46 shows a feedback delay patch.

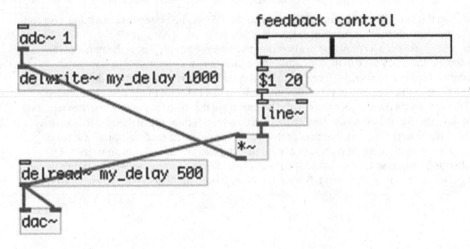

Figure 1-46. *Feedback delay line patch*

As I've already mentioned, the default range of Hslider is from 0 to 127. But multiplying the delayed sound by such a great number will greatly amplify it and distort it, which might not be so pleasant to your ears . What we did in a previous example with a slider was to divide its output by 127, so we get a range from 0 to 1. Another way to do that is to use its properties. Right-click it and select **Properties**. A window like the one shown in Figure 1-48 will show up. Go to the field named **output-rage**: and set the value in the **right:** field to 1. Click **Apply** and **OK**, or simply hit **Return**. This way, you can have any desired range. It might be advisable to place a comment next to a slider that has its range changed. When a slider is controlling the amplitude of a signal that's coming out through the speakers, it's rather obvious that its range is from 0 to 1. Mind that we send the output of the slider to the message [$1 20] and then to [line~]. This is to avoid clicks, as I mentioned in the "Control Domain vs. Signal Domain" section of this chapter. Turn on the DSP and start playing with the slider as you talk into the microphone.

[delread~] has one control inlet, which sets the delay time (it will override the second argument). Using it in real time will create clicks as we combine the control and the signal domain without using [line~] (it's not possible to use [line~] here, as the inlet itself is a control inlet). If we want to be able to change the delay time on the fly, we must use another delay object: [vd~]. "vd" stands for *variable delay*. Change [delread~] in your patch with [vd~ my_delay]. [vd~] takes one argument only, which is the name of the delay line. You can't set the delay time with an argument, but only with input in its inlet. Make your patch look like the one shown in Figure 1-47.

feedback control
adc~ 1
delwrite~ my_delay 1000
$1 20
line~
79
$1 20
*~
line~
vd~ my_delay
dac~

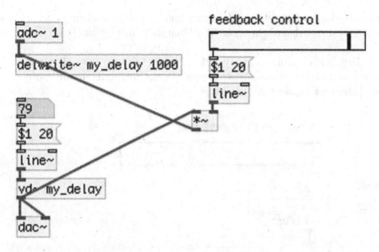

Figure 1-47. *Delay patch using [vd~] instead of [delread~]*

|hsl| Properties
--------dimensions(pix)(pix):--------
width: 128
height: 15
-----------output-range:-----------
left: 0
right: 1
lin
No init
Steady on click
Messages
Send symbol:
Receive symbol:
Label
X offset -2
Y offset -8
DejaVu Sans Mono
Size: 10
Colors
Background
Front
Label
Compose color
o=||=o
Test label
Cancel
Apply
OK

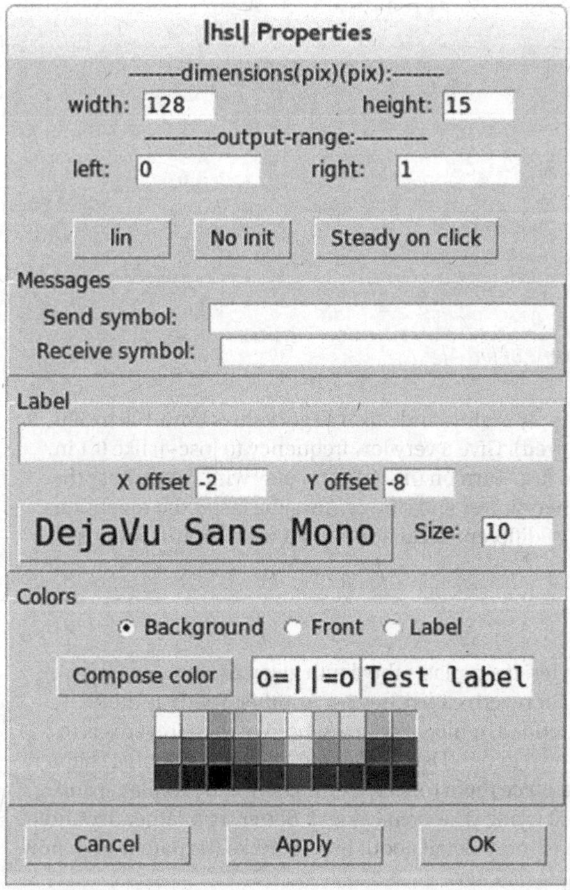

Figure 1-48. *Slider's properties window*

Now you can play with both the feedback amount and the delay time. Again, if you're not using headphones, be careful with the physical feedback that might occur. Mind that we're using [line~] with [vd~] too, in order to avoid clicks.

Before closing this section, let's make a final enhancement to our delay patch. Let's use an oscillator to control the delay time of [vd~] to see how sound generators can act as controllers. Our patch remains more or less the same, only the input to [vd~] changes. Figure 1-49 shows this.

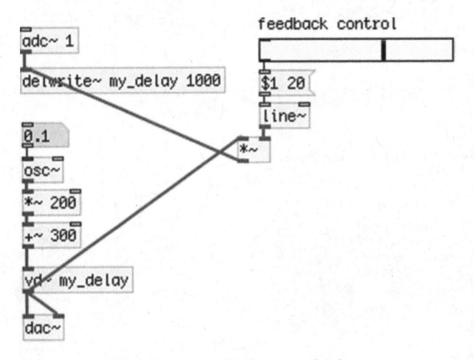

Figure 1-49. *Using an oscillator to control the delay time of [vd~]*

Since oscillators output values from –1 to 1, we apply scaling and offset to get values from 100 to 500 (think of simple math to understand how this is achieved). Give a very low frequency to [osc~], like 0.1 in order not to get changes in the delay time that are too fast. Turn on the DSP and play with. You'll hear the delayed audio being repeated faster and faster, and then slower and slower. Also, the faster the repetitions get, the higher their pitch becomes, and vice versa. Try different oscillators with it to see how it sounds.

Reverb

Reverb is another celebrated effect in lots of music styles. It simulates the depth given to sound in large rooms. In Pd-extended, there is a vanilla abstraction for reverb called "rev3~". Another option is the "freeverb~" external, which is also included in Pd-extended. A nice way to test both of these reverbs is to combine them with the audio file playback patch, shown in the "Using Tables in Pd" section in this chapter. Check their help patches to see how to use them, and place them (one at a time) between [tabplay~] and [dac~] (use the arguments of [rev3~] help patch before trying your own values). Better try a rather dry audio file to hear the full effect of the reverb. We won't go into more detail about how to make the patch, as by now you should have gained some fluency in making simple patches.

Filters

Pd-extended has a variety of raw filters and a few user-friendly ones. The raw filers are [rpole~], [rzero~], [rzero_rev~], [cpole~], [czero~], [czero_rev~], and [biquad~]. These are quite tough to handle because they require quite some knowledge on filter theory. Just for the information, "rpole" stands for *real pole*, and "cpole" stands for *complex pole*. You don't really need to worry about understanding all of this, as a few user-friendly filters are included as well. These are [lop~], [bp~], [hip~], and [vcf~], where "lop" stands for *low pass*, "bp" stands for *band pass*, "hip" stands for *high pass*, and "vcf" stands for *voltage controlled filter*.

[lop~] and [hip~] have one signal and one control inlet. The control inlet takes a frequency value (which can also be set as an argument), which is called the *cutoff frequency*. It is called "cutoff" because they will let all frequencies below or above that pass. The low pass will let the frequencies below the cutoff pass, and the high pass will let the frequencies above it, hence their names. The left inlet, which is the signal inlet, takes the signal to be filtered.

[bp~] and [vcf~] are somewhat different. They both have three inlets, the second of which receives the center frequency. It is called like that because they both let a band around that frequency pass (hence "band pass"). The right-most inlet takes the so-called Q, which is the width of the band of the frequencies that pass. Both of these values can be set as arguments in the case of [bp~]. [vcf~] takes one argument only, which is the Q. The difference between [bp~] and [vcf~] is that the latter can have its center frequency controlled by a signal (an oscillator for example); whereas [bp~] takes a signal only in its far-left inlet, which is the signal to be filtered, like with [lop~] and [hip~], and both the center frequency and the Q inlets are control inlets.

[biquad~] is not considered a raw filter, but it's still difficult to handle. It takes a list of five parameters, and can more or less take the form of any kind of filter (low pass, shelving filters, and others). You need to know how to calculate the five coefficients to design the desired filter. Since it is not very user-friendly, we won't be using it in this book.

An obvious way to use these filters is to filter out some high, low, or middle frequencies of a sound. Again, you can test them with a sound file, like you did with the reverb. Another way to use these filters is with oscillators to shape their wave forms. Figures 1-50 and 1-51 show a square wave and a triangle wave oscillator, respectively, passed through a low pass and a band pass filter, respectively. If you want to build these two patches, go to the properties of each array. Give each the appropriate name ("unfiltered" and "filtered") and change their size to 512. Don't have both patches open at the same time, as you will have each array twice, using the same name, and that will create a clash and warning messages in Pd's console. [metro] is an object that outputs bangs in time intervals provided via its argument (or its right inlet) in milliseconds. Check the help patch for more information.

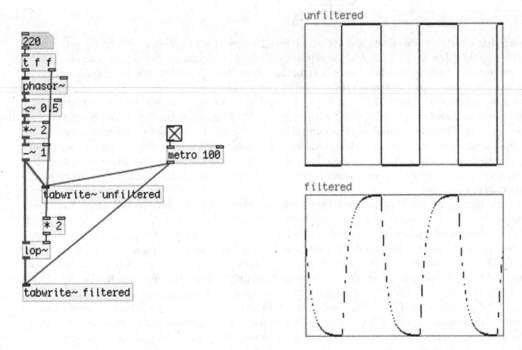

Figure 1-50. *A square wave oscillator passed through a low pass filter*

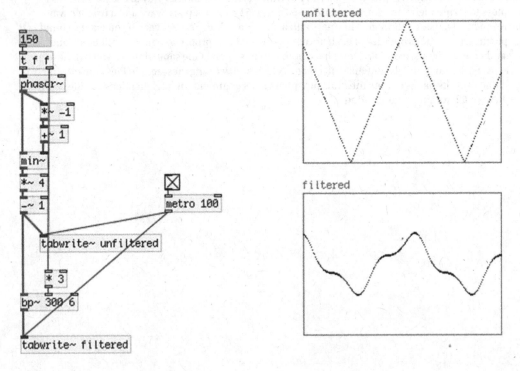

Figure 1-51. *A triangle wave oscillator passed through a band pass filter*

You can see how the shape of the oscillator changes drastically when filtered. This might be desirable for the immediate audio result, but also for using them as controllers in various techniques, like FM. Go back to the "Basic Electronic Music Techniques in Pd" section and try the patches shown there, with filtered oscillators. You'll see that the variety of sounds you can create will expand greatly.

Before we close this section, let's show [vcf~] in action and its advantage of being able to control its center frequency with signals. Build the patch shown in Figure 1-52. The [pd triangle~] subpatch contents are shown in Figure 1-53, although by now you should be able to tell without being shown.

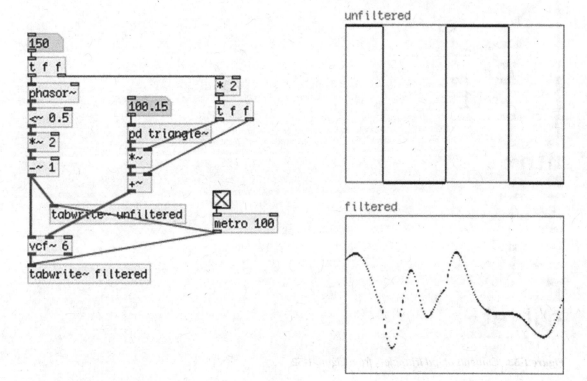

Figure 1-52. *[vcf~] in action*

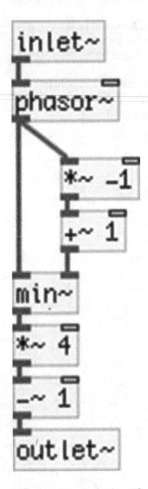

Figure 1-53. *Contents of [pd triangle~] from Figure 1-52*

The result of this specific patch is definitely for hearing as well, and not only for graphing its output. What happens is that the center frequency of the filter is being controlled by a triangle oscillator. In Figure 1-52, the values output from the [pd triangle] subpatch, go from 0 to 600, in a triangle fashion. The frequency of the triangle is a little bit above two-thirds of the frequency of the filtered square wave oscillator. This small difference creates a slowly evolving shift in the timbre. Put a [dac~] in your patch and listen to the result. Try different values for both oscillators. As you can imagine, this is a kind of modulation, which could be included in the modulation patches in the "Basic Electronic Music Techniques in Pd" section.

Making Wireless Connections

We've covered most of the basics of Pd, so now we can talk about more generic things. Here we'll talk about how to connect objects, messages, numbers, and so forth, wirelessly. Visual programming can be very intuitive, because of graphing the data flow in a program, but the more complex a patch becomes, the more difficult it is to read. Figure 1-54 shows a patch that's pretty messy.

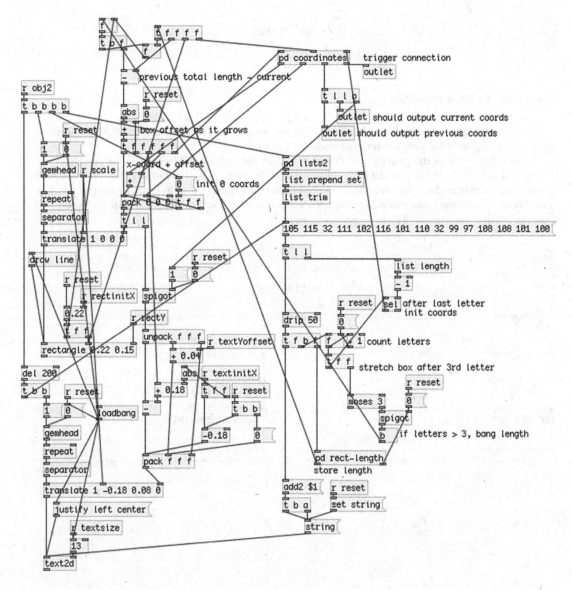

Figure 1-54. *A rather messy patch*

Thankfully, Pd provides a way to connect things without the connection cords, and this makes things a lot cleaner. In the control domain, we can connect things wirelessly using [send] and [receive], abbreviated [s] and [r]. These objects take one argument, which is the name to send to and to receive from. A [send my_send] will be heard by a [receive my_send], as shown in Figure 1-55.

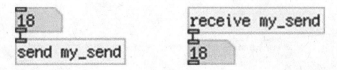

Figure 1-55. *A wireless connection*

You can have as many [send]s and [receive]s with the same name as you want. Mind that whenever we use these objects, we'll use their abbreviated aliases, [s] and [r].

In the signal domain, things are a bit different. There are the [send~] and [receive~] objects (also abbreviated [s~] and [r~]), but you can have only one [s~] with many [r~]s. If you want to send many signals to one destination (the [dac~] for example), you need to use the [throw~]/[catch~] pair, where you can have many [throw~]s with the same name, sending signals to one [catch~] with that name. Figures 1-56 and 1-57 show the [s~]/[r~] and [throw~]/[catch~] pairs, respectively.

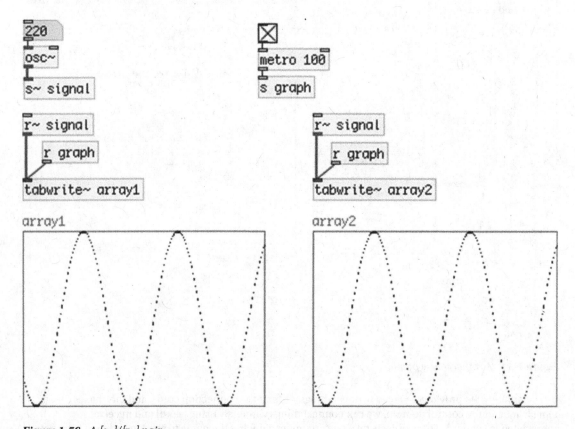

Figure 1-56. *A [s~]/[r~] pair*

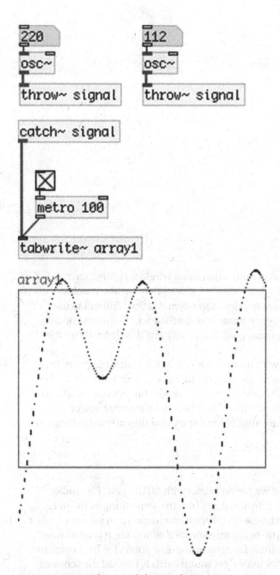

Figure 1-57. *A [throw~]/[catch~] pair*

As mentioned in the additive synthesis patch, signals sent to one destination are being added. You can see that in Figure 1-56, where the graph of the signals exceeds the limits of –1 and 1 (the frame of the graph). When using [throw~]/[catch~] pairs to send sound, make sure that you scale the sum of signals sent to [catch~] to keep them within limits. If the audio signal limits are exceeded, they will be clipped before they reach the sound card. So the graph shown in Figure 1-57 will actually look like the one in Figure 1-58, and the resulting sound will be distorted.

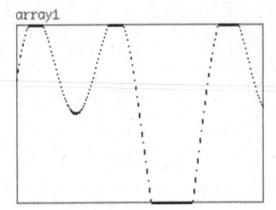

Figure 1-58. *A clipped audio signal*

Before we close this section, I'll note that you must be careful when using wireless connections, as a patch can become difficult to read because the connections are not immediately visible. Wireless connections are there to facilitate us when a patch becomes too dense or when it is very difficult to use wired connections for other reasons. Most of the time, they are much less than the wired ones. A good balance between wired and wireless connections is an optimal goal. But ultimately, this depends on the programming style of each person.

Another thing to note is that when using one [send] with many [receive]s in the control domain, there is no way to force the execution order, as the order the [receive]s were created doesn't have the same effect as the order of the connections we make (data will go to the first connection made, but not necessarily to the first [receive] created). Be very careful when using this pair this way. There is an external object that remedies this situation, [iemguts/oreceive] (iemguts is the name of the library that the external belongs to).

Audio and MIDI Settings

Until now, we have used the default settings for audio, but we haven't dealt with MIDI at all. The audio settings let you configure various settings, like the sample rate, the delay (not the same thing as the delay we saw previously), and the block size. They also let you choose an external sound card and set the number of channels both for input and output. In general, when you make music with a computer, it is advisable to have an external sound card. This is for better performance, for more noise-free audio, for having more than two channels for input and output, and so forth. A discussion on sound cards is beyond the scope of this book, so I'll presume you already have a sound card and take it from there. If you don't have one, it's not really a problem, but read this section because it is helpful for other things concerning your audio settings.

To change the audio settings in Pd, go to **Media ➤ Audio Settings...**. The Audio Settings window will open. There are a few things you can change in this window. The first one is the "Sample rate," which defaults to 44100. We'll leave that as is for now, but if you want to play back audio files that have been recorded in a different sample rate, then you should set that rate here. The next field is the Delay, which defaults to 20 milliseconds in OS X and Linux, and to 100 on Windows. This is a delay set to give time to the computer to do its audio calculations. If you have too much latency with your audio input (for example, the patch with [adc~ 1]), try a smaller number. You should probably set the smallest number that doesn't distort the sound.

To test your latency, go to **Help ➤ Pd Help Browser** and a window will open. In that window, go to **Pure Data ➤ 7.stuff ➤ tools**, and double-click **latency.pd**. A patch will open with instructions on how to use it. This patch calculates the latency created by Pd and reports interruptions and errors (if your system doesn't have enough time to do its calculations). The next field is **Block size**, which defaults to 64. This is the number of samples in each block that the signal objects receive and output. We won't really need a different block size in this book. Note, though, that the smaller the block size, the faster the audio, but the higher the CPU. We'll also leave the **Use callbacks** tick box unticked.

All this leads us to the **Input device 1:** field. If you have no sound card plugged in your computer, Pd will choose the built-in microphone, as shown in Figure 1-59. If you plug in a sound card, you'll need to restart Pd, as it's not going to see the sound card if it is already launched. If your sound card is not chosen by Pd automatically, click the **(0)Built-in Microph** menu and a pop-up menu will appear, where you can choose your sound card (make sure you have the necessary drivers for your sound card, if there are any. You must install them and make any configurations necessary before trying to use it with Pd).

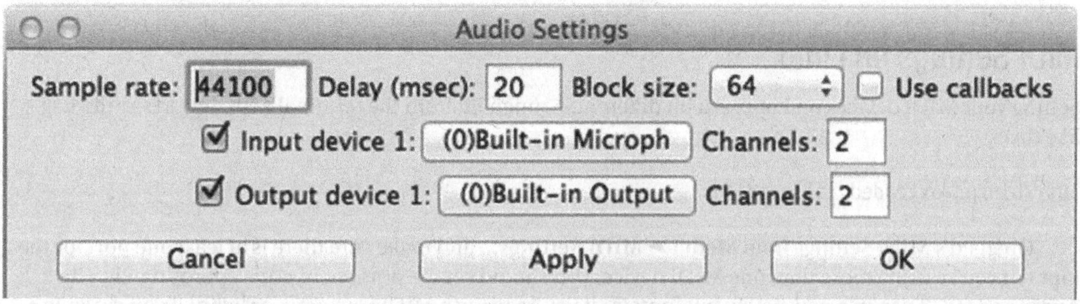

Figure 1-59. The Audio Settings window in Pd

The field below **Input device 1:** is the **Output device 1:**, which sets the output sound card. You can have one sound card for input and another for output, if, for example, you want to use the built-in microphone and use your external sound card's output.

The Channels fields set the number of channels for the input and the output. If you want to set up a quadraphonic system, this is where you'll set the number of channels. In all the projects of this book, we'll use a stereo setup, so we're not going to change these fields. When you make the setting you want, click **Apply** and then **OK**. It will take a few seconds, and the Audio Settings window will close. Your settings are now ready and you can use your sound card with Pd.

To set your MIDI devices, go to **Media ➤ MIDI Settings...** and the MIDI Settings window will open, as shown in Figure 1-60.

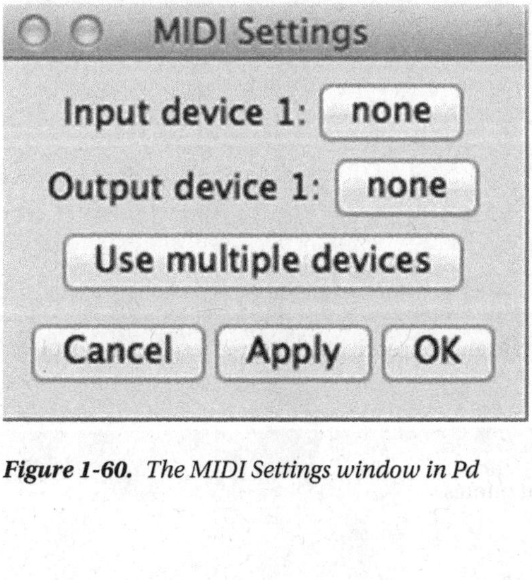

Figure 1-60. The MIDI Settings window in Pd

As with the audio settings, you need to plug in your MIDI devices before you launch Pd, otherwise it won't be able to see them. In the MIDI Settings window, you can choose your input and output devices. We'll barely use any MIDI, only in Chapter 5, where we'll build a MIDI keyboard synthesizer, but we're covering this part in this chapter since you're getting introduced to Pd. Usually, we use only input devices, like MIDI controllers, keyboards, and so forth, but you might also want to control a MIDI device from Pd (actually, Pd's predecessor, Max, was initially made—by Miller Puckette as well— to control a hardware MIDI synthesizer).

In the MIDI Settings window, again you have two fields, **Input device 1:** and **Output device 1:**, like in the Audio Settings. This is where you choose your device. If you have more than one device, first click the **Use multiple devices** button and you'll be able to choose more than one device both for input and output. When you select your devices, click **Apply** and then **OK** for your settings to be activated.

MIDI Settings on Linux

Setting your MIDI devices on Linux is a bit different. Launch Pd from the terminal with the -alsamidi flag, like this:

```
/usr/bin/pd-extended -alsamidi &
```

Open Pd's MIDI Settings from **Media ➤ MIDI Settings...** and make sure there is at least one port for the input. If you're using more than one MIDI device, you should set the number of ports appropriately. Click **Apply** and **OK**. Then in a new patch, put [notein] if you're using a MIDI keyboard, or [ctlin] if you're using a controller with potentiometers. Open its help patch and see if you get input from your device. If you don't get input, go back to the terminal and type:

```
aconnect -lio
```

This will print the available MIDI devices and software currently plugged in and running on your computer. In my computer I got the following:

```
client0:'System'[type=kernel]
    0'Timer            '
    1'Announce         '
client14:'MidiThrough'[type=kernel]
    0'MidiThroughPort-0'
    ConnectingTo:128:0
     Connected From: 128:1
client 20: 'nanoKEY' [type=kernel]
    0 'nanoKEY MIDI 1  '
    Connecting To: 128:0
    Connected From: 128:1
client 128: 'Pure Data' [type=user]
    0 'Pure Data Midi-In 1'
    Connected From: 14:0, 20:0
    1 'Pure Data Midi-Out 1'
    Connecting To: 14:0, 20:0
```

What I get is that I have a Korg nanoKEY sending to port 0, and Pd receiving in port 0 (it sends to port 1). The following example connects the nanoKEY to Pd:

```
aconnect nanoKEY:0 'Pure Data':0
```

Pd's name is two words, so you need to place it without quotes.

A Bit More on MIDI

Since we're talking about MIDI, let's talk a bit about how to receive input from various MIDI devices. Data from MIDI keyboards can be fetched with [notein]. If you provide no argument to it, it will have three outlets and no inlets. The outlets from left to right are the MIDI note number, the velocity, and the MIDI channel number. If you know the channel, you can set it via an argument, and [notein] will have only two outlets, for the first two. We'll see this object in more detail in Chapter 5.

Controllers with sliders and potentiometers send Control Change messages, which can be retrieved with [ctlin], which stands for *control in*. Open its help patch and use your already set controller to see the input it gives. Again this object has three outlets and no inlets, and the outlets give the controller value, the controller number, and the channel number. The last two can be set via arguments. Check the help patch for more information. Also, click the [pd Related_objects] subpatch to see all the MIDI objects available in Pd.

Additional Thoughts

Before we move on to the next chapter, I'd like to give a few tips concerning Pd programming. First of all, make sure that you save your patches in an organized way. Some people prefer to save files depending on the project; others prefer to have a tree structure of the files in a single program or programming environment. You might prefer a different way. The more you deal with Pd, the more files you'll save on your computer. Whether you are a hobbyist or you want to make things with Pd professionally, it's best that you find a way to organize your patches that suits you well.

Make sure that your patches are clean. Placing objects here and there will most likely create a chaotic patch. Try to have your objects aligned as much as you can. Make use of subpatches, and whenever applicable, abstractions. Try to make your patches self-explanatory (by giving describing names to subpatches, for example), and wherever this is not possible, use comments to describe what happens in that specific part of the patch. Even if you know what each object does in a patch, if there are not enough comments (or no comments at all), it might still be very difficult to understand what happened. This applies to your own patches too; when you come back to them after some time (even a couple of weeks can be enough, to make things confusing).

Use [trigger] whenever necessary. This means that wherever one piece of data goes to more than one destination, and the order matters, always use [trigger]. Not using it will very likely create bugs that are very hard to trace. Also, it makes the understanding of a patch very difficult, or even impossible. We haven't seen all features of this object yet, as apart from forcing the order of execution, it also converts data types (it can convert a float to a bang, for example), so in the chapters that follow, you'll see that it is a very important and helpful object.

Try to build simple things and to make them more complicated as you go. Trying to create something complex and beyond your skills will cause confusion and disappointment. With programming, you can create amazing things, but it takes time to handle a programming language. Take one step at a time, and try to take joy even with very simple programs that you build. As already mentioned, learning Pd is a matter of personal practice, and if you practice it frequently, you'll find yourself building rather complex programs before you expected.

I should mention that at the time of writing (August 2015), Pd-extended is not being maintained. Pd-vanilla, on the other hand, is being actively maintained by Miller Puckette, with new features added and new versions released frequently.

Pd's community is very active too, with lots of developers running and sharing their own projects. There has been a discussion among the community as to what route should be taken to either revive or replace Pd-extended. The prevailing idea for now seems to be one that centralizes external objects in a repository that you can pull from and add to your Pd-vanilla. Since Pd is open source, this kind of issue is likely to arise sometimes. But with an active community like the one around Pd, there's no fear that all issues will be solved. Still, the current version of Pd-extended is fully functional and can be used at both an amateur and a professional level. If you keep using Pd, it' good to stay up-to-date as to how it is going to be maintained in the future. You can place yourself on Pd's mailing list, which you can find on its web site, or you can sign up on the Pd forum at `http://forum.pdpatchrepo.info/`.

Conclusion

This concludes the first chapter and the introduction to Pd, a very flexible and powerful programming environment. In this chapter, you have been introduced to Pd, its philosophy, some of its features, capabilities, but also to some basic techniques of electronic music in general. Learning how to use Pd is a matter of personal practice, though.

Later on in this book, we're going to build more complex patches to make musical programs. What this book will try to provide is basic knowledge on the tools used, but also ways to research when you want to realize a personal project. The musical projects built here are limited, so they cannot meet every musician's needs. What they can do is give inspiration and insight to musicians so that they can realize original projects of their own. The main focus in this book is to combine the physical world with that of the computer to make musical interfaces. The basics of this communication will be covered to such an extent, that you will be able to use these tools in many different ways, much more than the ones shown here.

Next are the Arduino basics, where you will be introduced to its language and some simple circuits, and to the communication features between Arduino and Pd.

CHAPTER 2

■ ■ ■

Introduction to Arduino

In this chapter, we'll be introduced to the Arduino prototyping platform. As with the previous chapter, if you're already using Arduino, and you're programming it yourself, feel free to skip this chapter. Mind you, that apart from the Arduino language itself, we'll also focus on its serial communication capabilities, in combination with Pd. This means that we'll be using both ways of serial communication (`Serial.println()` and `Serial.write()`), and we'll analyze the way they work, their differences, as well as their advantages and disadvantages compared to one another.

By the end of this chapter, you'll be able to

- Write simple programs for the Arduino for your physical modeling projects

- Use looping mechanisms to facilitate your coding

- Understand the way serial communication is achieved, and which way to choose when

- Use the Arduino in combination with Pd and take advantage of each platforms capabilities

Arduino Jump Start

The Arduino board is a microcontroller that takes input from the physical world, using various sensors, and uses it in computer programs. It can be used as a stand-alone application, but also in combination with a computer to realize things that are more complex. The communication between the physical world and the computer goes also the other way round. The Arduino can give input to the physical world, by using LEDs, lights, motors, solenoids, and so forth. The Arduino is also a programming environment and a programming language. The language is built on C++, but has a great set of its own functions. The third element that comprises the Arduino in its entirety is its community. It has a large community of users, makers, developers, enthusiasts, that share work and projects between them. It is very similar to the Pd community, as they are both open source and widely used.

In contrast to Pd, Arduino is a textual programming language, but a very intuitive one. It also runs on all three major operating systems, like Pd, and its software is for free. Being open source, its hardware is open as well. All the schematics and circuit designs are open for anyone to use. So, if you have the facilities, you can build one yourself. That is a difficult task though, and you are encouraged to buy an original Arduino from your local reseller, or from their web site.

What makes the Arduino so special is not that it's a microcontroller that uses sensors, or that it can communicate with a computer to give input from or to the physical world, but the fact that it has been packaged with its software in such a way that it makes physical computing (the communication between the computer and the physical world) much easier than ever before. Microcontrollers are said to be a very difficult field in programming, but the Arduino is very simple to program. Also, the way it is build, facilitates prototyping to a great extent, where you can plug in a few sensors and start using them in a matter of a few minutes. The Arduino has actually revolutionized the way we use microcontrollers and the contributed in the expansion of the maker communities worldwide.

To follow this chapter and the rest of this book, you'll need to buy an Arduino board, and download the software. Go to its web site at www.arduino.cc, get the Arduino IDE (Integrated Development Environment, the Arduino software), and find your local distributor. At the time of writing (August 2015), there are issues within the Arduino team. This means that if you buy an Arduino outside the United States, it will be called Genuino. The name is the only thing that changes, the rest remain the same. This change applies for a few boards, the UNO, the MICRO, and the MEGA. The NANO and the PRO MINI, shouldn't be affected. These issue should actually be solved with the appearance of the Genuino. Still, we'll refer to it as Arduino, and we'll mean both boards, since they are essentially the same.

With Arduino, it is advisable to get the UNO, as it is designed for prototyping. In this chapter, this is the Arduino we'll use. In the chapters that follow, we'll use other types of Arduino, like the NANO and the PRO MINI, as they are much smaller, so not so good for prototyping, but perfect for being embedded in a project. Their cost is rather low, maybe the NANO is a little bit more expensive, so it shouldn't be very difficult to get one of each. Don't bother to buy them all yet, if you want to build a project in this book that requires another type that you don't have, then go get one.

Along with the Arduino, we'll be using some peripherals, like LEDs, switches, potentiometers, and so forth. Each project will needs its own peripherals, so you should probably get them as you go. In this chapter, we'll use peripherals for prototyping, so instead of potentiometers, we'll use trimmers (these are breadboard-friendly potentiometers). These prototyping peripherals will be helpful for many more projects, as when building an electronics project, we first prototype and them start building.

At this point, I should mention that there is a rather easy way to use the Arduino, if you're already using Pd. That is the Firmata library, which lets you program the Arduino through Pd (or other programming environments). Since we'll be using some built-in functions of the Arduino language, using Firmata here won't really help, so we're not going to use it at all. Instead, we're going to write our own small programs and restrict the Arduino to the few simple things we need to use. This way we'll get a better understanding of its language, the serial communication, and how it is combined with Pd.

Parts List

In this section, we'll review the parts you'll need to build all the projects of this chapter. Table 2-1 shows what each project will use. In addition to that, you'll need some jumper wire (make sure that you get a few), a breadboard (a half size will probably do, but a full size won't be bad, as it will prove useful for future projects too), and of course, an Arduino Uno and a USB cable.

Table 2-1. *Parts List*

Project	LEDs	Push buttons	Potentiometers	Resistors
1	1	0	0	0
2	0	1	0	$1 \times 10K\Omega$
3	1	1	0	$1 \times 220\Omega$
4	0	0	$1 \times 10K\Omega$	0
5	1	0	$1 \times 10K\Omega$	$1 \times 220\Omega$
6	0	0	$3 \times 10K\Omega$	0
7	0	3	$3 \times 10K\Omega$	0
8	3	0	0	$3 \times 220\Omega$

Make sure that the push buttons you get are breadboard-friendly (also called *tactile switches*), as well as the potentiometers (these one are also called *trimmers*). The resistors are counted in ohms, so a 10KΩ resistor is 10 kiloohms, and a 220Ω is a 220-ohm resistor.

The Blink Sketch

Before we start looking at Arduino code, make sure you have yourself an Arduino, preferably the UNO, so you can realize all the programs in this chapter. Figure 2-1 shows an Arduino UNO. The chip in the middle of the board is the actual microcontroller, an Atmel ATMEGA 328. We can also see a USB socket that we'll use to connect it to our computer. There's also a power JACK socket on the same side with the USB, but since we'll use the Arduino always in combination with a computer, we won't need that, as it will be powered through the USB. On the sides, we can see a few sockets with some indications on them. These are the pins to which we'll be attaching sensors, LEDs, and so forth. There are both analog and digital pins, for the corresponding sensors. On each project, there will be a diagram of the circuit, so it will be easy to follow.

Figure 2-1. *Arduino UNO*

When learning programming, usually the first task is to print "Hello, World!" to a monitor. In Pd, the first thing we did was to output a sine tone at 440 Hz (that is the usual case when learning audio programming). When learning how to program the Arduino, we usually make an LED blink. In electronics, making LEDs blink is the very basis. It is said that, if you can make an LED light up, you can do anything. So what we'll do first in this chapter is to make an LED blink. The Arduino IDE has a sketch (this is how we refer to Arduino code) that does exactly that.

Go ahead and launch the Arduino IDE. What you'll get at the beginning is a new sketch window, like the one in Figure 2-2. In contrast to Pd, this window is not totally empty. First of all, at the very top, it writes "sketch_aug07a | Arduino 1.6.5". This is a default sketch name given by the IDE. "aug" stands for August (all this is written in August), "08" stands for the eighth of the month, and "a" stands for the first sketch of the day. If you reach the limit of the Latin alphabet, you'll get a window saying, "You've reached the limit for auto naming of new sketches of the day. How about going for a walk instead?" and it won't let you create a new sketch. Just restart the application and it will work again. I'm pretty sure that you won't reach that limit unintentionally (I kept on creating new windows till I got to the end, just to see what happens). The second part of the top line is the version of the IDE you're using.

Figure 2-2. *A new sketch window*

On the very bottom of the window, on the left side you can see the line in your sketch where the cursor currently is. On the right side, you see the selected Arduino board and its serial port (can't really remember if there was any serial port when I first installed the IDE and opened a new window without having set a port yet). In the case in Figure 2-1, it's an Arduino Uno, on port /dev/cu.usbmodem411 (this is on OS X). Later on, we'll talk about all this in more detail.

In the window, we see a little bit of code. The first line reads void setup() {. This is a built-in function that runs once as soon as the Arduino is powered. We won't bother with the word void for now. setup is the name of the function. When we program in C++, we can create our own functions, and we have to give them a name. Think of it a bit like an abstraction in Pd. The parenthesis are obligatory when writing a function. They are there in case the function takes arguments, and even if it takes no arguments, you must still include them. After the parenthesis, there is an opening curly bracket. When we define a function, its code is included in curly brackets, and we can see the closing bracket in line 4. Whatever is written inside these brackets is the code of the function, which will be executed when we call that function (for the setup function, as soon as the Arduino is powered). By "define," I mean to write the code of the function. Both setup and loop are not defined by default, but are there for us to define them any way we want.

Line 2 has a comment. This is like the comments in Pd, they are there to give us information, and they don't affect the program at all. The compiler (the compiler is the program that turns code into an executable program) will ignore all comments when it will compile the code. This is a single line comment and it must start with two forward slashes. The comment reads "put your setup code here, to run once:". This actually tells us what really happens with this function, it runs only once, when the Arduino boots.

In line 7, we read void loop() {. This is another built-in function of the Arduino language and runs immediately after the setup function, over and over again, hence its name loop. Again we have the parenthesis, since it's a function, and the curly brackets, because it hasn't been defined yet. Inside it, we read the comment, "put your main code here, to run repeatedly:". This is where we'll be writing most of our code, and this will run for as long as the Arduino is powered.

The different colors for various keywords of functions and others, are there to facilitate the reading and writing of code. Most of IDEs have color highlighting for this reason. In the Arduino language, the blueish color of void is the color for data types (like integer, float, byte, etc.; you'll see them later on). The color of setup and loop is the color for these two functions and all control structures (if, for, while, and others). All comments are grey and all defined functions (not setup and loop) are orange. We'll see all this as we read further on.

Now go to **File ➤ Examples ➤ 01.Basics**, and click **Blink**. This should open a new window with the code in Listing 2-1 in it.

Listing 2-1. The Blink Sketch: the Equivalent to the "Hello World!" Program in Most Programming Languages

```
1.     /*
2.      Blink
3.      Turns on an LED on for one second, then off for one second, repeatedly.
4.
5.      Most Arduinos have an on-board LED you can control. On the Uno and
6.      Leonardo, it is attached to digital pin 13. If you're unsure what
7.      pin the on-board LED is connected to on your Arduino model, check
8.      the documentation at http://www.arduino.cc
9.
10.     This example code is in the public domain.
11.
12.     modified 8 May 2014
13.     by Scott Fitzgerald
14.    */
15.
16.
17.    // the setup function runs once when you press reset or power the board
18.    void setup() {
19.      // initialize digital pin 13 as an output.
20.      pinMode(13, OUTPUT);
21.    }
22.
23.    // the loop function runs over and over again forever
24.    void loop() {
25.      digitalWrite(13, HIGH);   // turn the LED on (HIGH is the voltage level)
26.      delay(1000);              // wait for a second
27.      digitalWrite(13, LOW);    // turn the LED off by making the voltage LOW
28.      delay(1000);              // wait for a second
29.    }
```

This is the first Arduino sketch we'll upload to our board. To get a grasp of how things work with the Arduino, we'll go through its code in detail, step by step. The code here has all lines numbered for convenience, but the Arduino IDE doesn't show these line numbers, only the number of the line where the cursor is, at the bottom of the window. The first fourteen lines of code are a multiline comment. To make a multiline comment in Arduino, start it with a forward slash and an asterisk, and end it with an asterisk and a forward slash. Whatever you write in between will be ignored by the compiler. This specific comment tells us what this sketch does, gives some information about the integrated LED on the Arduino and some other meta-data.

After the multiline comment, we have a single line comment giving information about when the setup function runs, and then we have the actual setup function. In line 19, we have a comment: "initialize digital pin 13 as an output". Indeed, line 20 does exactly that. Since we'll use an LED and we will be turning it on and off, the pin we'll use for this is a digital pin, because it has two possible states only, on or off (same as 1 or 0). Also, this pin will output voltage to the LED, so it must be an output pin. Digital pins can be either input or output, whereas analog pins are only input. This line also shows how intuitive the Arduino language is. The function to set a pin either as input or output—in other words, to set the mode of the pin—is called pinMode, and you can tell that it's a function by the parenthesis after its name. This function is part of the core of the Arduino language and has already been defined. That's why it has no curly brackets, like setup and loop, and you can't write code in it, but only use it as is. This is the first predefined function we encounter and we can see that it is color highlighted in orange, as mentioned earlier. pinMode takes two arguments, the pin to set the mode to, and the mode to set to that pin. The pin we'll use is pin 13, because that's the pin Arduino Uno has an integrated LED on. For the second argument, we use a keyword of the Arduino language, OUTPUT. This keyword (the language is case sensitive, so output won't work) tells pinMode to set the specified pin as an output. Mind the semicolon (;) at the end of the line. pinMode is a predefined function, so we must put a semicolon whenever we call it. All executable lines of code in C++ end with a semicolon. This tells the compiler that this is the end of the line and that this line must be executed. Functions that are being defined (like the setup and loop) don't take a semicolon, but all code written inside them does.

Our setup function consists of one line of code only (plus a single line comment). Line 21 closes the brackets of the setup function and we move on to the loop function. Notice how all code within setup is indented. The same happens with loop too. All functions and control structures have their code indented, for readability. You'll see that it is a very helpful feature. In the Arduino IDE, when you open a curly bracket and hit enter, the code automatically gets indented.

In our loop function, we can see four lines of code, where we call two functions, two times each. Both these functions are predefined, like pinMode. The first one, on line 25 is the digitalWrite function. As its name states, this function writes a value to a digital pin. Like pinMode, this one also takes two arguments, the pin to write a value to, and the value to write to that pin. In this case, we want to turn the LED on pin 13 on. Again, we provide the number 13 for the pin number. On can be indicated by the keyword HIGH. HIGH stands for high voltage. Sending voltage to that pin, will turn the LED on. Mind the semicolon after the function call.

We can see that after the semicolon there is a single line comment. The comments we've seen so far start at the beginning of the line, but single line comments can start at any point of a line. The compiler will compile the line up to the point of the two forward slashes, after which point it will ignore everything. This comment tells us what this line of code does, which is to "turn the LED on (HIGH is the voltage level)".

The next line reads delay(1000);. delay is another built-in, predefined function. What it does is delay the rest of the program by a specified amount of time, which we set via its argument (delay takes one argument only), in milliseconds. This line of code will delay the rest of the program for 1000 milliseconds, or for 1 second. After we call that function, we put a semicolon to let the compiler know we're done with that line, and then there is a comment saying that we'll wait for a second.

After delay, we call digitalWrite again, but this time we set the pin LOW, meaning we drop the voltage low, so the LED will turn off. The comment after the semicolon explains that as well. Lastly, we delay our program for another second by calling delay(1000); as we did before. And this concludes our first Arduino sketch. In line 29, we put the closing bracket for the loop function (no semicolon here as we are defining the function).

What you might have noticed is that in textual programming the code is being executed line by line, the same way we read it. In visual programming, we can see the data flow by the connections between the objects, in textual programming the data flow is being defined by the position of each line of code. Since I have explained the Blink sketch, let's upload the code to our Arduino board, to see it in action. Plug in your Arduino and go to **Tools ➤ Board:** and you'll get the menu shown in Figure 2-3.

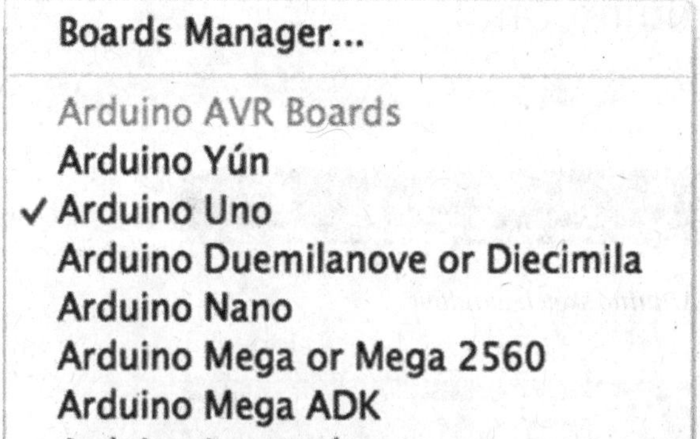

Boards Manager...

Arduino AVR Boards

Arduino Yún

✓ Arduino Uno

Arduino Duemilanove or Diecimila

Arduino Nano

Arduino Mega or Mega 2560

Arduino Mega ADK

Arduino Leonardo

Arduino Micro

Arduino Esplora

Arduino Mini

Arduino Ethernet

Arduino Fio

Arduino BT

LilyPad Arduino USB

LilyPad Arduino

Arduino Pro or Pro Mini

Arduino NG or older

Arduino Robot Control

Arduino Robot Motor

Arduino Gemma

Figure 2-3. *The Boards menu on the Arduino IDE*

You can see that there are many different boards supported by the IDE. If the Uno is not already selected, go ahead and click it (if you're using another board, click that). The menu will close, but the board will have been selected. Once you've done that, you must select your port. Go to **Tools ➤ Port** and a menu with all available ports will open. The Arduino port should have an indication like the one in Figure 2-4. On OS X the port is /dev/cu.usbmodemx,[1] on Linux it's /dev/ttyACMx, and on Windows it's COMx, where x is a number. Select your port (again, the menu will close, but the port will have been selected) and you're ready to upload the sketch to your board. On the top of the sketch window, there are a few icons, as shown in Figure 2-5. The icon with the arrow inside it is the **Upload** button. Click it and the IDE should start uploading the sketch to the board. Before it uploads the code to the board, it will first compile it, and on the bottom of the sketch window, you'll see the compilation process progress, shown in Figure 2-6. When the code has been compiled, the IDE will start uploading it to the board, and now you'll see the upload progress, shown in Figure 2-7.

---

[1]On OS X there's a /dev/tty.usbmodemx and a /dev/cu.usbmodemx. It used to be the tty. but the latest IDE hides it. Selecting the cu. one is exactly the same.

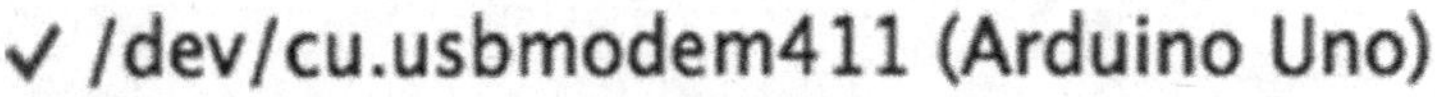

✓ /dev/cu.usbmodem411 (Arduino Uno)

Figure 2-4. *Indication of the Arduino Uno port*

Figure 2-5. *Verify, Upload, and other choices on the Arduino sketch window*

Figure 2-6. *Compilation process progress on the IDE window*

Figure 2-7. *Upload progress on the IDE window*

When the uploading has finished, on the bottom of the window you'll read: "Done uploading." On your Arduino board, you should see an LED blinking, turning on for one second, and off for another second, repeatedly. Congratulations! You've uploaded your first Arduino sketch!

Before we move on and start writing our own code, let's use an external LED with this sketch. Usually when we use LEDs, we also need to use resistors, because the voltage supplied by the Arduino is too much for an LED, and most likely it will be burned, if there's no resistor. Pin 13 on the Arduino Uno has an integrated resistor, so to make this sketch work with an external LED, there's no real circuit you need to build yourself. An LED has two legs, one long and one short. The long one takes voltage, while the short one connects to ground. By ground in electronic circuits, like the ones we'll build with the Arduino, we usually mean zero volts. Even though there's no connection with the real ground (the earth), we still call this ground. Figure 2-8 shows how you should connect the LED to your Arduino.

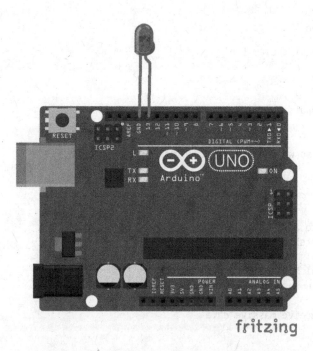

Figure 2-8. *LED connected to digital pin 13 and Ground*

It is very convenient that there is a ground pin next to digital pin 13, so we can insert the LED straight into the pin sockets (these are called *headers*, and that's how we'll refer to them from now on). Once you plug your LED into the Arduino, you should see it blinking along with the integrated LED that is already blinking. Now let's write our own code to the Arduino!

Digital Input

From this point on we'll start needing some components to build the circuit of each project. The parts for this project are shown in Table 2-2.

Table 2-2. *Project 2 Parts List*

Part	Quantity
Push buttons	1
Resistors	1 × 10KΩ

Since you saw how we give digital output, now we'll receive some digital input. To do this we'll need a switch that we can read from the Arduino (actually, we'll use a momentary switch, essentially a push button). We'll connect the switch to a digital pin, which we'll configure as an input pin, and we'll read whatever the switch gives in the Arduino serial monitor. Open a new sketch window using the same shortcut you opened a new window in Pd with: **Ctrl/Cmd+N**. Listing 2-2 shows the code you should write in the new window.

Listing 2-2. Receiving Digital Input in the Arduino

```
1.    // set a global variable for the pin of the switch
2.    int switch_pin = 2;
3.
4.    void setup() {
5.      // initialize the switch pin as input
6.      pinMode(switch_pin, INPUT);
7.
8.      // start the serial communication so we can see the readings in our computer
9.      Serial.begin(9600);
10.   }
11.
12.   void loop() {
13.     // store the state of the switch to a variable
14.     int switch_state = digitalRead(switch_pin);
15.
16.     // print the variable to the Serial Monitor
17.     Serial.println(switch_state);
18.
19.     // short delay so that we don't receive massive data
20.     delay(250);
21.   }
```

Defining Variables in Arduino

Line 1 says that we'll "set a global variable for the pin of the switch", and we do exactly that on line 2. This line sets a variable of type int, called switch_pin, and assigns the value 2 to it. An int in the Arduino language is an integer (a value with no decimal point), that is two bytes long (it can hold values from –32,768 up to 32,767). In Pd, we didn't deal with different data types because all numbers are actually floats. In Arduino (and C/C++ programming), we must always define the type of the data whenever we create a new variable. Sometimes we might need an integer, other time we might need a float, also we might need two bytes, or four, or one. Therefore, we must always define the data type. The syntax of line 2 is the one we use when we create a new variable, which is: *data type, identifier, value assignment.* The last part (where we assign a value to the variable) is not mandatory, but the first two are. In this specific sketch, using a variable is not really necessary, as with the Blink sketch, but it's good practice to use it, so we can get the hang of it. Also, this variable is called *global* because it is defined outside any function, at the top of the sketch, therefore is accessible by any function (note that this variable is called by both setup and loop). If it were defined inside one of the two functions, then the other function wouldn't have access to it, and it would be called a *local variable.* We'll see more of these as we write more code.

Further Explanation of the Code

Now that I've explained the first two lines of code, which are not specific to this sketch only, let's move further. Note that we use digital pin 2, and not 0 (pins start counting from 0), because digital pins 0 and 1 are used for receiving and transferring data, so we start using pins from 2 onward.

In line 4, we define our setup function. Like with the Blink sketch, we call the pinMode function to set the mode of the pin we'll use. This time, since we'll be receiving input from the Arduino, we set the pin as INPUT. With the sketch we'll also need to have serial communication, so the setup function goes on to set that as well. Line 8 contains the comment "start the serial communication so we can see the readings in our computer" and line 9 calls the begin function of the Serial class.

Classes in Arduino and the Serial Communication

We can tell that Serial is a class, because there is a dot between it and its function that we call, begin. The class name is also in bold letters (not on OS X), which declares that this is a class. A class in C++ is a set of functions and data that comprise a user-defined data type. They are there to make things easier when coding, as they are actually a package of methods that we often use. It is different than an abstraction, because the abstraction is a single function, we write once and use lots of times, whereas a class packs many functions together, along with its own data. It's not really necessary to grasp what a class is in C++, the details provided are there just to give some information. To come back to our code, line 9 begins the serial communication between the Arduino and the computer at the rate of 9600 bits per second, which is set via the argument of begin. The communication is called *serial*, because the bits come in the communication line is series, one at a time.

Further Explanation

Line 10 has the closing curly bracket of the setup function, which is there by default, when you open a new window in the Arduino IDE. Curly brackets are necessary for function definitions and control structures, which we will see later on. Forgetting to include one will create an error message at the bottom of the sketch window and the code will fail to compile. Luckily, when you put an opening curly bracket in the Arduino IDE, and hit return, it automatically inserts the corresponding closing bracket, so it's almost impossible to forget it and cause an error.

In line 12, we start our loop function. Line 14 creates a new variable of type int, called switch_state, and assigns to it the value returned by the digitalRead function. digitalRead is the counterpart of the digitalWrite function, and as its name states, it reads the value of a digital pin. This function takes one argument only, which is the pin to read from. Compare line 14 to line 2. They are very similar, only in the case of line 2, we assign a fixed value to our variable, whereas in the case of line 14 we assign a different value every time, the one read and returned by the digitalRead function. The argument we provide to digitalRead is the variable that holds the number of the pin we'll attach the switch to, defined in line 2. Using variables with names that make sense, make our code self-explanatory and easier to read. Line 14 should be fairly easy to understand without explanations.

Line 17 calls the println function of the Serial class. Like with the begin function, we must include the class name and place a dot between it and the name of the function, like this:

```
Serial.println(switch_state);
```

This function prints whatever is provided inside its parenthesis, to the serial port. In this case, it prints the value stored in the switch_state variable. For this sketch, we'll use the serial monitor of the Arduino IDE to see what the Arduino prints. Later on, we'll start receiving data in Pd, which is our goal.

Finally, in line 20 we delay our program by 250 milliseconds, so that we don't get massive amounts of data in the serial monitor. This delay is only for this reason, when we'll use Arduino with Pd, we won't be using these delays.

Building Circuits on a Breadboard

Before we check the circuit for this sketch, I'll explain what a breadboard is and how it works. A breadboard is a board that facilitates testing circuits a lot. It has small holes where jumper wires fit, and these holes are connected in a certain way to help connect the wires to other parts, like resistors, push buttons, LEDs, and so forth. Figure 2-9 shows the wiring of a small breadboard. On the top and the bottom, there is a blue and a red line. The black wire that goes along the blue line shows that all holes along the wire are connected with each other. So if you plug in a wire at one end of this line, and another wire at the other end, these wires will be connected. The same goes for the red wire along the red line, and this applies to both top and bottom. The green and yellow wires show how the holes are connected in the inside part of the breadboard. Up until the notch in the middle of the board, the lines are connected vertically, as we see the board in Figure 2-9. Building the circuit of this sketch will help you understand how the breadboard works.

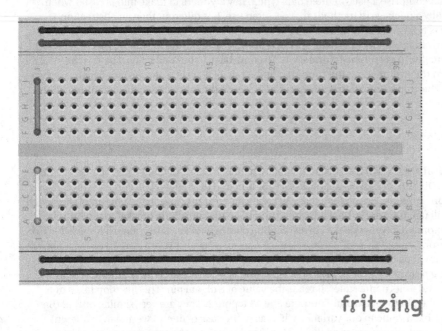

Figure 2-9. *Breadboard wiring*

The circuit is shown in Figure 2-10.

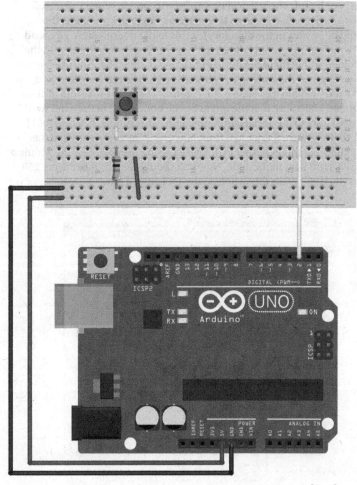

fritzing

Figure 2-10. *Digital input circuit*

The push button of the circuit has two pairs of two connected legs. The legs on the left side are connected between them, and the legs on the right side too. The two sides are not connected until we press the button. One of the two leg pairs connects to 5V (this means five volts), and the other leg pair connects to one leg of the 10kΩ resistor (it doesn't matter which one), and the other leg of the resistor connects to ground (anywhere along the bottom blue line). The resistor applies some resistance to an electrical current. The amount of resistance it applies is expressed in ohms. If we don't use a resistor in our circuit, as soon as we press the switch, we'll actually connect 5V to ground, creating a short circuit. The same leg pair of the button that connects to the resistor also connects to the digital pin 2 of the Arduino (it's the pin we've set as input and the one we're reading in the Arduino code). We could have connected the resistor straight to the ground pin of the Arduino (GND pin), and the right leg pair of the button straight to the 5V pin, but providing voltage and ground to the board is good practice for later projects where we'll have more components requiring voltage and ground. Also, traditionally, we use black wire for ground and red for 5V.

Go ahead and upload the sketch to your board. The IDE will prompt you to save it. When you install the Arduino IDE, it automatically creates a folder called Arduino to the Documents folder (on Linux the directory is called Sketchbook and should be in your home directory). If this folder doesn't exist, go ahead and create it. Save the sketch with the name Digital_input. Check if it's saved. You'll see a folder with the name Digital_input in your Arduino folder, and in there the file Digital_input.ino. The .ino extension is for files read by the Arduino IDE. Once you upload the code to your board, open the serial monitor. To open it, click the rightmost icon on top of the window, shown in Figure 2-5. Figure 2-11 shows the readings of the switch being printed onto the serial monitor. Make sure that the menu on the bottom left of the window reads "9600 baud". This is the baud rate we've set to the Serial communication with Serial.begin(9600); (baud rate is the rate of bits per second, that's how we'll refer to it from now on). Also, make sure that the menu next to it reads "Newline" (I'll explain what this is further on). These two are necessary for the Arduino to print to the monitor properly, since we've set them in the code. Now you should see a number every 250 milliseconds, which should be a 0 when you don't press the switch, and a 1 when you press it.

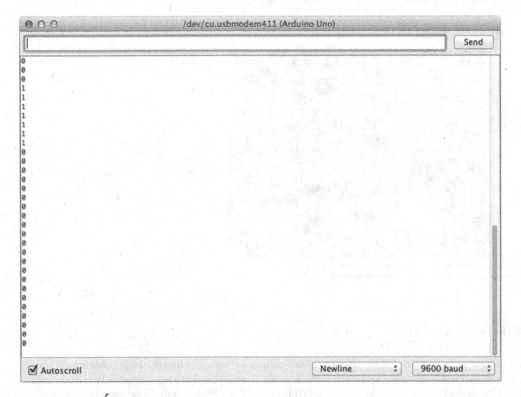

Figure 2-11. *Arduino's serial monitor*

Pull-up vs. Pull-down Resistors

There's one last thing I need to explain before we move on to the next sketch. The resistor used in this circuit is called a *pull-down resistor*, because it connects the switch to ground. If instead we reverse the connections, so the resistor connects to 5V (any hole along the red line), and the right leg pair of the push button connects to ground (any hole along the blue line), the resistor will be a pull-up resistor. This will create a reversal in the readings of the switch, meaning that the Arduino will print a 1 when you don't press the switch, and a 0 when you press. This is a bit counter intuitive, but it is said that pull-up resistors are more stable in a circuit, than pull-down. Apart from that, all pins in the Arduino have internal pull-up resistors, which are disabled. To enable a pull-up resistor in a pin, we must call the pinMode function. Open a new sketch window and copy the previous code to it (if you change the code in the previous sketch and upload it, the IDE will automatically save it). In the new sketch change line 6 to this:

```
pinMode(switch_pin, INPUT_PULLUP);
```

Also, change your circuit to the one in Figure 2-12. Using Arduino's internal pull-up resistors reduced the circuit we need to build a bit. This will come in handy when we'll start building circuits on a perforated board, as it will reduce the amount of soldering to a great extent. These resistors are 20kΩ, but they're still good for us to use with switches. Don't confuse them with the internal resistor on pin 13, which we used with the Blink sketch. That resistor connects pin 13 of the processor to the header where we attached the LED, whereas the pull-up resistors connect the pins of the processor to 5V. We first built the circuit with an external resistor, to clarify how the actual circuit works, because if we used the internal one straight away, you probably wouldn't understand the circuit the same way.

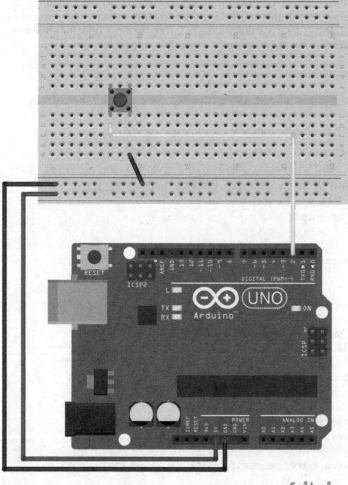

Figure 2-12. *Digital input with internal pull-up resistor enabled*

Both Digital Input and Output

A logical next step would be to combine the two sketches we've already analyzed. What we'll do is use the switch both for visualizing it in the Arduino serial monitor, but also to control an LED. Table 2-3 shows the parts needed to realize this project.

Table 2-3. *Project 3 Parts List*

Part	Quantity
Push buttons	1
LEDs	1
Resistors	$1 \times 220\Omega$

This time we'll use another pin for the LED, that doesn't have an internal resistor like pin 13 (unlike the pull-up resistors, only pin 13 has an internal resistor that can be used with an LED). Here we'll also see why it is good practice to store readings in variables, as we'll use the reading of the digital pin 2, both for projecting in to the serial monitor, but also for controlling the LED. Listing 2-3 shows the code you should write.

Listing 2-3. Digital Input and Output Sketch

```
1.   // set a global variable for the pin of the switch
2.   int switch_pin = 2;
3.   // set a global variable for the pin of the LED
4.   int led_pin = 8;
5.
6.   void setup() {
7.     // initialize the switch pin as input with the internal pull-up resistor
8.     pinMode(switch_pin, INPUT_PULLUP);
9.     // initialize the LED pin as output
10.    pinMode(led_pin, OUTPUT);
11.
12.    // start the serial communication so we can see the readings in our computer
13.    Serial.begin(9600);
14.  }
15.
16.  void loop() {
17.    // store the state of the switch to a variable
18.    int switch_state = digitalRead(switch_pin);
19.
20.    // write the reading of the switch to the LED
21.    digitalWrite(led_pin, switch_state);
22.
23.    // print the variable to the Serial Monitor
24.    Serial.println(switch_state);
25.
26.    // short delay so that we don't receive massive data
27.    delay(250);
28.  }
```

This code is very similar to the code in Listing 2-2. What's new is line 4, where we set a global variable for the pin of the LED, which is 8. In the setup function, we call the pinMode function for both pins, but we set the switch_pin as INPUT_PULLUP, and the led_pin as OUTPUT. Then in line 21, we use the value stored in the switch_state variable, to control the LED, by calling the digitalWrite function like this:

```
digitalWrite(led_pin, switch_state);
```

In line 24, we print the value of switch_state to the serial monitor. Instead of creating a variable for the switch readings, we could have called the digitalRead function twice. So line 21 could read:

```
digitalWrite(led_pin, digitalRead(switch_pin));
```

And line 24 could read:

```
Serial.println(digitalRead(switch_pin));
```

And line 18 could have been avoided altogether. We could have even avoided to create the switch_pin variable, and write the number 2 in its place instead. This whole approach is problematic for the following reasons. The Arduino takes some time to read a pin (especially the analog pins), and calling a function that reads a pin more than once is not very efficient. Calling that function once and storing its reading to a variable, and then calling that variable instead, is much faster, efficient, and easier to read and understand. Also, avoiding a variable for the pin number of the switch can cause some problems, if for some reason we decide to change that pin. If you use a variable, you'll have to change one line of code only, the variable declaration. If you're not using a variable, you'll have to change that pin number in any line of code where you use it. Figure 2-13 shows the circuit for this sketch.

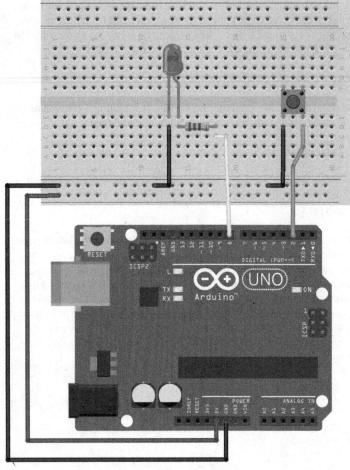

Figure 2-13. *Digital input and output circuit*

We could have use digital pin 13 for the LED, which has an internal resistor, but we prefer to use an external resistor, so you can see how a circuit using LEDs actually works. Build the circuit and upload your code. Don't save it yet when the IDE prompts you, just click **Cancel** and the code will be uploaded without being saved. Open the serial monitor too. Now whenever you press the switch the LED should go off, and whenever you release it, it should go on. But wait a minute, this should be the other way round, right? This inversion happens because of the pull-up resistor we have enabled in the switch pin. The LED should be aligned with the readings you see in the serial monitor. Whenever you press the switch, you should see 0s in the monitor, and the LED going off, and whenever you release it, you should see 1s and the LED going on. We can very easily reverse this whole process by adding a single character to our code. Go back to your code and change line 18 to this:

```
int switch_state = !digitalRead(switch_pin);
```

All we did was add an exclamation mark just before `digitalRead`. The exclamation mark in C/C++ when used before a value (`digitalRead` returns a value, so we should treat calling it like writing a value) means "the reverse of." Adding the exclamation mark to this line, should reverse the readings of the digital pin 2, so now whenever you press the switch you should see the LED going on, and 1s in the serial monitor, and the other way round. There should be a tiny bit of lag to the reaction of the LED, which is because of line 27:

```
delay(250);
```

This is used to avoid receiving massive amounts of data. We won't be using that when we build musical interfaces.

Analog Input

The next thing that we'll look at is getting input from the analog pins of the Arduino. Table 2-4 shows the parts for this sketch.

Table 2-4. *Project 4 Parts List*

Part	Quantity
Potentiometers	1 x 10KΩ

There are many sensors you can use with the analog pins, like proximity sensors, vibration sensors, accelerometers, and many more. For now, we'll just use a potentiometer, to see how to use the analog pins of the Arduino. Listing 2-4 shows the code.

Listing 2-4. Analog Input Sketch

```
1.    int analog_pin = 0;
2.
3.    void setup() {
4.      // begin the serial communication
5.      Serial.begin(9600);
6.    }
7.
8.    void loop() {
9.      // store value of potentiometer to a variable
10.     int pot_val = analogRead(analog_pin);
11.
12.     // print it to the Serial Monitor
13.     Serial.println(pot_val);
14.
15.     // short delay to avoid massive data
16.     delay(250);
17.   }
```

This code is also very similar to the code in Listing 2-2. In line 1, we define a variable for the analog pin number. If you look at your board, you'll see that the analog pins start from A0, up to A5. We can omit the letter A, which stands for analog, but if you like, you can include it and write this line like this:

```
int analog_pin = A0;
```

In the setup function, we're not calling pinMode anymore, because as already stated, the analog pins are input only, so we don't need to set their mode. We're just starting the serial communication with a 9600 baud rate. In the loop function, we create a variable to hold the value read by the potentiometer, but this time we call the analogRead function. This function is very similar to its digital counterpart, digitalRead. It takes one argument, which is the analog pin to read from, and returns the value read from that pin. In line 13, we print that value to the serial monitor, the same way we did before. Finally, we use the delay function, in order not to get a massive amount of data. Figure 2-14 shows the circuit for this code.

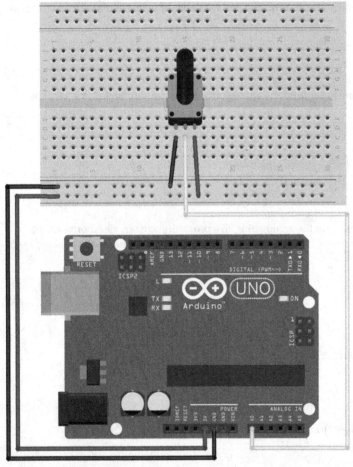

Figure 2-14. *Analog input circuit*

The potentiometer is actually a variable resistor. That's why the ohms are mentioned in the components of this circuit. It has three legs, where one of the side ones connects to ground, the other side leg connects to 5V, and the middle leg (called the *wiper*) connects to the analog pin of the Arduino. As you spin the potentiometer, the resistance it applies to the circuit varies. It doesn't really matter which of the side legs will go to ground and which to 5V, only the increasing/decreasing of the resistance will change direction. If you connect the left leg to ground and the right to 5V, then the resistance will drop as you spin the potentiometer clockwise, and the values you'll receive will increment. If you connect the legs the other way round, this process will be reversed. Most of the time, we want to have incrementing values as we spin the potentiometer clockwise, so you might want to connect its legs as shown in Figure 2-14.

Upload the sketch to your board and when prompted, save it as `Analog_input`. Open the serial monitor of the IDE and you should see something like the Figure 2-15. As you spin the potentiometer clockwise you should see the values increase (or decrease, depending on the way you built the circuit) and vice versa. The minimum value you get is 0 and the maximum is 1023. This is because the Arduino Uno has 10-bit resolution analog pins. This means that it can express the voltage it receives with 10 bits. In the decimal numeral system, this is expressed as 2 to the 10nth power, which is 1024. Since the number 0 is in that range, what we get is a range from 0 to 1023, which is in total 1024 values. In general, in a number stream that represents a signal, when starting from 0, the maximum value is always (2^bit-depth) – 1 (the ^ symbol raises 2 to the power of the bit-depth).

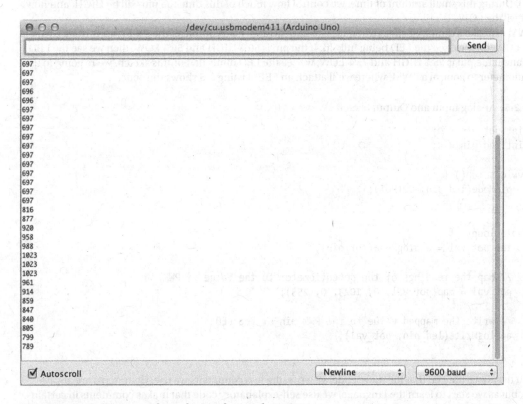

Figure 2-15. *Receiving analog values in the serial monitor*

Analog Input and Output

As with the digital pins, we'll now look at both input and output with the analog pins. Table 2-5 shows the components needed for this sketch.

Table 2-5. *Project 5 Parts List*

Part	Quantity
Potentiometers	1 x 10KΩ
LEDs	1
Res	1 × 220Ω

The name of this section might sound a bit strange, as I've already mentioned that the analog pins of the Arduino are input only. By "analog output" I don't really mean analog, but digital. Six of the digital pins of the Arduino have PWM capabilities. PWM stands for *pulse-width modulation*. This is similar to the duty cycle of the square wave oscillator we made in Pd. PWM essentially controls the amount of time a digital pin will be HIGH and LOW, during one period of a specified frequency. Quoting from the Arduino web site, "The frequency of the PWM signal on most pins is approximately 490 Hz. On the Uno and similar boards, pins 5 and 6 have a frequency of approximately 980 Hz." To make this a bit clearer, most PWM pins run at a 490 Hz frequency. When we control the width of the pulse, we control the percentage of the HIGH and LOW states of one period of this frequency, which lasts 1/490 seconds (*hertz* is a time unit of repetitions per second). During this small amount of time, we control how much of this time the pin will be HIGH, and how much it will be LOW.

PWM can fake a dimming effect when we use LEDs with it. For example, if the PWM pin is 100% HIGH and 0% LOW, then we see the LED being fully lit. If the pin is 50% HIGH and 50% LOW, then we see the LED half lit, and it the pin is 25% HIGH and 75% LOW, we see the LED dimly lit. For this sketch, we're going to use a potentiometer to control a PWM, where we'll attach an LED. Listing 2-5 shows the code.

Listing 2-5. Analog Input and Output Sketch

```
1.    int pot_pin = 0;
2.    int led_pin = 9;
3.
4.    void setup() {
5.      pinMode(led_pin, OUTPUT);
6.    }
7.
8.    void loop() {
9.      int pot_val = analogRead(pot_pin);
10.
11.     // map the readings of the potentiometer to the range of PWM
12.     pot_val = map(pot_val, 0, 1023, 0, 255);
13.
14.     // write the mapped value to the PWM pin of the LED
15.     analogWrite(led_pin, pot_val);
16.   }
```

You may have noticed that the more code we write, the less comments we use. Comments are always helpful, but as we start to learn the language, we use self-explanatory code that makes comments in certain cases unnecessary. For example, we don't use any comments in the first two lines, and by now you should understand what these two lines of code do. Notice that in the setup function, we don't start the serial communication, as we don't care to see the values of the potentiometer, since the LED will provide the necessary visual feedback (the brighter the LED, the greater the potentiometer value). We only call the pinMode function to set the mode of the LED pin.

In our loop function, we first store the potentiometer value to a variable, and then we call a new function, map. This function maps a specified range of values to another range. It takes five arguments, which are the variable that holds the range we want to map, the lowest value of the range we want to map, the highest value of the range we want to map, the lowest value of the desired range, and the highest value of the desired range. Notice that we're mapping the pot_val variable, but we're saving the value returned by map to the same variable, since we write:

```
pot_val = map(pot_val, 0, 1023, 0, 255);
```

This is legal and works as expected. What we see in this line of code is that we want to store to the pot_val variable, the value it holds, mapped from a range from 0 to 1023, to a range from 0 to 255. If the potentiometer has a value of 511, then this line will store the value 127 to the pot_val variable. This mapping is necessary because PWM in Arduino has an 8-bit resolution. Remember that a number stream starting from 0, will go up to (2^bit-depth) -1. 2 to the 8th power, yields 256, minus 1 yields 255.

The last line of the loop function calls the analogWrite function, which is very similar to its digital counterpart, digitalWrite. It takes two arguments, the pin to write a value to, and the value to write to that pin. Here we write to the led_pin the pot_val value. The mapped example of the previous paragraph (511 mapped to 127), will give approximately 50% (127 is almost half of 255), so the LED will look half lit.

Figure 2-16 shows the circuit. As you spin the potentiometer, you should see the LED dimming in and out, looking like a real analog output.

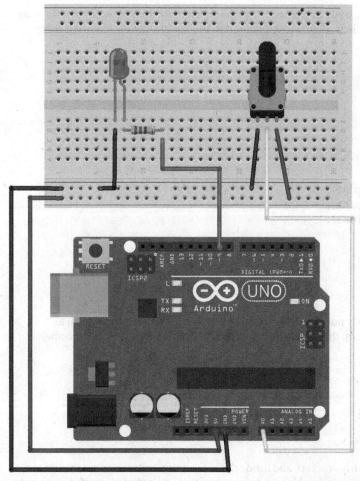

Figure 2-16. *Analog input and output circuit*

Reading More Than One Pin, Arrays, and the for Loop

We have covered quite a lot of the Arduino language and ways to use it both with analog and digital input and output. Now let's see how we can read more than one pin, in an efficient way. Table 2-6 shows the necessary components for this sketch.

Table 2-6. *Project 6 Parts List*

Part	Quantity
Potentiometers	3

This part may be a little bit tricky, so you might need to go through it more than once. Say that we want to use three potentiometers and print them all to the serial monitor. You could write the following code:

```
int pot_pin1 = 0;
int pot_pin2 = 1;
int pot_pin3 = 2;

void setup() {
  Serial.begin(9600);
}

void loop() {
  int pot_val1 = analogRead(pot_pin1);
  int pot_val2 = analogRead(pot_pin2);
  int pot_val3 = analogRead(pot_pin3);

  Serial.println(pot_val1);
  Serial.println(pot_val2);
  Serial.println(pot_val3);
}
```

But this way of writing code is really not efficient, as the more pins we add, the more we have to duplicate code, plus we can't really group the values we want to read. This is where the for loop comes in handy.

Explaining the for Loop

The for loop has the following syntax:

```
for(int i = 0; i < some_value; i++)
```

After this declaration, we insert curly brackets, and inside the brackets we write the code we want to have executed within the loop, much like we do when we define a function (the setup of loop function, only the curly brackets for these functions are there by default). What this loop does is create the variable i and assign it the value 0. Then it goes to the second field, which is a condition. If this condition is met (in this example, if i is less than some_value), the code inside the loop's curly brackets will be executed, and the loop will go to the last field, i++, which is a shortcut for incrementing i by 1. After that, the condition is tested again, and if it's true, again the loop's code will be executed. And this runs over and over, until i is not less than some_value anymore.

Using Arrays in Arduino

Before we apply the for loop to code that we'll write, I need to explain the array in the Arduino language. This is much like the array in Pd, only there's no graph of the table. An array can be of any data type (except void) and its declaration has the following syntax:

```
int pots[3];
```

This will create an array of three ints, called pots. We can access the elements of the array by means of indexing, much like we did in Pd. So, to write a value to the first element of pots, we must do the following:

```
pots[0] = some_value;
```

Applying these two features to our code, we can now write the code in Listing 2-6.

Listing 2-6. Using the for Loop and Arrays

```
1.    // create an array to store the values of the potentiometers
2.    int pots[3];
3.
4.    void setup() {
5.      Serial.begin(9600);
6.    }
7.
8.    void loop() {
9.      for(int i = 0; i < 3; i++){
10.       pots[i] = analogRead(i);
11.     }
12.
13.     Serial.print("Pot values: ");
14.     Serial.print(pots[0]);
15.     Serial.print(" ");
16.     Serial.print(pots[1]);
17.     Serial.print(" ");
18.     Serial.println(pots[2]);
19.
20.     delay(500);
21.   }
```

What happens in line 9 is that the for loop will run for as long as i is less than 3, which will happen three times (mind, not less than or equal to three, but only less than 3), as many as the potentiometers we're using. Note that we use the variable i both for indexing the pots array, but also as the argument to the analogRead function. All this is legal, since i will take the values 0, 1, and 2 sequentially, which are the indexes of the pots array, and the analog pins we want to read. We could have use a for loop for printing the values too, but that would make things a bit more complicated, so we'll leave it for later. One important thing here is that i is a local variable to the for loop, and as soon as the loop is finished, i won't exist anymore, until the next the loop will run. The advantage of this is that a local variable is faster to access, and frees the memory it allocates when it is destroyed (when the function defined inside it exits).

Line 13 calls the print function of the Serial class. Its difference to the println function of the same class is that it will print whatever is inside its parenthesis, but anything printed afterward will be printed to the same line. println causes the serial monitor to go one line below after it prints, like hitting the Return key on your keyboard (ln stands for *newline*). We print white spaces in between the values to get a clearer print on the monitor. Running this sketch and opening the serial monitor, you should get something like what's shown in Figure 2-17.

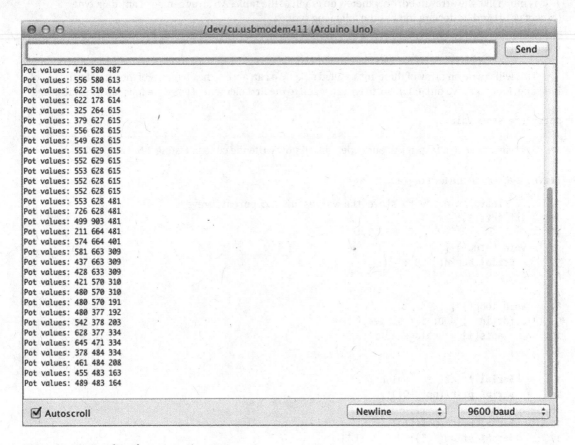

Figure 2-17. *Reading three potentiometers*

Although it is probably rather obvious, Figure 2-18 shows the circuit for this sketch.

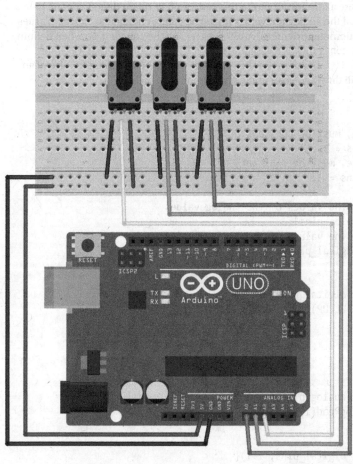

Figure 2-18. *Three potentiometer circuit*

Analog and Digital Input

Now that we've seen the for loop in action, let's write some code that utilizes both potentiometers and push buttons. Table 2-7 shows the necessary components.

Table 2-7. *Project 7 Parts List*

Part	Quantity
Potentiometers	3
Push buttons	3

This time we'll make our code even more efficient, by applying the loop to the printing functions as well. We'll keep the three potentiometers we used in the previous example, and we're going to add three push buttons to our circuit and code.

77

Listing 2-7 shows the code. There are a few new things in this code, so I'll explain them in detail. Line 2 defines the size of the array that will hold the analog pin values (the array name changed to analog_values to be more generic, and not only potentiometer oriented). Array sizes need to be constant, so when defining its size we must use the const keyword. const makes a variable read-only, which means that we cannot modify it anywhere else in our program. If you omit to use the const keyword when defining the size of an array via a variable, the Arduino IDE will throw an error and won't compile the code.

Listing 2-7. Analog and Digital Input

```
1.    // analog values array size, must be constant
2.    const int num_of_analog_pins = 3;
3.    // digital values array size, must be constant
4.    const int num_of_digital_pins = 3;
5.
6.    // create an array to store the values of the analog values
7.    int analog_values[num_of_analog_pins];
8.    // create an array to store the values of the digital values
9.    int digital_values[num_of_digital_pins];
10.
11.   void setup() {
12.     for(int i = 0; i < num_of_digital_pins; i++){
13.       pinMode((i + 2), INPUT_PULLUP);
14.     }
15.     Serial.begin(9600);
16.   }
17.
18.   void loop() {
19.     for(int i = 0; i < num_of_analog_pins; i++){
20.       analog_values[i] = analogRead(i);
21.     }
22.
23.     for(int i = 0; i < num_of_digital_pins; i++){
24.       digital_values[i] = !digitalRead(i + 2);
25.     }
26.
27.     Serial.print("Analog values: ");
28.     for(int i = 0; i < num_of_analog_pins; i++){
29.       Serial.print(analog_values[i]);
30.       Serial.print(" ");
31.     }
32.
33.     Serial.print("Digital values: ");
34.     for(int i = 0; i < (num_of_digital_pins - 1); i++){
35.       Serial.print(digital_values[i]);
36.       Serial.print(" ");
37.     }
38.     Serial.println(digital_values[num_of_digital_pins - 1]);
39.
40.     delay(500);
41.   }
```

Line 4 defines the size of the array that will hold the values of the digital pins, the same way line 2 did for the analog ones. We could have initialized both arrays by writing their size as a number inside their square brackets, but we need these two values in more places in our code, so it's more efficient to initialize the arrays this way. Lines 7 and 9 initialize the two arrays using the preceding two constant values.

In the setup function we use the for loop to set the mode of the digital pins. This way we only need to write the for loop header (the header of the loop is this (int i = 0; i < num_of_digital_pins; i++)), and one line of code to set the mode of all pins we're using. In this case, without using the loop, we would write three lines of code, since we use three digital pins, and now we have written two lines, which is not so much less. Imagine if we used all 12 available digital pins. Then we would have saved quite some coding. In general we prefer to use the for loop in many cases of repetition. As stated earlier, we use the num_of_digital_pins constant in the condition test of the for loop. This should make it clear why we prefer to initialize arrays with constants, rather than hard-code their size in their declaration square brackets.

Another thing to mention is that when we define a control structure, like the for loop, if the code of the loop is only one line (like our case, where the code is only pinMode((i + 2), INPUT_PULLUP);), we can omit the curly brackets; we can even write that line on the same line with the control structure's header. So lines 12 to 14 can also be written like this:

```
for(int i = 0; i < num_of_digital_pins; i++) pinMode((i + 2), INPUT_PULLUP);
```

This syntax is perfectly legal. Before we move on to the rest of the code, notice that we add 2 to the i variable inside the pinMode function. This is because i has been initialized to 0 (so we can combine it with the num_of_digital_pins constant, and the for loop can run three times), but we use digital pins from 2 onward. We can see here that a variable of this type is a numeric value and we can apply math operations to it.

In line 19 we run a for loop to read and store the values from the analog pins. Again, we could have omitted the curly brackets. In line 23, we run another for loop to read and store the values from the digital pins we're using. Again we're adding 2 to the i variable, as we need to initialize it to 0, so the loop can run properly, combined with the num_of_digital_pins constant.

In line 27, we print an indication that the following values are from the analog pins. The text inside the quotation marks is called a *string*. A string is essentially an array of characters, used to display text. After we print the string, we use a for loop to print the values of the analog pins. This time we cannot omit the curly brackets, because the code of the loop is two lines. The first line prints the value that is stored in the analog_values array (which we access via the index held in the i variable), and the second line prints a white space, so that the values are separated and easy to read.

Line 33 prints an indication that the following values are from the digital pins, and below that, we run another for loop to print the digital pin values. This time we run the loop for as many times as the number of digital pins we're using, minus one. This is because we want to print the last value using the println function, so that the values printed in the next loop, will be printed one line below. If we had used only print, then all text and values would be printed in one line, like in Figure 2-19.

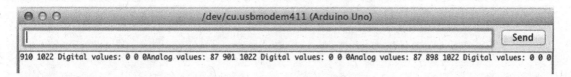

Figure 2-19. *Printing all data in a single line*

Using `println` for the last value makes things a lot clearer, as one line will contain each value only once. Figure 2-20 shows the serial monitor using `println` for the last value. Also, the index we use to access the last digital pin is `num_of_digital_pins` - 1, because `num_of_digital_pins` is 3, but array indexes start counting from 0, so the last index is 2, and not 3.

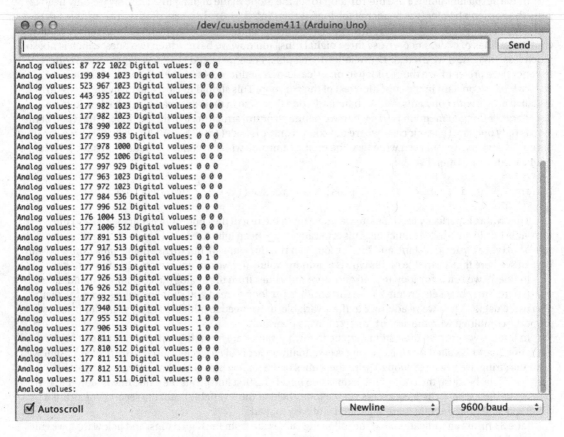

Figure 2-20. *Using println for the last value*

The circuit for this code combines the circuit of the previous code, with the circuit of the push button, in the "Digital Input" section of this book; it is shown in Figure 2-21.

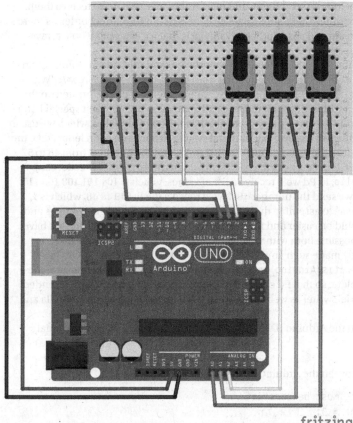

Figure 2-21. *Circuit for analog and digital input*

Communicating with Pd

Since we've covered some basic concepts of Arduino programming, we can now combine it with Pd. As you can imagine, this is possible with the serial communication capabilities of the Arduino. On the Pure Data side, we use [comport], an external object for serial communication. The Serial class of the Arduino language has three functions to send data to the serial line: print, println, and write. We've already seen the first two with the Arduino IDE's serial monitor, but I must explain their differences and how and when to use which of the two when we combine it with Pd.

I've already explained the difference between print and println, which is that println adds the newline character at the end, causing the serial monitor to go one line lower. So, for a while we'll talk about these two functions as if they were one, and we'll call them simply print. Before I explain this function, let's talk about write. write writes a single byte, or an array of bytes, to the serial line. This should be straightforward. If we type

Serial.write(100);

we should receive the value 100 in the serial line. This has both advantages and disadvantages. The advantage is that we receive a value as is, but a disadvantage is that we can only receive bytes, and not values longer than that (remember, an int for example is two bytes long, and can't pass the serial line as is). To send and receive values greater than a byte (an analog pin has a 10-bit resolution and a byte is 8-bit)

81

we must somehow disassemble them before we send them, and reassemble them when we receive them. Another disadvantage of write is that we must make sure we receive the bytes in the correct order, as there's no default for beginning or ending a stream of data. Another advantage is that we can send whole arrays, which reduces the code we write to a great extent.

print sends anything as ASCII characters. ASCII stands for American Standard Code for Information Interchange, and it's a 7-bit code (now extended to 8-bit), quoting from http://ascii-code.com/: "Where every single bit represents a unique character." This means that every character that can be printed on the screen by a computer, takes a unique bit within this 7-bit range, so all characters (except from special Latin characters) get a unique value from 0 to 127. What we most care about is the Latin alphabet, which we use to pass strings over serial, and the 10 numbers of the decimal numeral system. Uppercase Latin letters take the values 65 to 90 (A–Z), lowercase letters take the values from 97 to 122 (a-z), and the numbers from 48 to 57 (0–9) (if you want to check the whole ASCII table, check the ASCII URL earlier, http://ascii-code.com). So, if we use print to pass the string hello, in Pd we'll receive a list of values, which is 104 101 108 108 111, where 104 is h, 101 is e, and so forth. If we send the number 100, we'll receive the list 49 48 48, which is 1, 0, and 0 in ASCII. Taken that there is a way to assemble these ASCII values back to the original strings and numbers, using the print function should be easier and more intuitive than using write. Pd doesn't have an object to do the assembly, still it is possible both with vanilla and extended objects to create an abstraction that does that. I have already made such an abstraction, which you can find in my GitHub page, https://github.com/alexdrymonitis/Arduino_Pd. Download the .zip file, unzip it, and save the abstractions to your "abstractions" folder, so that Pd can find them. We'll use the [serial_print_extended] abstraction now, but later we'll use [serial_write] as well (mind, the [serial_print] abstraction is vanilla and won't work with extended).

Open the serial_print.ino file in the Arduino IDE. Listing 2-8 shows the code without the initial multiline comment.

Listing 2-8. Serial Communication Between the Arduino and Pd Using Serial.print

```
1.   // analog values array size, must be constant
2.   const int num_of_analog_pins = 3;
3.   // digital values array size, must be constant
4.   const int num_of_digital_pins = 3;
5.
6.   // create an array to store the values of the analog pins
7.   int analog_values[num_of_analog_pins];
8.   // create an array to store the values of the digital pins
9.   int digital_values[num_of_digital_pins];
10.
11.  void setup() {
12.    for(int i = 0; i < num_of_digital_pins; i++) pinMode((i + 2), INPUT_PULLUP);
13.
14.    Serial.begin(9600);
15.  }
16.
17.  void loop() {
18.    for(int i = 0; i < num_of_analog_pins; i++) analog_values[i] = analogRead(i);
19.
20.    for(int i = 0; i < num_of_digital_pins; i++) digital_values[i] = !digitalRead(i + 2);
21.
```

```
22.    Serial.print("Analog_values: ");
23.    for(int i = 0; i < (num_of_analog_pins); i++){
24.      Serial.print(analog_values[i]);
25.      Serial.print(" ");
26.    }
27.    // print last value of the group with Serial.println()
28.    Serial.println(analog_values[num_of_analog_pins - 1]);
29.
30.    Serial.print("Digital_values: ");
31.    for(int i = 0; i < (num_of_digital_pins - 1); i++){
32.      Serial.print(digital_values[i]);
33.      Serial.print(" ");
34.    }
35.    // print last value of the group with Serial.println()
36.    Serial.println(digital_values[num_of_digital_pins - 1]);
37.  }
```

We have used the code in Listing 2-7 with some minor modifications. In lines 12, 18 and 20 we have three for loops without curly brackets, as we already saw that it is legal and works. This time we print both analog and digital values, all but last with a for loop, and we print the last outside the loop because we have to use println instead of print. This is because of the way the Pd abstraction is made. It receives values in groups, which groups can be recognized by a string, which is used as a tag (here Analog_values: and Digital_values:). Each value group must end with println, because this function adds the newline character at the end (it actually adds the carriage return, and the newline character, which have ASCII 13 and 10 respectively), and this way we can tell that a data stream of a value group has finished. Notice also that we have slightly modified the strings, as we separate their two words with an underscore, instead of a white space. This is because the Pd abstraction uses the white space to separate the string of a group from its values, but also to separate the values of a group between them. Take a close look to each string, they both have a white space at the end. This way we can tell Pd that this is the tag of a value group, and now we'll start receiving the values. We also print a white space after every value, but the last, which we print with println. Last modification is that we're not using a delay at the end of our code, since we don't need it to visualize the values in Pd. Figure 2-22 shows part the help patch of the Pd abstraction.

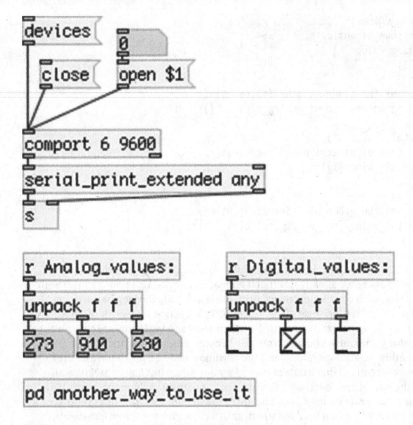

Figure 2-22. *The [serial_print_extended] abstraction help patch*

In this patch, we use two new objects along with the abstraction. These are [comport] and [unpack f f f]. [comport] is an object for serial communication with devices like the Arduino. It optionally takes two arguments, which are the port number and the baud rate. If no argument is provided, it will try to open port number 0, with a 9600 default baud. When you put this object in a patch, you might get an error message saying that it cannot open the port. This is nothing to worry about. Send the message "devices" to [comport] and you'll get a list of available serial devices connected to your computer, along with their port numbers, like Figure 2-23.

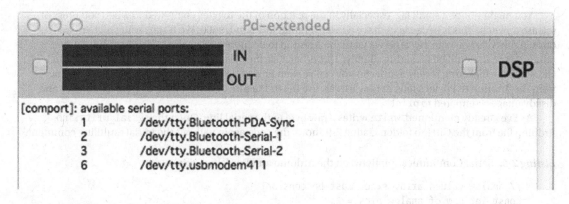

Figure 2-23. *List of available ports in Pd*

Figure 2-23 shows a list with four serial ports, where we can see that the Arduino port is number 6 (in this case it's /dev/tty.usbmodem411, depending on your platform, you should see something equivalent to the port you open in the Arduino IDE). Lock your patch and type the port number in the number atom connected to [open $1(, and [comport] will open that port and print a message declaring that in Pd's console. Taken that you have already uploaded the code in Listing 2-8 to you Arduino, and that you still have the circuit in Figure 2-21 patched, you should immediately see the potentiometer values in the three number atoms below [r Analog_values:], and the toggles below [r Digital_values:] should go on whenever you press a switch.

I'll explain how this works. [serial_print_extended] takes in ASCII values from [comport]. The first thing that comes in is a string, which is used as a tag. This string is being assembled to a symbol (in Pd, strings can be displayed via the symbol atom, the fourth element of the Put menu—Ctrl/Cmd+4—or stored in [symbol], an object that stores strings) using the [bytes2any] external object of the moocow library (in vanilla, this is achieved with the new feature of the [list] object, [list tosymbol], which is a bit more efficient). When [serial_ print_extended] receives a white space, it knows that the ASCII values of the tag are finished and it sends the assembled tag out its right outlet to the right inlet of [s]. This way you can dynamically set the destination of [send]. Afterward, the values of the first group arrive one by one, in ASCII. [serial_print_extended] assembles them to their original values and stores them in a list, again using white spaces to separate the values from each other. When the abstraction receives the newline character (ASCII 10) which is sent with the last potentiometer value from the Arduino, it sends the list of values to the left inlet of [s], which sends that to the corresponding [r]. In our case, we first print the values of the analog pins, and our tag is Analog_values:. So [r Analog_values:] (we don't include the white space to the argument of [r] here, it is used as a delimiter) will receive this list, and will send it to [unpack f f f]. This object takes in a list and unpacks its elements. It will create as many outlets as its arguments (in this case three). "f" stands for *float*, as you already know, so this specific object unpacks a list of three floats, which we can see in the three number atoms below it.

Once we have received and assembled all analog values, we'll start receiving the digital values. Receiving the newline character denotes the end of a value group, so [serial_print_extended] knows that we're done with it and will start assembling the next value group, starting from the string tag. This time the string tag is Digital_values:, so we'll retrieve these values with [r Digital_values:]. [unpack f f f] below [r Digital_values:] is connected to three toggles, and this is because the digital pins give only 0s and 1s, so a toggle is a nice way to visualize this. The same assembling and listing technique applies here as well, and we can visualize our switches with the three toggles, which correspond to their respective switches in our circuit, from left to right.

The argument of [serial_print_extended] sets the delimiter character, which can be either a white space, a comma, or a tab. Here, "any" means that all three characters will be used as delimiters. So in the Arduino code, we could have used any of the three, in case we didn't want to use a white space.

You might notice a small lag, especially in the reaction of the toggles compared to when you press or release a switch. This is because we have chosen a rather low baud, 9600. We'll start using higher baud rates in the code we write, because we want the Arduino to be more responsive. Also, the values in the number atoms should be flickering slightly. This is because there's a little bit of noise in the circuit, but it's OK, since the flickering is only a little. So now we've seen how to use the print function to receive data from the Arduino to Pd, let's look at how we can use the write function, and what are the advantages and disadvantages compared to print.

As I've already mentioned, write writes a raw byte to the serial line. Open the serial_write.ino Arduino file from the GitHub folder. Listing 2-9 shows the code again without the initial multiline comment.

Listing 2-9. Serial Communication Between the Arduino and Pd Using Serial.write

```
1.    // analog values array size, must be constant
2.    const int num_of_analog_pins = 3;
3.    // digital values array size, must be constant
4.    const int num_of_digital_pins = 3;
5.
6.    // assemble number of bytes we need
7.    // analog values are being split in two, so their number times 2
8.    // and we need a unique byte to denote the beginning of the data stream
9.    const int num_of_bytes = (num_of_analog_pins * 2) + num_of_digital_pins + 1;
10.
11.   // array to store all bytes
12.   byte transfer_array[num_of_bytes] = { 192 };
13.
14.   void setup() {
15.     for(int i = 0; i < num_of_digital_pins; i++) pinMode((i + 2), INPUT_PULLUP);
16.
17.     Serial.begin(57600);
18.   }
19.
20.   void loop() {
21.     int index = 1; // index offset
22.
23.     // store the analog values to the array
24.     for(int i = 0; i < num_of_analog_pins; i++){
25.       int analog_val = analogRead(i);
26.       // split analog values so they can retain their 10-bit resolution
27.       transfer_array[index++] = analog_val & 0x007f;
28.       transfer_array[index++] = analog_val >> 7;
29.     }
30.
31.     // store the digital values to the array
32.     for(int i = 0; i < num_of_digital_pins; i++)
33.       transfer_array[index++] = !digitalRead(i + 2);
34.
35.     // transfer bytes over serial
36.     Serial.write(transfer_array, num_of_bytes);
37.   }
```

Lines 6, 7, and 8 are comments explaining how many bytes the array we'll transfer over serial must have. We'll now store all values, analog and digital, to one array, and we'll transfer that array with one function call to Pd. As I've already mentioned, the analog pins have 10-bit resolution, but we can only pas bytes through the serial line, and a byte is 8-bit. For this reason, we must split the analog values to two, which we will assemble in Pd. Therefore, we need two bytes for every analog pin. The digital pins will give either a 0 or a 1, and that fits in a byte, so we use one byte only for each pin. Lastly, we need a unique byte at the beginning of the data stream, to denote that we're starting to receive a new package. Using three potentiometers and three switches, makes 10 bytes in total. Line 9 does the appropriate calculations to get the necessary byte number, using the constants of line 2 and 4. This value is a constant as well, since it will be the size of an array.

In line 12, we define the array that will be transferred to Pd. This array is of the type byte, since we can only transfer bytes over serial. We're also initializing the array's first value to 192. This value will denote the beginning of the data stream. This value must be unique, and since we're sending bytes, we must make sure it will be in a range no other value will ever reach. This range is in the 8th bit.

To understand this, it's better to visualize it. Think of binary numbers. In an 8-bit binary number, 0 is

00000000

Number 127 is

01111111

This number has the first seven bits on, and the 8th bit off. Number 128 is

10000000

So from 128 onward, we begin to utilize the 8th bit. Any value above that (until 255, which can be represented by 8 bits) will have the 8th bit on, while all values below it (from 0 to 127) will have the 8th bit off. If we make sure that all other values are restricted to a 7-bit range (0 – 127), and assign to the first byte a value between 128 and 255, then this byte will be unique. I got this technique from the rePatcher project by Open Music Labs, where they used 192 as the unique byte, so this is what I use as well. Though, any value between 128 and 255 will do. We'll also see that in practice along with the Pd patch.

Our setup function is the same as before, only this time we start the serial communication with a much higher baud rate, 57,600. And then we move on to the loop function. Line 21 creates a new variable and assigns the value 1 to it. This variable is called index and it will be used as the index to access elements of the transfer_array. We assign the value 1 to it, because we have already stored the value 0xc0 to the first element of the array, which has index 0.

In line 24, we run the for loop to read and store the values from the potentiometers. We first create a variable of type int, called analog_val, and assign it with the value returned by the analogRead function. In line 27, we read:

```
transfer_array[index++] = analog_val & 0x007f;
```

The index++ inside the curly brackets of the transfer_array is called the *post-increment technique*. This will first use the current value of the index variable, which is 1. So in this line we're accessing the element with index 1 of the transfer_array. When the whole line has been executed, the index variable will be incremented by 1 (with the double plus sign, ++), so for the next line of code, it will hold 2. The part of this line after the equals sign takes the value stored in the analog_val variable and wraps it around a 7-bit range. This means that when this value is between 0 and 127, it stays as is, but when it goes to 128, it wraps back to 0. As the analog_val value rises up to 255, the result of analog_val & 0x007f will go up to 127, and when analog_val reaches 256, analog_val & 0x007f will again wrap back to zero, and this will happen over and over, for the whole range of analog_val. This happens because of 0x007f, which is the hexadecimal version of the number 127. Hexadecimal values are useful here because we can express the size of a value without needing to define it as a specific data type (an for example). Since analog_val is an , when we apply operations to it, we need to use the same data type, or a value with the same length (two bytes in this case). A hexadecimal number can express 256 values with two digits, from 00, which is 0, to ff, which is 255. This is the range of one byte. Using four digits, we can express a value in the range of two bytes, which is the same as an int. The 0x prefix is the C++ prefix to indicate a hexadecimal value. So, without having declared a variable of the type int, we can use the hexadecimal value here to indicate the same length of the int data type, and be sure that our operations will have correct results.

Line 28 again uses the post-increment technique, where index now holds 2, so we're accessing the third element of the transfer_array. The second half of the line reads:

```
analog_val >> 7;
```

What this does is shift the bits of analog_val by 7 positions to the right. The result of this is a number that increments by one, whenever the line wraps back to 0. By restricting both values to a 7-bit, we can be sure that no analog value will ever reach the first byte of the array, which is 192, in the 8-bit range. All this might sound very strange, but when we look at the Pd patch, where we reassemble these values, it will start making sense.

In line 32, we read and store the value of the digital pins to transfer_array, again using the post-increment technique we used in the loop. This time the code of the for loop is one line below, without curly brackets. This syntax is also legal, since the executable code is only one line. We indent it manually (the IDE won't do it automatically if there are no curly brackets) to make the code more readable.

Lastly, in line 36 we call the write function, of the Serial class, to write our array to the serial line. When we write an array using this function, we must provide two arguments, the name of the array, and the number of bytes we want to transfer. Usually we want to transfer the whole array, so we're using the constant that sets the size of it, num_of_bytes. Notice that we're using names that explain what each variable or array does, so we can restrict our comments to a minimum, making our code self-explanatory.

Now let's check the help patch of the [serial_write] Pd abstraction, part of which is shown in Figure 2-24.

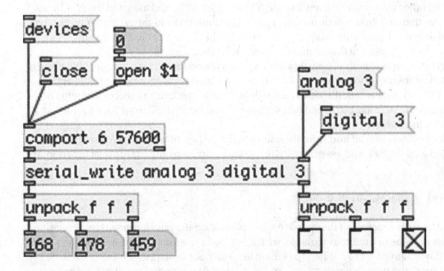

Figure 2-24. *Pd [serial_write] abstraction to read bytes sent from Arduino with Serial.write()*

The first thing to notice is that we are using arguments with [comport], which are the port number and the baud rate. From Figure 2-23, we've seen that the Arduino port on my computer is number 6, so I'm using this number for the first argument. You should use the number of your serial port, which is very likely to be different. If you don't remember which one it is, send the "devices" message to [comport] and it will print all available serial ports to Pd's console. This time we're also using a higher baud rate, which is 57,600, so we use that as the second argument (the comma is used only in the text of this book. When programming, use the number 57600 (as in Figure 2-24).

Again, it can be quite cumbersome to receive and assemble all the data we receive from the Arduino, so I have created another abstraction for this purpose, [serial_write], which is included in the GitHub link I posted earlier. This abstraction takes two to four arguments. The first argument is the type of pins we're reading (analog or digital), and the second is the number of these pins that we'll be reading (in this case 3). We use the third argument is in case we're reading both types of pins, analog and digital, so depending on the first argument, this will set the other type of pins (in this case we first read the analog pins, so the third argument is "digital"). The last argument is the number of pins of the second type that we're reading (again 3, as we have three potentiometers and three switches). The first and third arguments cannot be set any way we want, but must be aligned with the Arduino code. In our code, we first store the analog values, and then the digital, so we must type our arguments in this order.

[serial_write] will output the values of the first argument (the analog values) out its left outlet, and the values of the third argument (the digital ones) out its right outlet. We're again using [unpack f f f] to unpack the lists we're receiving so we can see these values in the number atoms and the toggles.

It's not necessary to know the workings of [serial_write], but to make the Arduino code a bit clearer, I must explain a couple of things. As already mentioned, an abstraction is clickable. So in a locked patch, if you click it, the abstraction window will open. In there, click **[pd $0-route_list]**, then click **[pd specify_analog]**, and then click one of the two **[pd assemble_analog]** subpatches. The patch in Figure 2-25 will open.

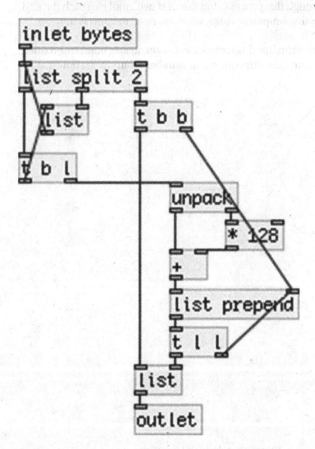

Figure 2-25. *The assemble_analog subpatch of [serial_write]*

This subpatch receives the list of the analog values. Since we're splitting the analog values to two, this subpatch will receive a list with elements two times the number of analog pins we're using. It then goes on to split this list and separate the first two elements, using [list split 2], which comprise the first analog value. These two values are sent to [unpack] (no arguments to this object is equivalent to [unpack f f]), which unpacks the two element list and makes the operations below it. Bearing in mind the right-to-left execution order in Pd, the second element will come out first, and the first element will come out second. In Pd, apart from right-to-left, the depth-first execution order applies as well. This means that the right outlet of [unpack] will go to [* 128] and then to the right inlet of [+]. Since the right inlet is cold, all execution will stop there, and only then will the left outlet of [unpack] spit its value.

Now remember the operations we did in the Arduino code. We split each analog value to two, and wrapped the first one to a range between 0 and 127, so whenever the analog pin would read 128, we'd get 0 and start incrementing anew, up to 127, and then wrap back to 0 again. The second element of this list is a number that increments by 1 (starting from 0) whenever the first element wraps back to 0. So, for the first 127 values, the second element will be 0, which will be multiplied by 128, which will again be 0, and will be stored to the right inlet of [+]. So as the first element goes from 0 to 127, it will be added to 0, so it will stay intact.

When the first element wraps back to 0 for the first time, the second element will be $1 \times 128 = 128$, which will be added to the first element, so we'll get $0 + 128 = 128$, the next number after 127. As the first element again increments to 127, we'll get $127 + 128 = 255$, and then it will wrap back to 0, and the second element will be $2 \times 128 = 256$, the next value after 255. And this goes on until we get 1023, which is the maximum value of the 10-bit resolution (2 to the 10th power, minus 1). After the first two elements of the analog values list is assembled, the next two elements will go through this process, and the next two, until we reach the end of the list. Don't worry so much about how exactly this subpatch works, what you need to know is how we combine the Arduino code with the Pd patch.

Figure 2-26 shows a simple Pd patch that helps visualize this technique. Go ahead and build it and drag the values of the top number atom, from 0 to 1023 and check the other four number atoms to see how it really works.

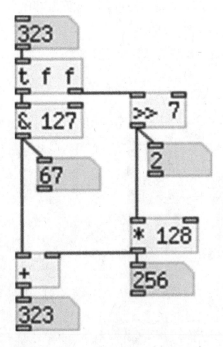

Figure 2-26. *Visualization of the Arduino-Pd technique for the Serial.write() function*

We've covered both ways of communicating the Arduino to Pd. Each has its own advantages and disadvantages, plus they might seem a bit difficult to use. The advantage of both is that we can more or less use them the way we've showed them here, for any number of pins we're using, doing only minor changes to both Arduino and Pd. So we have a tool to help us combine the two platforms, without needing to write all the code from scratch every time. This shows the great advantage of using abstractions, where one patch can be used many times, without needing to create it anew. When to use which of the two techniques depends on the project. We'll use both in the projects of this book, but not at the same time, as in one project it might be better to use one technique, and in another it might be better to use the other technique.

Sending Data from Pd to Arduino

Let's build the final project of this chapter where we'll send data from Pd to the Arduino. Table 2-8 shows the components needed for this sketch.

Table 2-8. *Final Project Parts List*

Part	Quantity
LEDs	3
Resistors	3 × 220Ω

Until now, we've covered the communication between Arduino and Pd, where the Arduino was sending data to Pd. Now let's look at this communication the other way round, we'll send data from Pd to the Arduino. The simplest form the Arduino code can take is shown in Listing 2-10.

Listing 2-10. Sending One Byte from Pd to the Arduino

```
1.    int led_pin = 13;
2.
3.    void setup() {
4.      pinMode(13, OUTPUT);
5.
6.      Serial.begin(57600);
7.    }
8.
9.    void loop() {
10.     if(Serial.available()){
11.       byte in_byte = Serial.read();
12.       digitalWrite(led_pin, in_byte);
13.     }
14.   }
```

Up until the loop function, there should be nothing new. In line 10, there is something we haven't encountered yet. This is the if control structure. Like the for loop, if has code of its own, written inside its curly brackets. What if does is test the statement inside its parenthesis. The following line of code demonstrates how this test works:

```
if(some_variable){
  // code to be executed if the test above is true
}
```

The code inside the curly brackets (here it's only a comment) will be executed only if some_variable is true. True in the Arduino language is any non-zero value, and this concept is often used in control structures like if. If some_variable is not true (if it's zero), then the whole code inside the curly brackets will be skipped and we'll move further down to our program.

Inside the parenthesis of the if control structure in Listing 2-10, we call the available function of the Serial class. Quoting from the Arduino web site, this function will "get the number of bytes (characters) available for reading from the serial port". This means that if there are any bytes available in the serial port, then available will return a non-zero value, so the statement inside the parenthesis will be true, and the code inside if's curly brackets will be executed. This means that whenever we send something from Pd to the Arduino, Serial.available() will be true.

Inside the curly brackets of the if control structure, we define a variable of type byte, and assign it the value returned by the read function, of the Serial class. This function will return the first byte of the incoming serial data, so it will assign to in_byte whatever value we send from Pd. Then we use this variable to set the state of the led_pin, in line 12. The Pd patch that works with this code is extremely simple, and it's shown in Figure 2-27.

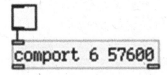

Figure 2-27. *Simple patch to control an LED of the Arduino*

All we need to do is send either a 1 or a 0 to [comport], and the integrated LED of the Arduino board will turn on and off accordingly. What if we want to control more than one pin on the Arduino? This technique is not efficient because we cannot really separate bytes we send to the Arduino, so we'll have to use another technique. Listing 2-11 shows a way to control various pins of the Arduino from Pd.

Listing 2-11. Sending More Than One Byte from Pd to the Arduino

```
1.    const int num_of_leds = 3;
2.
3.    int led_pins[num_of_leds] = { 3, 5, 6};
4.
5.    void setup() {
6.      for(int i = 0; i < num_of_leds; i++) pinMode(led_pins[i], OUTPUT);
7.
8.      Serial.begin(57600);
9.    }
10.
11.   void loop() {
12.     if(Serial.available()){
13.       static int temp_val;
14.       byte in_byte = Serial.read();
15.       if((in_byte >= '0') && (in_byte <= '9'))
16.         temp_val = temp_val * 10 + in_byte - '0';
17.       else if((in_byte >= 'a') && (in_byte <= 'z')){
18.         int which_pin = in_byte - 'a';
```

```
19.          analogWrite(led_pins[which_pin], temp_val);
20.          temp_val = 0;
21.       }
22.    }
23. }
```

Now we'll use an array to store the pins we want to control. In line 1, we set the size of the array, and in line 3, we define the array and initialize it. Initializing an array, means to provide values for it, and the syntax is the one used in line 3, we use curly brackets to include the values of the array, which are separated by commas. Here we use the pins 3, 5, and 6, because they are PWM pins, and we want to control LEDs with PWM. In the setup function we set the mode of the pins as OUTPUT with a for loop, where the variable i is used as the index to access the elements of the led_pins array. Then we start the serial communication.

In the loop function, we use the same if statement as before, but its code is very different. Line 13 defines a variable of type int, which is static. A static variable means that it is a local variable, but it retains its value even when the function, which it belongs to has ended. We must define the variable as static, otherwise the code won't work. Line 14 stores the incoming bytes, one by one. In line 15 we use another if statement, this time using a Boolean AND (the double ampersand, &&). All Boolean operations utilize the concept of truth and falsity. The Boolean AND tests both sides, and only if both sides are true, it will be true, otherwise it will be false. So the if statement in line 15 will be true only if both (in_byte >='0') and (in_byte <= '9') are true. The numbers inside single quotes represent the ASCII values of these characters, which are 48 and 57, respectively. This tells us that the data we'll be sending from Pd to the Arduino will be in ASCII form. If this statement is true, the code of line 16 will be executed.

Notice that the code of this if statement is one line only, so we don't need to use curly brackets, still we're writing it one line below, and not in the same line with the declaration of the control structure. This is legal as well, and we indent this line to make the code more readable (this time, the IDE won't indent the cursor automatically, you need to do it manually).

The executable code of the if control structure reads:

```
temp_val = temp_val * 10 + in_byte - '0';
```

These simple operations will assemble a value passed as ASCII. Let's say that we send the value 152 from Pd, in ASCII. A list with the values 49, 53, and 50 will arrive to the serial line of the Arduino. The preceding line of code will run three times, because the if statement in which it is included, will be true for all the three values. The first time temp_val hold nothing, which is equivalent to zero. Multiplying this by 10 will give zero, adding in_byte , which will hold the first value of the three, will give 49, and subtracting ASCII 0 (0 in single quotes), which is 48, will give 1. The second time temp_val will hold $1 \times 10 = 10 + 53$ (the value in_byte will now hold) $= 63 - 48$ ('0') $= 15$. The third time temp_val will hold $15 \times 10 = 150 + 50 = 200 - 48 = 152$. This way we can assemble values arriving in ASCII. This is the same technique used in the [serial_print] abstraction.

Line 17 calls the else if statement. This is an optional statement when we use the if control structure. The else if statement will be tested only if the if statement it is false. In Pd, we send messages of the type "200a", where 200 is the value for the PWM, and "a" the PWM pin to use. After the numeric value has been assembled, the if statement of line 15 won't be true and the program will move on to the else if. The serial line will now have the ASCII a, which is 97. Indeed, line 17 tests if the current byte from the serial line is between ASCII a and ASCII z, which are 97 and 122, respectively. So this line will be true and its code (now inside curly brackets, as it's more than one line) will be executed. This statement's code creates a variable called which_pin, and assign to it the value in the serial line, minus ASCII a, which is 97. If the byte in the serial line is indeed ASCII a, line 18 will assign the value 0 to which_pin. Line 19 calls the analogWrite function, and passes as arguments the led_pins array using which_pin as the index, and temp_val as the value. Since which_pin is 0, led_pins[which_pin] accesses the first element of the led_pins[which_pin] array, which is the value 3 (check line 3). So line 19 will write the value 200 to digital pin 3.

When we're done writing to the digital pin, using the analogWrite function, we assign 0 to temp_val, so that we can start assembling value from scratch (temp_val is static, so it will retain its value, therefore we must zero it). Figure 2-28 shows the circuit for this code, and Figure 2-29 shows the Pd patch.

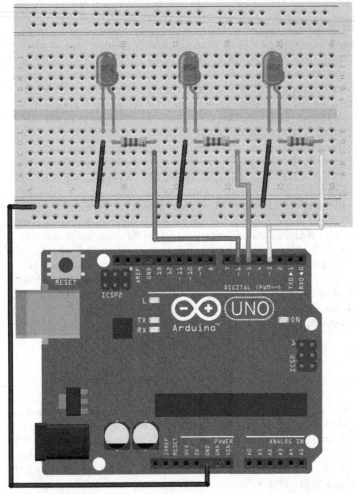

Figure 2-28. *PWM from Pd*

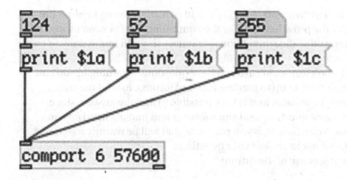

Figure 2-29. *Pd patch for controlling PWM pins in Arduino*

Again, this is a very simple patch, where we send messages to [comport] of the type "print $1a". The word print in the message converts the elements of the message to their ASCII values, which is necessary to manipulate the data in the Arduino code. $1 will take the first value of a list that arrives to the inlet of the message, and since we send one value only, $1 will take that value. So in the patch in Figure 2-29, the three messages are actually "print 124a", "print 52b", and "print 255c". Going back to the Arduino code, we see that the letters are converted to indexes in order to access the led_pins array. So the message "print $1a" will control the LED on pin 3, the message "print $1b" will control the LED on pin 5, and the message "print $1c" will control the LED on pin 6 (these are the three elements of the led_pins array, check line 3 of the Arduino sketch). Upload the code to the Arduino. ([comport] must have its port closed to upload code to the Arduino using its IDE, so send the message "close" to [comport] and when you've uploaded the code, send the message "open port number", where port number is the number of the Arduino port—in Figure 2-29 it is 6.) Open the Arduino port in the Pd patch and start playing with the number atoms. You'll see the LEDs dimming in and out, as you send various values between 0 and 255. This concludes the communication between Arduino and Pd, and the whole chapter.

Conclusion

We have gone through the basics of the Arduino language, and some very simple circuits. What has been covered in this chapter will be sufficient for most of the projects we'll be building in this book. Even if we encounter something new in one of the projects, we'll be able to easily understand it, since we have been introduced to the philosophy of the Arduino language and electronics. For all code in this chapter, we used push buttons for the digital pins, and potentiometers for the analog ones, but further on we'll use different kinds of sensors, depending on the project. So even though we used the same elements to build all the circuits, we have gained the foundation to comprehend and build other kinds of circuits, using the Arduino.

All circuits in this chapter were built using a breadboard, since they were testing circuits, and not a final version. In the building projects of this book, we'll use perforated boards, which make it possible to build more steady circuits, compared to the breadboard. To build circuits using a perforated board, you must be able to solder. Explaining how to solder is beyond the scope of this book, but there are many tutorials on the web, so it shouldn't be too difficult to learn how to solder yourself.

The last part of this chapter covered the communication between the Arduino and Pd, both directions. You should be able to see the potential of this, since being able to use various kinds of sensors and receive their input in Pd, gives infinite possibilities as to what you can build. There are many kinds of different sensors, proximity sensors, accelerometers, tilt sensors, gyroscopes, even humidity sensors, and so forth. The only limitation as to what you can realize is your own imagination. Being able to also give input from Pd to Arduino, expands the possibilities one has using these platforms. The Arduino can give input to the physical world, not only with LEDs, but also with motors, solenoids, and so forth. Bearing in mind the simplicity of the Arduino language, along with the provided tools for the communication between it and Pd, you should just let your imagination take over, and realize things you might have thought impossible, or even come up with things that never crossed you mind.

Lastly, you might want to sign up to Arduino's forum, in case you have questions regarding it. You can find that on Arduino's web site. Take care when signing up to mailing lists and forums, to read the "rules" before posting. Search the web before you post a question, as it is very possible to find the answer, since someone else before you might have had the same question, and since forums and mailing lists keep an archive, it's rather easy to search for solutions. Nevertheless, if you play nice, you will be warmly welcomed to these communities, and you'll find people willing to answer your questions.

Now on to embedded computers and wireless use of the Arduino.

CHAPTER 3

■ ■ ■

Embedded Computers and Going Wireless

In this chapter, we're going to cover two topics, which are very popular with multimedia programming and creative coding. The first topic is the embedded computers. These are small and low-cost computers that can be embedded into projects, hence their name. There are quite some embedded computers around, and we'll talk about which to choose for your needs.

The second topic of this chapter is the wireless use of the Arduino. This is possible with two ways. One way is using a Bluetooth module with the Arduino, which is a quick and low-cost solution. The Bluetooth is widely used, and there are dedicated modules that work with the Arduino. The other way is using the XBee transmitter/receiver modules, which work nicely with the Arduino. These modules are not so expensive, low-power, and easy to configure radios. There are many boards that help combine them with various Arduino types (usually these boards are called *shields*), so you can easily find a board type that fits your project needs.

In this chapter, you'll learn how to

- log in an embedded computer remotely, using your laptop (or desktop)

- navigate through the Linux operating system

- find and install software on a Linux computer

- launch software on a Linux computer

- do basic editing in text files on Linux

- transfer files between your computer and an embedded computer

- change the IP address of an embedded computer

- shut down an embedded computer using a script

- configure the XBee transmitter/receiver modules

- use the XBee modules to make the Arduino wireless

Before You Begin

Usually these are open source hardware computers, which run on Linux. Therefore, we're going to cover some Linux—navigating through the operating system, installing software, and launching software on a Linux computer. In some of this book's projects, where we're going to use embedded computers, we're going to write some small scripts as well, to make the computer plug-and-play.

The goal of using such a device is to use the capabilities of a computer, without using the interface we're familiar with, the screen, the keyboard, and the mouse. Since we will be building interfaces with the Arduino, we're going to use these devices headless, meaning without a monitor to display information. To do that we need to configure a computer in such a way that it will launch all software we want on boot, and quit all software when we want to shut it down. Using scripts enables these features and helps us finalize our projects.

Even though we're not going to deal with scripts in this chapter, you'll learn how to use a text editor on Linux, as this will prove very helpful with writing scripts and debugging certain parts of an interface that might occur (like crashes or immediate shutdown as soon as a patch loads), and configuring certain settings of the embedded computer. We're going to do all this using our own laptop or desktop, without needing an extra screen, keyboard, and mouse, to use the embedded computer. We'll log in the embedded computer using an Ethernet cable and simple Unix commands, and we'll use our laptop's/desktop's screen, keyboard, and mouse to interact with the embedded computer.

In some projects in this book, we're going to use the XBee because it frees the performer from having a bunch of cables hanging out. Usually, we need to use the XBee (or Bluetooth) when we're not using embedded computers, as with the latter, it's probably pointless to make the project wireless, since the computer will be built-it the project's interface. So this chapter will cover two sort of complementary topics, which will prove very handy as we build the interfaces of this book's projects.

Parts List

Table 3-1 lists the parts that you'll need for this chapter.

Table 3-1. *Parts List*

Part	Quantity
Raspberry Pi or other embedded computer	1
SD card	1 microSD 8GB, class 10
Charger	1 5V 1A maximum. A smartphone charger will do.
Ethernet cable	1
XBee radios	2 (series 1 or 2)
Arduino XBee shield	1
XBee USB Explorer	1
Arduino Uno	1
Bluetooth module (HC-06)	1

Why Use Embedded Computers?

When building a project and putting it in an enclosure, it's very likely we don't want to carry a laptop to use the interface we've built. An embedded computer helps us build a stand-alone interface, which can also be plug-and-play. Think of digital synthesizers. Being digital means that they function with a computer, which does all the calculations to produce the sound. Imagine a digital synthesizer that is only a controller (the keyboard and a few knobs), and you need to bring along a laptop for it to work. This scenario doesn't sound very appealing. Using a computer to make electronic music has gained so much ground, that the laptop has become a standard piece of equipment.

In the past few years, the revolution of the Arduino was followed by the revolution of the embedded computers. Of course, they are nothing new because they have been around for years (many devices that you use every day have an embedded computer). Over the past few years, new, easy-to-use, open source, and low-cost embedded computers have made their appearance, making them more easily accessible. Consequently, communities have been created around each embedded computer, where people share work and knowledge and help each other with problems that may occur.

All this have brought the DIY communities to a very high level, as it is easier than ever before to create lots of things, from hobbyist to professional level. The Maker movement has grown to a great extent. You see people realizing DIY projects all over the world. This book is a result of the appearance of such tools, as it realizes projects that the reader is encouraged to try at home.

Which Embedded Computer?

Again, there are quite a lot of embedded computers around, but I'll mention the most popular. The computer with the wider use is by far the Raspberry Pi, https://www.raspberrypi.org/. This is a small computer (well, all embedded computers we'll talk about are small), used by a very wide community, and gets most of the attention when it comes to embedded computing. It runs on a Linux flavor based on Debian, called *Raspbian*, but other Linux distributions are available. The latest version has a special Windows 10 version, especially for the Raspberry. It is the only embedded computer, which can run on an operating system other than Linux or Android.

Another very popular computer is the BeagleBone Black, http://beagleboard.org/black. Till recently, it was quite more powerful than the Pi, but the latest Pi version has a processor, which runs a little bit slower than the BeagleBone Black, but has four cores, whereas the BeagleBone Black has only one. Still, this is a very nice low-cost embedded computer, which you should bear in mind. One disadvantage I find with the BeagleBone Black is that it lack an audio output jack, and can play audio only via its HDMI, whereas the Pi (and the other boards I'll mention here) does have an onboard audio jack, and that makes things quite easier.

The third embedded computer in our list is the Udoo, http://www.udoo.org/. This one is rather special because it is a computer with an embedded Arduino Due (a 32-bit Arduino board)! At the time of writing (August 2015), a new version is being released with an Arduino Uno pinout, and a few on-board sensors. The two pre-existing versions are a dual and a quad core computer, with an on-board Arduino Due. This is a powerful computer that integrates an Arduino, so it can make things very flexible. It comes at a higher price though (the Arduino is included in this price), and it has quite a bigger size (it's still small), so sometimes it ca prove not very appropriate as being a one-board machine, reduces casing flexibility.

The last embedded computer we're going to look at is the Odroid, http://www.hardkernel.com/main/main.php. This is the most powerful of all four computers. There are a few different versions available, where the lowest performance one has a higher performance than all the computers mentioned here. There are other powerful versions of the Odroid, but it's not really necessary to look at them at this point. This computer has the smallest size as well, so one might really consider one of these boards for a project. The latest versions of the Odroid don't include an onboard audio jack, like the BeagleBone Black, but even an older version provides very poor audio quality. You should either use the HDMI for audio (there are some shields for that), or an external sound card.

Having covered shortly these four embedded computers, we'll stick to the first one, the Raspberry Pi, for a few reasons. It might be the one with the lowest performance, but still it's powerful enough to host some of this book's projects. Price is one reason, even though the Odroid-C1+ can compete with it. The latter is missing the audio jack feature, so the Pi wins this round, as we'll be dealing with audio throughout this book. Another reason is its community. The other boards do have an active community, but the one of the Pi outnumbers them all, and this is a good reason for one to choose one board over another. The Pi is so popular that even Miller Puckette, the maker of Pd, has a compiled version of the latest Pd-vanilla for the Pi. He has compiled one for the Udoo as well, but with the previous things in mind, again the Pi wins this round over the Udoo. One last reason is that, apart from the on-board audio jack (which provides a restricted quality audio) there dedicated external sound cards, which mount on top of the Pi, increasing its performance, and making it quite compact, easy to fit in an enclosure. We're not going to cover the external sound card topic in this book, but it's good to know in case you want to extend one of your projects.

All these reasons lead us to the Raspberry Pi, which we'll use throughout this chapter. If you want to follow this chapter, it's better to get an embedded computer, as only by reading the instructions, you're not going to get as familiar with all processes described here. If you want to use another embedded computer, you're free to do so. The steps we're going to take are similar for all boards. You'd better use a Debian Linux flavor, however, as we'll be working with the Debian flavor for the Raspberry Pi: Raspbian. Still, even if you use Ubuntu Linux, it's OK, since Ubuntu is Debian-based. If you don't want to get an embedded computer yet, and leave it for later, you can skip this chapter and come back to it when you have one. Not all projects of this book will use the Raspberry Pi, so you'll be able to build quite some of the projects without one. You can buy a Raspberry Pi (or whichever computer you want) from its web site, or ask your local electronics store whether they distribute it.

Getting Started with the Pi

Now that you have your new Raspberry Pi, let's start using it. The first thing you'll need to do is install the Linux image to an SD card (image is the way to refer to the OS distributions. The SD card should be at least 8GB, preferably class 10). Go to the DOWNLOADS section of Raspberry's web site, and click the Raspbian icon (not the NOOBS one). As of September 2015, there are two Raspbian images: a Jessie and a Wheezy. I recommend downloading Jessie, which is more up-to-date, since Wheezy is the previous Debian OS. On the Jessie page, you can get instructions for writing the image to your SD card. It's beyond the scope of this book to get you through the steps required, but the instructions provided by the web site should be sufficient. If you encounter problems, you can sign up the Raspberry's forum and search the archives, or ask for help. I'll now assume that you have written the Raspbian image to your SD card, and we'll take it from there.

Since we're not going to use an external monitor, a keyboard, and a mouse to use the Raspberry, we'll have to log in remotely, using our laptop/desktop. You'll need an Ethernet cable to connect the Raspberry to your computer. If you have a Mac without an Ethernet port, you should get an Ethernet converter, to be able to log in your Pi. You'll also need a changer (a smartphone charger will do, as the Pi uses a micro USB for power, and 5V/500mA will be enough power), and access to a router with Internet connection. Connect the Pi to the router with the Ethernet cable and power it up. Make sure your laptop/desktop is connected to the same router and that you don't have any firewalls enabled that could block the communication with the Pi.

First, you have to find the IP address of the pi. The simplest way to do this is to log in the router. Most Internet services give access to the IP addresses of devices connected to them, through the router. To log in the router open an Internet browser and type the router's IP address to the URL field. This should be something like 192.168.1.1, where the last two fields will be either a 1 or a 2 (most likely the very last will be 1 anyway). Try all four combinations until you get connected to the router. When you get in the router's home page, you'll be asked for the router's password, and maybe the username as well. Usually they are both *admin*, so if you're lucky enough, you'll be logged in. If it's not *admin*, it's very likely that the password for the device is written in the back of the router, so use that. On the router's home page, browse to the list of connected devices; it could be called Status or something similar—routers are all different. There you should see the IP address of the Pi. You'll know it's the Pi because its name will be displayed. In case your router doesn't provide this feature, you'll have to use another way. There should be quite some different ways to achieve that, depending on your platform (Linux, OS X, Windows). Nmap is one solution, which runs on all three platforms. Go to Nmap's web site (http://nmap.org/download.html), and install it on your computer. The fastest way is to type the following in your computer's terminal:

```
nmap raspberrypi.local
```

This prints some information about the Pi, including its IP address. On Windows, use the Zenmap GUI provided by Nmap, and use the same command on Zenmap's window.

If you're using another embedded computer, I don't really know how to get its IP from its name, but typing the following will give all the active hosts in your network:

```
nmap 192.168.2.0/24
```

The first three fields should be the same for your computer's IP (and the same with your router). This will print a few different IPs, one of which should be the IP of your embedded computer.

If all this seems a bit overwhelming, there's a simple, not very efficient way to find the IP. Check your computer's IP (if you're connected wirelessly, check the wireless IP, otherwise check the Ethernet IP of your computer). If you're at home, there shouldn't be many devices connected to the network, so if your IP is for example, 192.168.1.3, then it's very likely that the Pi's IP will be 192.168.1.4, or something similar. In general, the first three fields will be the same, since the two devices are connected to the same network, and the last field will be different, where it should be a sort of an incrementing number. Since your laptop was connected to the router prior to the Pi, the Pi should have a greater number in that field. Once you have the IP, it's time to connect to the Pi (if you're trying to guess the IP, just repeat the following process until you are logged in). On OS X and Linux, the process is the same, but on Windows, it differs, so we'll go through it immediately afterward.

Getting Your Computer's IP

Since we'll need to know our computer's IP, let's see how we can find it. On Linux, you can get it from the terminal by typing:

```
ifconfig | grep "inet " | grep -v 127.0.0.1 | awk '{print $2}'
```

This will print the IP address of the Wi-Fi, and the Ethernet port. In this case, we care about the Wi-Fi. On OS X, you can get it from the terminal by typing:

```
ipconfig getifaddr en1
```

Another way is to open your System Preferences by clicking the apple icon on the top-left part of your screen and choosing **System Preferences…**. Select **Network** under **Internet & Wireless**, and then on the **Wi-Fi** tab on the left side, and you'll get your IP address.

On Windows, Go to **Control Panel** ➤ **Network Connections** ➤ **Wireless Network Connection**, and under the **Support** tab, you'll find your IP address. This is tested on Windows XP, which is now obsolete, and Windows versions tend to change quite a lot of things in their OS upgrades, so the process might be a bit different on your Windows machine. Still, this information should give some insight as to where you should look for your IP address.

Logging in the Pi from OS X and Linux

On OS X make sure you have XQuartz installed, `http://xquartz.macosforge.org/landing/`. Open a terminal window (on OS X go to **Applications ➤ Utilities** and launch the **Terminal.app**; on Ubuntu press **Ctrl+Alt+T**, or search your files) and type the following:

```
ssh -X pi@192.168.x.x
```

`ssh` stands for *Secure Shell*, which is a protocol for getting a secure access to a remote computer, like the Pi. Replace the x's in the two last fields of the IP address with the ones of the IP of the Pi. The `-X` character (called a *flag*, which is recognized by the hyphen symbol) means that we will be using X11 (XQuartz on OS X) so that the Pi can display its GUI. Depending on the version of the OS X you're using, you might need to launch XQuartz prior to typing this command in the terminal. Hit **Return** and your computer will connect to the Pi. The first time that you connect, you'll get a warning concerning the authenticity of the host (the Pi) and you'll be asked whether you're sure that you want to continue. Type **yes** and hit **Return**. Then you'll be asked for the password of the Pi, which is *raspberry*. As you type it, you won't see anything in the terminal (like the top of Figure 3-1, there's nothing seen after password), but the password will actually be typed. Once you enter the password, hit **Return** and you'll be logged in the Pi. Figure 3-1 shows the prompt of your first log in to the Pi.

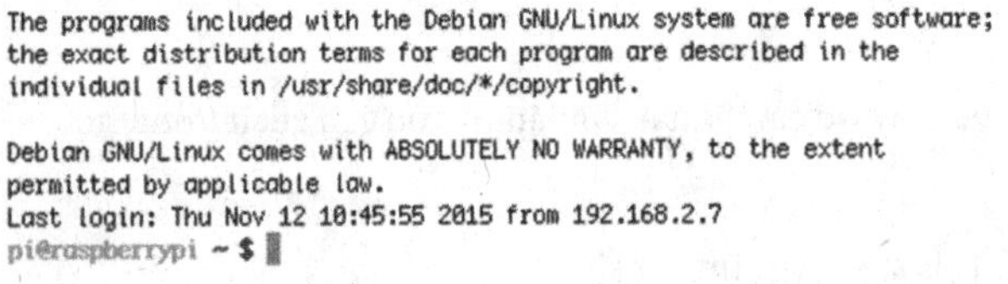

```
The programs included with the Debian GNU/Linux system are free software;
the exact distribution terms for each program are described in the
individual files in /usr/share/doc/*/copyright.

Debian GNU/Linux comes with ABSOLUTELY NO WARRANTY, to the extent
permitted by applicable law.
Last login: Thu Nov 12 10:45:55 2015 from 192.168.2.7
pi@raspberrypi ~ $ ▊
```

Figure 3-1. First log in to Pi

Logging in from Windows

In Windows, you'll need a couple of things before you log in your Pi. First, you'll need an X server, which the Pi will use to display its GUI on your computer. There are several X servers around, including MobaXterm, `http://mobaxterm.mobatek.net/`, Xming, `http://www.straightrunning.com/XmingNotes/`, and Cygwin, `http://x.cygwin.com/`. Choose whichever you like and install it on your computer. You'll also need PuTTY, `http://putty.org/`. Choose the first link on its download page, putty.exe. Next, launch the X server you have installed and run `putty.exe`. (I use Xming, and when launched, there's a small icon on bottom left of the screen, no window opens.) Figure 3-2 shows the PuTTY window.

Figure 3-2. *PuTTY window*

Type the Pi's IP in the **Host Name** (or IP address) field. (In Figure 3-2, I typed the IP of my Pi. You should use the IP of your Pi.) The **SSH** selection underneath should be selected by default, which we want. SSH stands for *Secure Shell*, which is a protocol for getting a secure access to a remote computer, like the Pi. Before you click **Open**, go to the right-side menu and click **SSH**. Next, click **X11**. Figure 3-3 shows this menu. Make sure that you select the **Enable X11 forwarding** tick box. Once you've done that, click **Session** on the left-side menu, and you'll get back to the login window, as shown in Figure 3-2. In this window, you can save your session, but don't do it just yet, because we're going to change in IP of the Pi, so you'll have to be a bit patient.

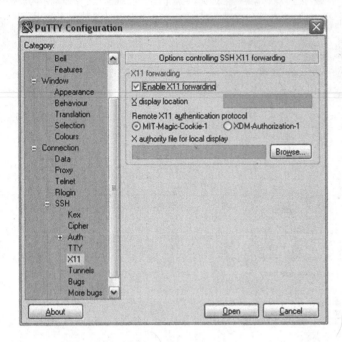

Figure 3-3. *X11 menu in PuTTY*

Once you have all this done, click **Open**. You'll get a window telling you that the host is not registered in your computer, as shown in Figure 3-4.

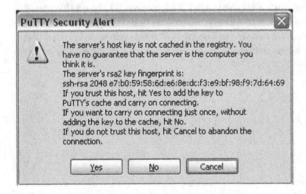

Figure 3-4. *PuTTY host registration window*

Click **Yes** and you'll get to the login window, shown in Figure 3-5. In the beginning there will be the first line only, which reads **Login as:**. Type **pi** and hit **Return**. You'll be asked for Pi's password, like in Figure 3-5 (don't mind the IP that appears in the Figure). Type **raspberry**. As shown in Figure 3-5, the password won't be visible as you type it, but the computer will receive it. In Unix terminals (Linux and OS X), passwords are hidden from the monitor. If you typed the password correctly, you'll be logged in. Congratulations!

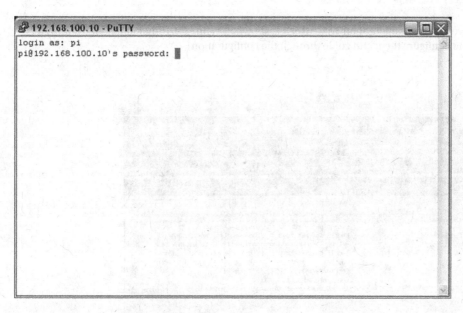

Figure 3-5. PuTTY login window

Configure the Pi

The steps from now on are common for all three operating systems (Linux, OS X, and Windows), since we're now logged in the Pi, and no matter what OS you use, we're now in Linux. The first time you log in you'll need to configure your Pi. To do this run the following command:

```
sudo raspi-config
```

This will lead you to the Configuration Tool, which is shown in Figure 3-6 (other embedded computers might prompt you to configure them and guide through the configuration).

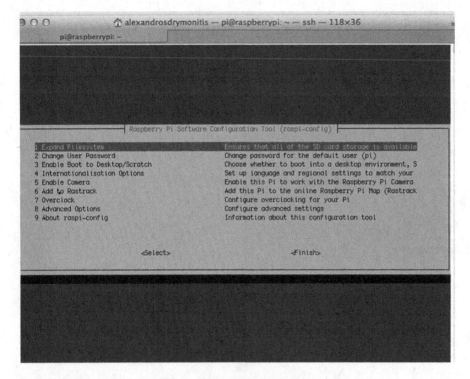

Figure 3-6. *Raspberry Pi's software configuration tool*

To navigate in this tool, you have to use the arrow keys. The up and down arrow keys will navigate through the options on the left side of Figure 3-2, which are numbered. The right arrow key will get you to <Select>, and if you hit it once more, it will get you to <Finish>. Clicking the left arrow key will get you back to the configuration options. Hit **Return** on the first option, **1 Expand Filesystem**. This will take just a moment. Hit **Return** to get back to the options, and navigate to **7 Overclock**. Hit **Return** and in the options, choose **High 950Hz**. Hit **Return** to enable it and go back to the main menu. Click the right arrow key and navigate to <Finish>. You'll be asked if you want to reboot now. Click **yes** and the Pi will reboot. This means that the connection between the Pi and your computer will be lost. Wait until the Pi reboots and log in again using the ssh command for OS X and Linux, and PuTTY for Windows, as before.

Navigating Through the Linux system

Now that we've logged in the Pi, let's navigate through its system. Since we're logged in the Pi remotely from another computer, we'll do all navigation, folder creation, file management, and so forth through the terminal. Figure 3-7 shows the prompt of the Pi's terminal.

Figure 3-7. *Raspberry Pi's command-line prompt*

In green, you see the name of the user, which is *pi*, logged in to the "raspberrypi" device. The dollar sign is the user sign, called the *prompt*. If you're not a super user (called a *root*), but a plain user, you get this prompt. If you're root, you get the # prompt. Go ahead and type the following: ($ or # will be included in all commands from now on to clarify the user. You shouldn't write it yourself in the terminal because it is already there.)

```
$ whoami
```

You should get pi. This is a Unix program telling you what kind of user is currently using the computer. Pi's default user is called a *pi*. Now type the following:

```
$ ls
```

This is a shortcut for list. "ls" is a program that prints to the monitor the contents of the current directory (by directory, we refer to what is usually called a *folder*). It should read python_games. To see what kind of file this is, type the following:

```
$ file python_games
```

Note that all commands are separated from their arguments with a white space. The first word (or combination of few letters) is the command, and afterward the arguments follow, don't mix them in one long word. The preceding command should print python_games/: directory. We can see that this is a directory by default on the Raspbian image. Directories appear is in blue to distinguish them from other types of files. Now type the following:

```
$ pwd
```

This stands for *present working directory*, which is a program that prints the path to the directory you currently are. This should give you /home/pi. This is the home directory of the user "pi". Now let's create a new directory in our home directory. Type the following:

```
$ mkdir pd_patches
```

mkdir stands for *make directory*, which is a program that creates directories. pd_patches is the name of the directory we want to create, which we pass as an argument to mkdir. If you type **ls** again, you'll now see two directories, one called python_games and another one called pd_patches. To go into the new directory, type the following:

```
$ cd pd_patches
```

This stands for "*change directory* and we provide the directory we want to go to as an argument (here pd_patches). Now if you type **pwd** again, you'll get /home/pi/pd_patches. Notice that Linux has a tree structure, where / is the hard drive of the computer, and subsequent directories are being separated by forward slashes. The new directory we created is in the pi directory, which is in the home directory, which is in the / directory, which is the hard drive of the Pi. Now let's go back to our home directory. Type the following:

```
$ cd ..
```

The double dot means one directory up. This will get us back to /home/pi. Now create a directory called pd_abstractions, inside the pi directory, the same way you did with the pd_patches directory. Go to this directory using cd, but do the following. Type the following:

```
$ cd pd_a
```

Hit the **Tab** key. You'll see the whole name of the directory appear on screen. The Linux command line has tab completion, which means that if enough letters of a directory are provided, hitting Tab will give the rest of the name (by "enough" I mean that there's no name clash. In /home/pi there's only one directory starting with pd_a, the newly created pd_abstractions). This is a good way to use the terminal, not only because it saves you some typing, but it's also less error-prone since the computer finds the name of a directory, so there's no way you'll make a typing mistake.

Let's go back to our home directory (by "home" I mean the /home/pi directory, not /home) and create another directory. Now we want to create a directory inside the pd_abstractions directory. We could have created it while we were in that directory, but we came back to the home directory for practice and to better understand the Linux tree structure. We'll create a directory to store the abstractions that deal with the Arduino, so we'll call it arduino_abs. In your home directory, type the following:

```
$ mkdir pd_abstractions/arduino_abs
```

Now go to that directory using cd. Type the following:

```
$ cd pd_a
```

Hit the **Tab** key, and you'll get cd pd_abstraction/. Type **a** and hit **Tab**. You'll get cd pd_abstractions/arduino_abs/. Hit **Return** and you'll be taken to the newly created directory. To go back to the home directory, type the following:

```
$ cd ../..
```

This will take you two directories up, where the home directory should be. Another way to navigate back to home, from whichever directory you are (as user "pi") is this:

```
$ cd ~
```

The tilde character means "the home directory of the current user."

Editing Text Files in Linux

Now that we've navigated a bit through the system, let's do a bit of text editing. Raspbian, and many other Linux distributions, has a text editor preinstalled, which runs in the terminal. This editor is called *nano*. There is also the *vi* text editor, but I prefer to use nano because it's very simple and easy to use, so we'll use that in this book. In your home directory, type the following:

```
$ nano test_text.txt
```

The nano editor will open, as shown in Figure 3-8.

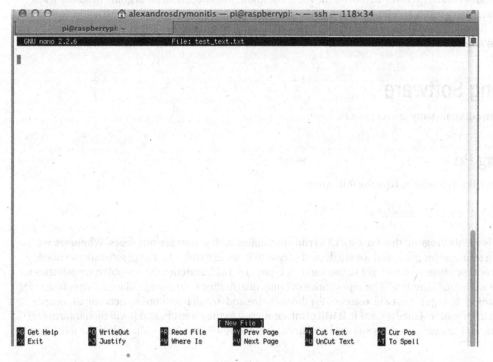

Figure 3-8. *The nano text editor*

Go ahead and type something; for example, "This is a test text using the nano text editor." Press **Ctrl+O** (if you're on OS X, still use Ctrl, as now we're actually on Linux) and at the bottom you'll see the prompt: File Name to Write: test_text.txt. Hit **Return**. At the bottom, in the place of [New File] (shown in Figure 3-8), you'll see [Wrote 1 line] (or as many lines of text as you entered). Press **Ctrl+X** to get out of nano and back to the Unix prompt. Now type **ls** to get a list of the present directory's contents, including the three directories (python_games, pd_patches, and pd_abstractions). You'll see the new text file, test_text.txt. To see the file's contents, type (use tab completion):

```
$ cat test_text.txt
```

And you'll get the text that you just wrote in nano. "cat" is short for *concatenate*, which is a program to concatenate files. Using it this way, we can just see the contents of a file, without entering an editor. We'll be using nano, mostly for writing some simple scripts to tell the Pi to launch certain software on boot or to shut it down, but also in case we need to debug a Pd patch that might cause a crash or an automatic shutdown. This is in case we cannot edit it in Pd itself, but only in its textual form.

Since the text file we wrote is only a test to see how nano works, let's delete it. To do this type (remember to use tab completion):

```
$ rm test_text.txt
```

rm stands for *remove*. When you use rm, the file is completely deleted, instead of going to some "trash" directory from where you can delete it later on. Also, the Pi doesn't ask you if you're sure whether you want to delete the file or not, so be very careful when you use it, as you might delete a file you don't really want to, and you won't be able to retrieve it.

Installing Software

It's time to install some software on your Pi.

Installing Pd

Let's start with Pd. To install it, type the following:

```
$ sudo apt-get install puredata
```

"sudo" is a Unix program that enables us to run commands with superuser privileges. Whenever we want to use a terminal program, but we are denied access to it, we use sudo. To install software, we need superuser (root) privileges, so we have to use sudo. apt-get install searches the Raspbian repositories for the software we want, "puredata". The repositories of Linux distributions are storage places with software for each distribution. apt-get install searches for the software and install it without us needing to compile the software from source. Once it finds it, it will print some data to the monitor, and it will then inform you about the amount storage place this software will need, and ask you if want to continue, like the following text:

```
Do you want to continue [Y/n]?
```

Type **y** and the installation will go on. The terminal will type quite some data about the installation process, and when it's done you'll get back to the $ user prompt. Installing Pd this way will actually install Pd-vanilla, so you'll need to install some external libraries. To find which ones are available in the Raspbian repository, type the following:

```
$ apt-cache search "^pd-"
```

We don't need to use sudo for apt-cache search, since it only searches the repositories, but doesn't install any software to the computer. This will print a list with all available external libraries in the Raspbian repositories. To enhance our Pd, we'll install some of them but not all. Go ahead and type the following:

```
$ sudo apt-get install pd-arraysize pd-comport pd-freeverb pd-ggee pd-iemmatrix pd-list-abs
pd-mapping pd-pan pd-pdstring pd-purepd pd-zexy
```

Again, type **y** for *yes* when asked if you want to continue with the installation. This will install all the libraries to a directory created by the Pd installation, which is /usr/lid/pd/extra. After the installation is done, type the following:

```
$ ls /usr/lib/pd/extra
```

This will print the directories of the installed external libraries:

```
Gem arraysize comport cyclone freeverb~ ggee iemmatrix libdir list-abs mapping
maxlib pan pddp pdstring pix_drum pix_fiducialtrack pix_mano purepd zexy
```

There are few more packages that were installed when we installed Pd, such Gem, pix_fiducialtrack, and the ones we installed ourselves.

Launching Pd

To launch Pd, you must type the following command in Pi's terminal:

```
$ /usr/bin/pd
```

What we're doing here is call an executable file, called pd, which is in the /usr/bin/ directory. This file launches Pd (you can launch Pd in a similar way on OS X too, since it's also a Unix system). The first time you launch Pd, it might take a while, and you'll see the "watchdog signaling pd..." message in the terminal. Don't worry, Pd will launch and these messages will stop appearing on your screen. When Pd launches, the first thing to do is test if the audio is working properly (we didn't do this in Chapter 1 because we started building patches from the beginning, and the first patch did essentially the same thing). Go to **Media ➤ Test Audio and MIDI...**, and the patch shown in Figure 3-9 will open.

As the comment on top of the patch states, this patch is made for testing the audio and MIDI connections in Pd. Plug a pair of headphones into your Pi and click **80** on the left radio underneath **TEST TONES**. This should create a (low in volume) sine tone at 440 Hz. To make it louder, type **100** in the number atom underneath the radio. You can also listen to Pd's white noise by selecting **noise**. If you have a MIDI device, you can plug it in the Pi and relaunch Pd to see if that is also working, as we'll use a MIDI keyboard in one of this book's projects.

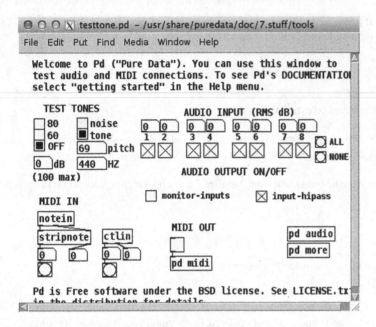

Figure 3-9. *Test Audio and MIDI... Pd patch*

Setting up External Libraries in Pd

Once you've tested audio in Pd, you should set the external libraries paths to Pd's search path. Go to **Media ➤ Preferences ➤ Path...** and you'll get the Path window. Click **New** and the window to add a new patch will open, shown in Figure 3-10. Click the **Directory:** to get the pop-up menu, which should display three directories, /, /home, and /home/pi. Click **/** and navigate to /usr/lib/pd/extra. From there, choose the libraries you've already installed, one by one, all but zexy and iemmatrix.

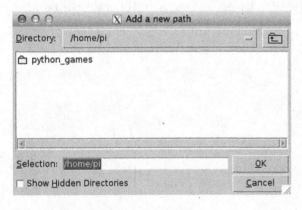

Figure 3-10. *Add new path window*

Once you've selected the libraries, you search path window should look like Figure 3-11. Click **Apply** and then **OK** so that the new settings will be enabled.

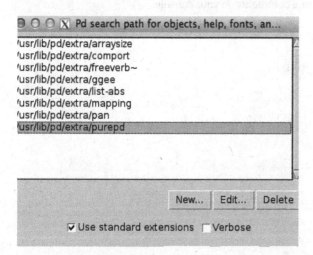

Figure 3-11. *Selected libraries in Pd's search path window*

For the other two libraries, zexy and iemmatrix, we need to do something else to add them to Pd. Go to **Media ➤ Preferences ➤ Startup...** and the **Pd libraries to load on startup** window will appear, as shown in Figure 3-12. In the **Startup flags:** field type **-lib zexy -lib iemmatrix**, like in Figure 3-12. Click **Apply** and then **OK** to set these features as well.

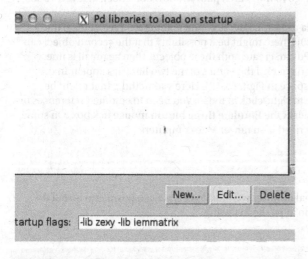

Figure 3-12. *Startup flags Pd window*

Now, to test the new setting, you must quit and relaunch Pd. Go to File ➤ Quit, or press **Ctrl+Q**. Go back to the terminal and relaunch Pd, by typing **/usr/bin/pd**. Figure 3-13 shows the Pd console with the zexy and iemmatrix libraries loaded. You should have the same comments on your console.

Figure 3-13. *Pd's console with the "zexy" and "iemmatrix" libraries loaded*

Test a few objects that we will very likely need for this book's project. Open a new patch and try to create the following objects (OS X users, all shortcuts that use Cmd on OS X, should be replaced by Ctrl, as this is Linux): [arraysize], [comport], [shell], [list-drip], [uzi]. If Pd can create all the objects, then everything is fine. If there are any objects that cannot be created, go to the search path window and check that you've set everything as suggested.

You should also test the two libraries we set via a startup flag: zexy and iemmatrix. Try to create the following two objects: [limiter~], and [mtx_*~ 4 4 20] (there might be a possibility that the second object can be created like this [mtx_mul~ 4 4 20], instead). If Pd can create both these objects, then again all is fine. If there's any problem, go to the startup window and check if the names of the two libraries appear in the window (not the Startup flags: field, but the white space in Figure 3-12). Here's something that might be useful to some users of OS X computers: to be able to right-click in the Pi, you need to edit the Preferences of XQuartz. Under the **Input** tab, you might need to select the **Emulate three button mouse** tick box. On some computers, it might not be necessary, but you may need it on newer Mac computers.

Installing Arduino

Now that we've tested Pd's audio and all the external libraries we want to use, we can go on and install the Arduino IDE to our Pi. As with Pd, to install Arduino, type the following:

```
$ sudo apt-get install arduino
```

This will take you through the same process as with Pd, and it will install the Arduino IDE to the Pi. Installing the Arduino IDE with apt-get creates a directory called sketchbook on your home directory (/home/pi) to save your sketches. When the installation is finished, type the following and the Arduino IDE will launch.

```
$ /usr/bin/arduino
```

You'll get a few warning messages in the terminal, but that's nothing to worry about. It will take a bit of time to launch, but eventually you'll see a new sketch window on your screen. It doesn't look exactly like the new sketch window of your computer's IDE, because it doesn't have the void setup() and void loop() functions already there. This is because the IDE we've installed is an older version than the one you have on your computer, if you installed it recently. Let's first test if the Arduino IDE works properly, and then we'll talk about versions. Go to **File ➤ Examples ➤ 01.Basics** and click **Blink** (the color highlighting is a bit different as well, because of the different version). This sketch, apart from being the first sketch to write/read when learning Arduino programming, also serves as a test sketch, to see that the IDE works fine with your board. Go to **Tools ➤ Board** and select your Arduino (if it's the Uno, it will most likely be already chosen), and then go to **Tools ➤ Serial Port** to choose the Arduino port. You'll probably get only one available port, /dev/ttyACM0, which is the Arduino port. Choose it and upload the sketch to your board. If all goes well, the integrated LED of the Arduino board should blink.

When launching software this way, we can't use the terminal anymore, until we quit the software. Still, as we create Pd patches, or as we write Arduino sketches, we might want to change things in the Pi, and since the terminal is the only way, we need to launch software in such a way that enables us to use the terminal. This is done using the ampersand (&) at the end of the command. So, to launch Arduino, it's best to use this:

```
$ /usr/bin/arduino &
```

This way, you'll be able to go back to the terminal and do whatever you want, while the Arduino IDE is running. The same applies to Pd, of course.

Let's Talk About Versions

Now let talk about software versions. The Arduino IDE version (which is on top of a sketch window) is 1.0.5, whereas the one we've installed on our computer from the Arduino web site is 1.6.5. That's quite a version difference. To see which Pd version was installed, run this command:

```
$ /usr/bin/pd -version
```

Using the -version flag will output the version of Pd and will not launch the software. This should produce the following output:

```
Pd-0.46.2 ("") compiled 07:48:11 Oct 29 2014
```

The latest vanilla version is 0.46-7, while we got 0.46.2 The version difference is not big, but it's worth mentioning. If you're using Raspbian Wheezy, then the version difference will be quite big, since you'll get Pd-0.43.2 instead.

When we install software using apt-get, the program will search the repositories of the Linux distribution, and install the software it will find, that fits its argument ("puredata" and "arduino"). The repositories are not always up-to-date, so even though you'll find the latest version of the software on its own web site, this won't apply to a Linux repository as well. If you really want to get the latest version of software, you'll need to download its source code and compile it yourself. This is beyond the scope of this book, so I won't explain the process. Still, the software versions from the Raspbian Jessie repositories are good enough.

Exchanging Files Between Your Computer and the Pi

We've already created a directory called arduino_abs inside the pd_abstractions directory, to store the Pd abstractions that deal with the communication with the Arduino. We could launch Pd and make those patches again, but since we have downloaded them from the Internet, and there are quite some subpatches there, it will be a bit difficult to make them from scratch, plus it will be rather error prone, which could create bugs difficult to trace. A solution to this would be to either install an Internet browser on the Pi and go the web page where these abstractions are and download them, or transfer the already downloaded files form your computer to the Pi. The second solution proves to be the best, because once you know how to transfer files, you can do that for any kind of file, and not only for stuff you've downloaded from the Internet. On OS X and Linux, this is done the same way, but on Windows it's different, so we'll go through the OS X/Linux first and then we'll talk about Windows.

Transfer Files from OS X and Linux to the Pi, and Vice Versa

Although there are some GUI programs to do this, like Cyberduck, a WinSCP Mac port, and some others, we'll use the good old terminal to do this, since it's rather straight forward and fast. In a new terminal window on your computer (not Pi's terminal), type the following command, taking care of the white spaces between the scp command and the two arguments, the path to the files we'll be transferring, and the destination computer:

```
$ scp /path/to/arduino/abstractions/serial*.pd pi@192.168.x.x:/home/pi/pd_abstractions/
arduino_abs
```

scp stands for *secure copy protocol*, which is a Unix program/protocol that copies files over an SSH connection. Replace /path/to/arduino/abstractions/ with the actual path, which, for example, could be ~/Documents/pd_patches/abstractions (on OS X, you can drag and drop the folder you want to navigate to in the terminal, and its path will automatically print). The two x's in the Pi's IP, with the actual values of the Pi's IP address (tab completion will work for the path of your computer, but not for the path of the Pi). Notice the asterisk after the word serial. This is called a *wild card* and it means anything starting with serial and ending with .pd, because we want to copy both [serial_print] (actually [serial_print_extended]) and [serial_write], plus their help patches, which are called serial_print-help.pd and serial_write-help.pd, respectively. So instead of writing this command four times, we write it only once. Hit Return and you'll be asked for Pi's password (which is *raspberry*). Type it (remember, you won't see the text as you type) and you should get the following in your terminal:

```
serial_print-help.pd            100% 4408    4.3KB/s   00:00
serial_print.pd                 100% 8576    8.4KB/s   00:00
serial_print_extended-help.pd   100% 4446    4.3KB/s   00:00
serial_print_extended.pd        100% 8672    8.5KB/s   00:00
serial_write-help.pd            100% 2997    2.9KB/s   00:00
serial_write.pd                 100% 9606    9.4KB/s   00:00
```

This informs us that all the files we wanted to transfer, have been transferred successfully.

We also want to copy the Arduino sketches we have that work with these abstractions. Remember that all Arduino sketches must be included in folders sharing the same name with the actual sketch. Using scp this way won't work with directories (folders). We need to include a flag to do this. Go ahead and type the following:

```
$ scp -r /path/to/arduino/abstractions/serial*/ pi@192.168.x.x:/home/pi/pd_abstractions/
arduino_abs
```

Here we're using the -r flag, which means recursive. This way the computer will copy not only the directory, but also its contents. This is necessary when we want to copy whole directories using the terminal. The fields of the command now are four, the command name, the -r flag, the path to the directories to copy, and the path to the copy destination. This means that we must insert a white space between each of these. Type Pi's password when asked to. Now in the arduino_abs directory on your Pi, you should have two new directories: serial_print and serial_write. Each contains the corresponding Arduino sketch.

What if you want to transfer a file from the Pi to our computer? We use the same program, the other way round. Go to Pi's terminal and create a test text file, like we did earlier, using nano. Type the following:

```
$ nano test_text.txt
```

Type some dummy text in there. Hit **Ctrl+O** and **Return** to save the contents, and then hit **Ctrl+X** to exit nano. Now, in your computer's terminal (not the Pi's; don't confuse the two), type the following:

```
$ scp pi@192.168.x.x:/home/pi/test_text.txt  ~/Documents
```

Again, you'll be asked for Pi's password, so type it and then go to your Documents directory, in your computer, and you should see the dummy text file in there. Check its contents to make sure it's exactly what you typed in the Pi. This way we can exchange files between our computer and the Pi, in a fast and easy way. Mind though, that this way you could make Pd patches on your computer and then transfer them to the Pi, but that's not a very good way to go, because your computer is much more powerful, concerning CPU. It's easy to make a heavy patch that the Pi will find very difficult, or impossible, to handle, and launch its CPU sky high. It's always advisable to develop your Pi project in the Pi itself.

Transfer Files from Windows to the Pi and Vice Versa

In Windows, you'll need to install some software to be able to transfer files to your Pi. I found WinSCP (https://winscp.net/eng/index.php), which is open source and a rather user-friendly software for that. There are more solutions, but this is rather straightforward that seems to work out of the box, with GUI, so we'll use that. Install it onto your computer and launch it. Figure 3-14 shows WinSCP's login window. "SCP" stands for "Secure Copy Protocol", which is a Unix program/protocol for transferring files from one computer to another. WinSCP is an SCP program for Windows.

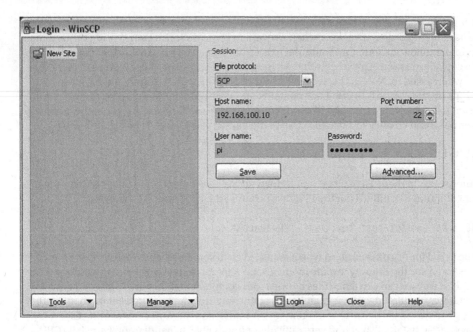

Figure 3-14. *WinSCP login window*

Click the **File protocol:** tab and select **SCP**. In the **Host name:** field type the IP of your Pi (not the IP you see in Figure 3-14). In the **User name:** field type **pi**, and in the **Password:** field type **raspberry**. Your session should look like Figure 3-14, except from the IP address. Hit **Login** and you'll get connected to the Pi. Mind that you don't need to log out from PuTTY to use WinSCP; you can be logged in from both programs simultaneously.

The first time you log in, you'll get a warning window saying that, "The server's host key was not found in the cache." Click **yes** to log in. Figure 3-15 is a WinSCP connected to a Pi session. Don't save this session yet, as you did with PuTTY, because we're going to change the IP of the Pi later on.

On the left side, you have your computer directories, and on the right side, you have the Pi directories. Transferring files from one computer to another is super easy, as it's done with a single drag and drop. Navigate to the Pd abstractions that deal with the communication with the Arduino, and drop these files to the arduino_abs directory on your Pi.

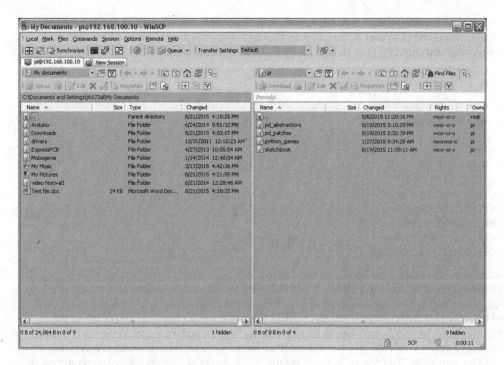

Figure 3-15. WinSCP connected to Raspberry Pi session

Changing the IP of the Pi

Up until now, we've used the Pi through a network router, but that's not always very convenient. We'll need to use the Internet sometimes, if we want to install software from the Raspbian repositories, but in general, when we want to work with the Pi, it's best that it is connected straight to our computer, instead of the router. If we use the Pi as is, it will automatically get an IP using the DHCP protocol (this is the protocol used by your computer to get an IP when logging in to a network). We can use tools like Nmap to get the IP, but this is not very efficient, as we want to log in pretty fast and start developing our projects. The best solution to this is to set the Pi to have a static IP, which we'll know. The way to set an IP address is via a file, which lies in /etc/network/ in the Pi. cd to that directory (by "cd" I mean to use the cd command to go to that directory) and type the following:

```
$ sudo nano interfaces
```

Don't forget to use tab completion. Nano will open with the `interfaces` file content. We need to use sudo because we're using the Pi as user pi, but we're not in our home directory, so we need root privileges. Listing 3-1 shows what you should type in that file.

Listing 3-1. The /etc/network/interfaces File

```
1.    auto lo
2.    iface lo inet loopback
3.
4.    auto eth0
5.    allow-hotplug eth0
6.    #iface eth0 inet dhcp
7.    iface eth0 inet static
8.    address 192.168.100.10
9.    netmask 255.255.255.0
10.    network 192.168.100.0
11.    broadcast 192.168.100.255
```

The "interfaces" file might have some more text in it, like "auto wlan0", "allow-hotplug wlan0", and so forth, but we don't mind it. Line 6 in Listing 3-1, which begins with #, is a comment. Well, it's not really a comment, but an executable line that has been commented out. By "commented out" I mean that we want this line to be there, but we don't want it to be executed, so we turn it into a comment. Of course, in the actual file, the lines won't be numbered, they're numbered only here for convenience. All we really need to care about are lines 7 and 8. Line 7 defines that a static IP address will be set. Line 6, which is commented out, defines that an automatic IP address will be set, via the DHCP protocol, but since it's commented out, it won't actually do it. Line 8 sets a static IP address. The first two fields (192 and 168) are standard, so we'll just use these. The last two fields are the ones we set manually. The third field, which is set to 100, is called the *subgroup*. This is a group without the network. We set it to 100, because usually DHCP sets small numbers like 1 and 2, so setting a high number will make it unlikely to get a clash with a DHCP address. The last field is the ID of the device. This can be anything between 1 and 254, but it must be unique within the subgroup. Don't worry about lines 9 to 11, just use them as they show in Listing 3-1.

For these setting to take effect, we must reboot the Pi, but we must also change the IP of our computer, otherwise they won't belong to the same subgroup, and they won't be able to see each other and get connected. In your Pi's terminal, type the following:

```
$ sudo reboot
```

Unplug it from your router (don't unplug the power). On Windows, PuTTY might warn you that the "Server unexpectedly closed network connection," but it's of no importance; just close the window. The Pi will now reboot, and the new, static IP will be set. Connect the Pi to your computer via the Ethernet cable, and set a static IP to your computer. Your computer's IP must have the first three fields the same as the Pi, and in the last field any number between 1 and 254, but not the same as the Pi, because this is the ID of the device, and the two devices must not have the same ID in the same subgroup (the third field is the subgroup). Each operating system has its own way to set a static IP, so we'll cover them separately.

Setting a Static IP on Linux

In your computer's terminal, type the following:

```
$ ifconfig eth0
```

The output of this should be something like the following:

```
eth0      Link encap:Ethernet  HWaddr 08:00:27:e3:ef:78
          inet addr:10.0.2.15  Bcast:10.0.2.255  Mask:255.255.255.0
          inet6 addr: fe80::a00:27ff:fee3:ef78/64 Scope:Link
          UP BROADCAST RUNNING MULTICAST  MTU:1500  Metric:1
          RX packets:1120 errors:0 dropped:0 overruns:0 frame:0
          TX packets:852 errors:0 dropped:0 overruns:0 carrier:0
          collisions:0 txqueuelen:1000
          RX bytes:800971 (800.9 KB)  TX bytes:454335 (454.3 KB)
```

This is information about the Ethernet connection of your computer. The second line reads inet addr:10.0.2.15, which is the IP of the Ethernet connection of your computer (yours is very likely to be different). To change it, type the following (don't forget to include sudo):

```
$ sudo ifconfig eth0 192.168.100.20
```

The last field can be any number between 1 and 254, but not the same with the Pi, which in Listing 3-1 is 10. Now type **sudo ifconfig eth0** again and check the **inet addr:** field, and make sure that it includes the IP you just set. This technique changes the IP of your computer only for this session. As soon as you reboot it, the IP will again be set via DHCP. To set a permanent static IP you have to do what you did with your Pi, since that's also a Linux system. Mind, though, to change only the Ethernet IP, not the Wi-Fi.

Setting a Static IP on OS X

On OS X, you can set a static IP in a very similar way to Linux, or via the System Preferences. For the command-line version, in your computer's terminal, type the following:

```
$ ifconfig en0
```

This will give you information about your Ethernet connection, which will look like this:

```
en0: flags=8863<UP,BROADCAST,SMART,RUNNING,SIMPLEX,MULTICAST> mtu 1500
        options=27<RXCSUM,TXCSUM,VLAN_MTU,TSO4>
        ether 00:25:00:a0:1a:b2
        inet6 fe80::225:ff:fea0:1ab2%en0 prefixlen 64 scopeid 0x4
        inet 169.254.15.79 netmask 0xffff0000 broadcast 169.254.255.255
        media: autoselect (100baseTX <full-duplex,flow-control>)
        status: active
```

What we care about is the fifth line, which reads inet 169.254.15.79. This is the IP set via DHCP (yours will probably be different). To change it, type (don't forget to include sudo) the following:

```
$ sudo ifconfig en0 192.168.100.20
```

As with the Linux version, the first three fields must be the same with the Pi, but the last must be a number between 1 and 254, not the same with the Pi, which is 10 in Listing 3-1. Now type **sudo ifconfig en0** again and check the line of the IP, to make sure that it's the one you've set.

Instead of the terminal, you can also use System Preferences. Go to **Applications ➤ System Preferences**, and click **Network**, in the Internet and Wireless section. Figure 3-16 shows the Network window of OS X's System Preferences. Click the **Configure IPv4** tab and select **Manually**. This will enable you to type in the following three fields: IP address, Subnet Mask, and Router. In the **IP address** field type the static IP for your computer, which should be 192.168.100.x, where x must be between 1 and 254, but not the same as the Pi's, which is 10. In the **Subnet Mask** field, type **255.255.255.0**, and leave the Router field blank. The **Apply** button on bottom right of the window will have become clickable, so when you type the two addresses, click it. Now your computer has a static IP for Ethernet connections. Using the System Preferences will make this IP permanent, whereas using the terminal will only change the IP for the current session, and as soon as you reboot your computer, the IP will be set via DHCP again. There are ways to set a permanent IP apart from using the System Preferences, but they are a bit complicated, and since the System Preferences allow us to do this so easily, we won't deal with that here.

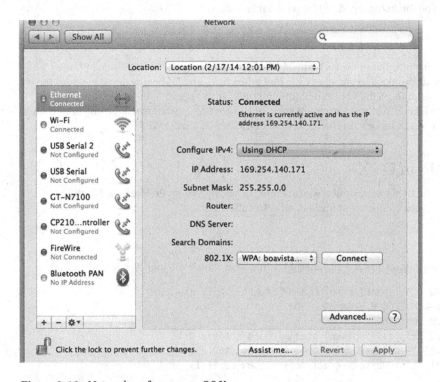

Figure 3-16. *Network preferences on OS X*

Setting a Static IP on Windows

The description here applies to Windows XP, but other Windows versions should be rather similar. Go to **Control Panel ➤ Network Connections ➤ Local Area Connections**, click **Properties**. In the Properties window, select **Internet Protocol (TCP/IP)** and click **Properties** again. The window shown in Figure 3-17 will open. The **Obtain IP automatically** option will be chosen by default. Click **Use the following IP address** and the fields that follow will become editable.

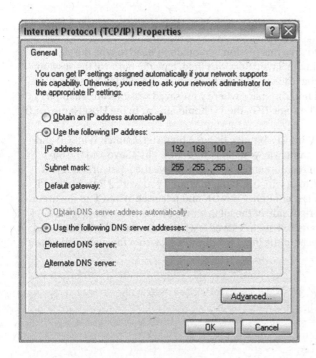

Figure 3-17. *IP set window on Windows XP*

In the IP field, type your IP , which should be **192.168.100.x**, where x must be between 1 and 254, but not the same as the Pi's last field of its IP, which is 10. In the **Subnet mask:** field type **255.255.255.0** and leave the rest blank. Click **OK** and the new, static IP will be set.

Log in to the Pi Without a Router

Now that both the Pi and your computer have a static IP and they are both in the same subgroup, you can connect to the Pi straight from your computer, without using your Internet router. Your Pi should already be connected to your computer's Ethernet, so go ahead and connect to it like you did before, only this time type the new IP where required. On Linux and OS X, in a terminal, type the following:

```
$ ssh -X pi@192.168.100.10
```

Then on Windows launch PuTTY (I assume that your X server is still running), put the new IP in the **Host Name (or IP address)** field, and click **Open**. If you are refused the connection for a few times when you try to reconnect, wait a bit as the Pi might be still booting. If you continue to get refused, insist a bit and eventually you'll be asked for Pi's password. If all goes well you should be logged in your Pi like before, only now the Pi has no Internet connection.

Whenever you want to connect the Pi to the Internet (to install software with apt-get, for example), you'll need to modify the /etc/network/interfaces file, shown in Listing 3-1. All you need to do is remove the comment sign from line 6 and comment out lines 7 to 11 (put a # at the beginning of each line). Then you'll need to find the IP of the Pi, as you did in the beginning (log in to your router, or use Nmap, for example) and log in with that IP. When your Internet session is done, go back to /etc/network/interfaces and do the reverse, comment out line 6, and remove the comment sign from lines 7 to 11, and reboot the Pi. For the use of the Pi in this book, we won't need any Internet, so it's best if you always reset the static IP whenever you log in the Internet.

Save Login Sessions on Windows

Since your Pi now has a static IP, you might want to save this session, so you don't need to type the static IP every time you want to log in the Pi. In PuTTY, in the **Saved Sessions** field, type the name of the session, which could be *raspberry*, and click **Save**. To test that the session has been saved, close PuTTY and open it again. Underneath the Saved Sessions field, you'll find the space where your saved sessions are. **raspberry** should be in there—click it and then click **Load**. The static IP of the Pi should appear in the Host Name (or IP address) field. Click **Open** and log in to the Pi.

You might also want to save the Pi session to WinSCP, to make things a bit faster. Launch WinSCP and type the new IP along with the rest of the data, the same way you did before. Now click **Save** and the **Save session** window will open, shown in Figure 3-18. It recommends that you not save the password, but you can do as you like. It can also create a desktop shortcut, which you might want if you work a lot with the Pi. If you have a shortcut of WinSCP however, it's still fast enough. Click **OK** and relaunch WinSCP to make sure your session has been saved. It will appear on the right side of the login window, shown back in Figure 3-14 (this figure doesn't include the saved session, because we hadn't saved it when we first used WinSCP), and is probable already selected. The Pi's data will appear in the login window (except from the password, if you didn't save it). Lastly, log in to test if all goes well.

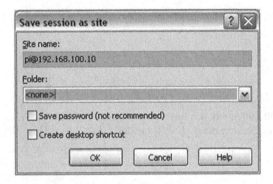

Figure 3-18. WinSCP "Save session" window

Shutting Down the Pi

Like all computers, the Pi needs to be shut down via a command (clicking buttons is the same: it generates commands), and not only by powering it off. To shut it down properly you must type the following in the Pi terminal:

```
$ sudo poweroff
```

You'll get a message in your computer's terminal saying that the connection is now closed (maybe not on Windows). To reboot it, unplug the power cable and plug it back in and log in as before. Note that on other embedded computers the command might be sudo halt instead. What we'll do now is write a script to shut the Pi down, because we will need this when we" be using the Pi without a screen and keyboard. cd to /etc, create a new directory called my_scripts, and cd to that. To do this, type the following:

```
$ cd /etc
$ sudo mkdir my_scripts
$ cd my_scripts
```

You'll need sudo only for the command that creates the new directory. Now we're in the new directory, where we'll write our script to shut the Pi down. We'll use nano to edit the script and afterward one command to make it executable. Type the following:

```
$ sudo nano shut_down.sh
```

We need sudo because we're not in the home directory, and we're not root. The .sh extension is the extension for shell scripts. In nano, type the following:

```
sudo poweroff
```

Hit **Ctrl+O** and **Return**, to write the text to the file, and then **Ctrl+X** to exit nano. To test if it's working, go back to your home directory and type the following:

```
$ cd ~
$ sh /etc/my_scripts/shut_down.sh
```

This will properly shut down the Pi . You might be wondering what the point of this is since you can just type **sudo halt** to shut down the Pi. When we're building projects using the Pi headless, we won't have the ability to type anything to the Pi's terminal, so we need to have a way to quit all programs and run sudo halt so that the Pi can shut down properly. Now go back to edit the script. From your home directory, type the following:

```
$ sudo nano /etc/my_scripts/shut_down.sh
```

In nano, type the following:

```
sudo killall pd
sleep 3
sudo poweroff
```

The last line was already there; you just need to type the other two lines above it. The first line will quit Pd (we assume that we'll be running Pd), then the Pi will wait for three seconds (with "sleep 3"), and eventually the halt command will run. Save it (**Ctrl+O** and **Return**) and exit nano. Now launch Pd and open a new window. Figure 3-19 shows the patch you should build. [shell] from the "ggee" library enables us to talk to the shell, therefore to run scripts, apart from other things. Since we've already wrote our script, all we need to do is call it, the same way we would call it from a terminal. Save the patch (I've named it shut_down), lock it and click the message. Pd will quit, and the Pi will shut down.

This is a very helpful feature, since we can use the Arduino to run this script and shut down the Pi properly, even if there's no keyboard or screen attached to it. Don't try yet to use the patch in Figure 3-19 with the Arduino, because it's very likely to cause a crash, since the Arduino will be sending its data at the specified baud rate, so this message will be called several times. We'll see in some of this book's projects how to implement this in a patch.

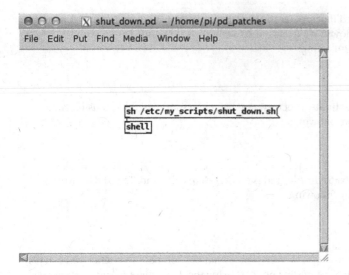

Figure 3-19. *Pd patch to shut the Pi down*

This concludes our Pi overview. It is a very useful tool that we'll use in some of this book's projects. Hopefully it will be useful to you in more projects that you build yourself. Now let's see how we can use the Arduino wirelessly.

Going Wireless

There will be some occasions where we'll rather use the Arduino wirelessly. This can be the case when building an interface for a performer that doesn't want to have cables hanging out of him/her. Being able to receive and send data between the Arduino and Pd wirelessly, expands the possibilities of interfaces a lot, so we'll go through that in this section. To have wireless communication, we'll need some more equipment. There are various options, depending on the actual interface we want to build. Here we'll cover the two most popular, the Bluetooth, and the very popular XBee.

Bluetooth vs. XBee

The Bluetooth solution is the cheaper of the two. The Bluetooth modules used with the Arduino are just that, Bluetooth modules, which can communicate with the computer if the computer's Bluetooth is activated. Then we can see it in our serial ports list and use the Arduino as we do with a USB cable. All you need to use Bluetooth with the Arduino is a Bluetooth module. Then you can immediately send and receive data between your computer and the Arduino.

XBee is a radio transmitter/receiver (also called a transceiver) produced by Digi International, which can be configured in any of the three OSes (Linux, OS X, Windows), and replaces the USB cable we've been using so far. The XBee is a more expensive solution. First of all, you'll need at least two XBees: one connected to the Arduino and another one connected to your computer. Apart from that, you'll also need some breakout boards to be able to configure them. The XBee has a non-breadboard friendly design so you can't build a circuit on a breadboard/perforated board to use it, you'll need to buy special breakout boards (called XBee Explorer) to make your life easier. If you want to use the XBee with an Arduino Uno, you'll also need a shield to mount on top of the Arduino, on which you'll mount the XBee so it can act like a transmitter/

receiver for the Arduino. If you want to use it with another kind of Arduino (the Pro Mini, for example), you'll need another kind of XBee Explorer for that.

All this might sound a bit expensive, but with a moderate budget, you can build a very flexible and powerful interface. The XBee has some advantages compared to the Bluetooth. First, it has a greater range than the Bluetooth, which might be very useful in a case of a concert where the performer is on stage and the computer is next to the mixing console (there can always be reasons for such a setup). In this case, the Bluetooth might not be able to cope with the demands of the setup. Other occasions are very likely to prove the XBee a more suitable solution, compared to the Bluetooth. Another advantage is that with the XBee, you can build more complex networks, like start networks (where many XBees communicate with a central XBee), mesh networks (where all XBees cooperate in the data distribution), and others. These are beyond the scope of this book, so I'm not going to cover them here; they're mentioned only as general information. Figures 3-20 and 3-21 show a Bluetooth module and an XBee transceiver respectively.

Figure 3-20. *HC-06 Bluetooth module*

Figure 3-21. *XBee series 1 with trace antenna*

Using a Bluetooth Module with Arduino

We'll use the HC-06 Bluetooth module to communicate with the Arduino. This is an inexpensive module that works out of the box. The only thing you need is a breadboard and some jumper wires. We'll first use it with its default settings, and then we'll configure it to change some things, like its name, its baud rate, and its passcode pin. Most likely this module comes with headers (the pins on its bottom in Figure 3-20) soldered on its communication pins, so you don't need to do it yourself. If there're no headers soldered, you should use angled male headers and solder them yourself. The pins we're going to use are VCC, GND, TXD, and RXD, which are indicated in the back side of the module.Go ahead and upload serial_print.ino sketch found in the GitHub page of the Pd abstractions [serial_print] and [serial_write]. Make the appropriate circuit (the three potentiometers and three switches circuit from Chapter 2). Once you do that, unplug the USB of the Arduino and make the following connections with it and the Bluetooth module:

HC-06 VCC connects to Arduino 5V.

HC-06 GND connects to Arduino GND.

HC-06 TXD connects to Arduino Rx (digital pin 0).

HC-06 RXD connects to Arduino Tx (digital pin 1).

We see here that we're using the first two digital pins of the Arduino, which we haven't used before. Chapter 2 mentioned that these pins are used for serial communication. Since we'll communicate serially with the Arduino via the Bluetooth module, we have to use these pins for this reason. Tx (or TXD) is the transfer pin, and Rx (or RXD) is the receive pin. So the transfer pin of the Bluetooth module should connect to Arduino's receive pin, and the receive pin of the Bluetooth module should connect to Arduino's transfer pin, so that whatever the Bluetooth module transfers is received by the Arduino, and the other way round. You can use the USB to power up the Arduino, or an external power supply, like a 9V battery (the external battery is preferable because there's no physical connection between the Arduino and the computer, so we can clearly see how the Bluetooth module works wirelessly). Figure 3-28 shows a battery connected to the Arduino with the XBee shield mounted on top of it. Connect the positive pole of the battery to the VIN pin of the Arduino (the first pin of the header next to the analog pins header), and its negative pole to the GND pin of the Arduino, which is right next to the VIN pin. With the Bluetooth module you just don't have the XBee shield on top as in Figure 3-28, so just connect the battery poles to the corresponding pins of the Arduino.

When you power up the Arduino you should see the a red LED blinking, on the Bluetooth module. This means that it is powered, but it's not connected to any device yet. You must first pair your computer with the Bluetooth, like you would pair it with any Bluetooth device (every OS has its own Bluetooth parity tool, but I'm not going to cover that here). If you're asked for a passcode key, use 1234. When you pair it, the red LED on the Bluetooth module will stop blinking and will stay lit. Open the help patch of the [serial_print] Pd abstraction and click the "devices" message. You'll get the available serial ports on Pd's console, and the Bluetooth module should be there. It will appear under the name /dev/tty.HC-06-DevB, where /dev/tty. might differ, according to your OS. If [comport] has already connected to another port, click the "close" message and then type the Bluetooth port number on the number atom connected to the "open $1" message. If all goes well, you should see the values of the potentiometers and the switches in the Pd patch.

Mind that we used the serial_print.ino sketch because it runs on a 9600 baud rate. If you have changed the baud rate of the sketch, it won't work, so set it back to 9600 and try again. This is because that's the default baud rate of the Bluetooth module. There might be some cases where we'll need another baud rate, so we'll have to configure the module for that. To do that we'll need to connect to it serially and type some AT commands (AT commands are commands for embedded computers). These commands are pretty easy, and we could use them with any terminal emulator program, like CoolTerm, or PuTTY. There is a drawback with HC-06, though. Commands don't get through with the newline feed (hitting Return), but the device waits for one second once you start typing. When that second is over, it will check its input buffer and use whatever is there. If a command is not full, it will just ignore it. This means that we need to type quite fast, and this can prove to be quite difficult. There is a work around to that issue, using the Arduino itself to configure the Bluetooth module.

Close the Bluetooth port of your computer, and unplug the Arduino. Disconnect the jumper wires from the potentiometer/switch circuit and make the following connections:

HC-06 VCC -> Arduino 5V

HC-06 GND -> Arduino GVD

HC-06 TXD -> Arduino D10 (digital pin 10)

HC-06 RXD -> Arduino D11 (digital pin 11)

Listing 3-2 shows code I found in a post on plasticbots.com, which I have slightly modified to fit the needs of this chapter. Write it in a new sketch window in the Arduino IDE.

Listing 3-2. Sketch to Configure the HC-06 Bluetooth Module

```
1.    #include <SoftwareSerial.h>
2.
3.    // RX, TX pins of the SoftwareSerial object
4.    // they connect to the TX and RX pins of the other serial device, respectively
5.    SoftwareSerial bt_serial(10, 11);
6.
7.    void setup()
8.    {
9.       Serial.begin(9600);
10.      //Serial.begin(57600);
11.      Serial.println("Set up HC-06 Bluetooth module!");
12.      bt_serial.begin(9600);
13.      //bt_serial.begin(57600);
14.      delay(1500);
15.
16.      // check connection, should receive OK
17.      bt_serial.print("AT");
18.      // wait for a second and half because HC-06 responds to commands after one second
19.      delay(1500);
20.      // check firmware version
21.      bt_serial.print("AT+VERSION");
22.      delay(1500);
23.      // set pin to 6666
24.      bt_serial.print("AT+PIN6666");
25.      delay(1500);
26.      // change name
27.      bt_serial.print("AT+NAMEmyBTmodule"); // Set the name to myBTmodule
28.      delay(1500);
29.      // set baud rate either to 9600 or 57600
30.      // bt_serial.print("AT+BAUD4");
31.      bt_serial.print("AT+BAUD7");
32.      delay(1500);
33.   }
34.
35.   void loop() {
36.      // loop does nothing
37.   }
```

Line 1 imports the **SoftwareSerial** library, which enables us to connect serially to other devices, while leaving the default Arduino serial pins, digital pins 0 and 1, free. This way we're able to upload code to the Arduino board, communicate with it via the IDE's serial monitor, and at the same time connect to the Bluetooth module. Line 5 defines an instance of the SoftwareSerial library, called bt_serial, which stands for *Bluetooth Serial*. An instance of a class is called an object, so bt_serial is considered an object of the SoftwareSerial class. This approach to programming helps us use functions and other data of a given class in a different way for each instance of the class. For example, if we defined two objects of the SoftwareSerial class, bt_serial1 and bt_serial2, we could use different pins for each, like 10 and 11 for bt_serial1, and 8 and 9 for bt_serial2. It's not really important to grasp the concept of objects in programming, this information is provided only for those interested.

Line 9 begins the serial communication between the computer and Arduino, using the Serial class, and line 12 begins the serial communication between the Arduino and the Bluetooth module, using the SoftwareSerial class, via the bt_serial object. Don't mind the commented lines for now, I'll explain them in a bit. Line 17 sends the AT command, which does nothing really, it only returns an OK if we're connected to the module. Note that all commands are sent with print and not with println, because we don't want to send the newline character.

Line 21 send the AT+VERSION command, which returns the firmware version of the module. It's not really necessary here, but it's provided just for the information. Line 24 sends the AT+PINxxxx, where x is a number, and it's used to change the passcode pin of the module. Again, you don't really need this, but you might want to change the default pin.

Line 27 sends the AT+NAME command, which is used to change the name of the module. Like with the AT+PIN command, there's no space between the command and its argument (the name you want to set to your module). This is a helpful command since you might want to use more than one module at the same time, and naming them will make things a bit easier, as it will be much simpler to tell which serial port communicates with which module. In this case, we're renaming it myBTmodule.

Finally, line 31 changes the baud rate of the module to 57600. The available baud rates are shown in Listing 3-3.

Listing 3-3. Availalbe Baud Rates for the HC-06

```
1 = 1200
2 = 2400
3 = 4800
4 = 9600
5 = 19200
6 = 38400
7 = 57600
8 = 115200
9 = 230400
A = 460800
B = 921600
C = 1382400
```

All the delays after each command are there to make sure that the module will have enough time to receive the commands and make the appropriate configurations. When you upload this code to your Arduino, open the serial monitor from the Arduino IDE. The first thing that will be printed is "Set up HC-06 Bluetooth module!" and then you'll receive the replies to each command from the module, which will be printed in the same line. Make sure you set the correct baud rate to the serial monitor. If you get the following line in the serial monitor, all will be fine:

```
OKOKfirmwareversionOKsetpinOKsetnameOKsetbaud
```

"firmwareversion" will be replaced by the firmware version of your Bluetooth module, of course. Now I'll explain what the commented lines do. We've now changed the baud rate of the module, if we want to reconnect to it, we must make sure we use the correct baud rate for both the Arduino and the Bluetooth module. Our baud rate is now 57600, so if you want to run this sketch again, you have to comment out lines 9 and 12, and remove the comment sign from line 10 and 13. This will start the serial communication between the computer and the Arduino, and between the Arduino and the Bluetooth module at the correct baud rate. The rest of the commands depend on what you want to do whenever you upload this code to the Arduino. The AT and AT+VERSION commands are not so important, but the commands on lines 24, 27, and 31 are.

If you want to set a different passcode pin to your module, set that in line 24, to change the name set it in line 27, and to change the baud rate do it in line 31. The 9600 baud is there in line 30 and it's commented out. If you want to set another baud rate, use the numbers from Listing 3-3. Note that if you want to configure another module, which hasn't been configured yet, you should use the begin functions in lines 9 and 12 (exactly like they are in Listing 3-2), and not the ones in lines 10 and 13, because this Bluetooth modules if configured to a 9600 baud by default.

Using this sketch enables us to easily configure the HC-06, since we don't need to type its commands in real time, and stress ourselves with the timing issue. Save it with a name that helps you understand what it does, so you can use it in the future.

Now unplug the Arduino and again build the potentiometer and switch circuit, and disconnect the Bluetooth module. Connect the Arduino to your computer and upload the serial_write.ino sketch to it (this sketch uses a 57600 baud rate). Unplug the Arduino and connect the Bluetooth module like you did with the serial_print.ino sketch, where the TXD pin of the module goes to the Rx pin of the Arduino (digital pin 0), and the RXD pin of the module goes to the Tx pin of the Arduino (digital pin 1). Power up the Arduino (again it's better to use an external battery), open your computer's Bluetooth and pair it to the Bluetooth module, which will now appear as myBTmodule. Now open the help patch of the [serial_write] abstraction and check the serial port number of the Bluetooth module. The port will be now called something like /dev/tty. myBTmodule-DevB (on OS X, replace /dev/tty. with your OS's port name). Open the port and you should get the values of the potentiometers and the switches. Congratulations, you're now free of wires!

Using the XBee with Arduino

To use the XBee, we'll use an XBee Arduino shield (shown in Figure 3-22) and an XBee Explorer USB (shown in Figure 3-23). Make sure that you've mounted the XBee on the Explorer properly, otherwise you could damage it. There are two lines on the Explorer and the shield indicating how the device must be mounted, by outlining the shape of the XBee. Figure 3-24 shows that. XBee has its own software called X-CTU, which runs on Windows only. This software provides a lot of features that we won't really need, but luckily there is free software for all three operating systems, which we can use to configure the XBee. On Linux and OS X, we can use the same software, CoolTerm (http://freeware.the-meiers.org/) to configure them. On Windows, we're going to use PuTTY to do this. There is also Minicom for all three platforms (on Windows it's available through Cygwin) and other solutions, but we're not going to use any of it here.

Connecting to the XBee from Linux and OS X

To connect wirelessly with an XBee (actually two XBees) we'll need an XBee shield, like the one shown in Figure 3-22, and an XBee Explorer USB, like the one in Figure 3-23. Launch CoolTerm with the XBee Explorer USB connected to your computer. In CoolTerm go to **Connections ➤ Options...** (or click the **Options** icon on top of the window) and the window in Figure 3-25 will open.

On the left side, Serial Port is chosen, which is what we want. Click the **Port:** tab to select the serial port of the XBee. This should be something like "usbserial-xxxxx" where x is either a letter or a number. The **Baud rate:** will be at 9600 by default, which is what we want for now, because the XBee is configured at this baud rate by default. Leave the rest of the fields as they are. Click **Terminal** on the left side menu. In Terminal, make sure the **Local echo** tick box is selected, otherwise what you type in the CoolTerm window won't appear on screen—much like when typing passwords in a Unix terminal, and this is not very convenient. Click **OK** and you'll be back to the CoolTerm session. On top of the window, click the Connect icon to get connected to the XBee.

Figure 3-22. *XBee Arduino shield*

Figure 3-23. *XBee Explorer USB*

Figure 3-24. *XBee Explorer indication for proper mounting*

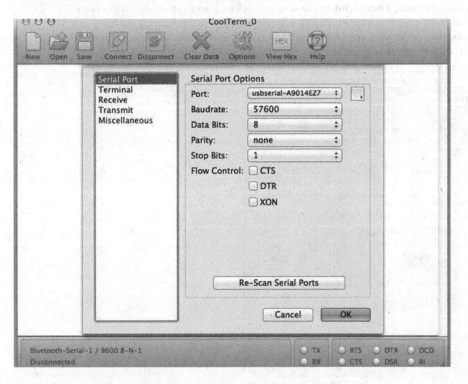

Figure 3-25. *CoolTerm's connections options window*

Connecting to the XBee from Windows

We'll use PuTTY to configure the XBee. First, check which serial port your XBee is at. On Windows XP, click **Start**, right-click **My Computer** and click **Properties**. On the Properties window click the **Hardware** tab and there click the **Device Manager** button. In the Device Manager, you'll find the serial port of the XBee under Ports, which should be a USB Serial Port, referred to as COMx, where x is a number. On other Windows versions, the procedure should be similar. If you don't get a port number, you might need to install drivers for the XBee. A simple online search should give sufficient results for this.

Now launch PuTTY, but instead of an SSH session, choose **Serial**, as shown in Figure 3-26. In the **Serial line** field, type the port name of the XBee. The Speed should be set to 9600, which is the default baud rate of PuTTY, but also the baud rate the XBees have by default. Before opening the connection, click **Terminal** on the left side menu, and under **Local echo**, make sure you tick on **Force on**, as shown in Figure 3-27. This will enable displaying whatever you write on PuTTY's terminal as you're configuring the XBee. If it's left unticked, then you won't be seeing what you type, which is not very convenient. Now click **Open** and PuTTY's terminal window will open, where you can start configuring your XBee.

Figure 3-26. *PuTTY Serial session*

Figure 3-27. Enabling Local echo in PuTTY's terminal

Configuring the XBee

Now that we're connected to the XBee, we're ready to type some commands to configure it. The following steps are common for all three OSes. I'll first show the commands and then explain what each does. Once you're connected and you start typing in the CoolTerm or PuTTY window. Type the following:

+++

Don't hit Return after this command, only wait until you receive an OK from the XBee. Once you receive the OK, go ahead and type the rest, now hitting Return with every command. Mind that once you get the OK you have about 30 seconds to type each command, so don't take your time doing it, otherwise the commands won't go through. You can tell that a command was set successfully by the OK respond from the XBee. If you don't get it, you must type the three plus signs again and start the whole configuration (not connection) procedure from the beginning. Listing 3-4 shows the rest of the commands. On PuTTY, all commands will appear on the same line, which is a bit confusing.

Listing 3-4. The Five AT Commands Used with the XBee

```
ATID1234

ATMY1

ATDL2

ATBD6

ATWR
```

Again we're using AT commands (this time with no plus sign), so we'll use only the last two letters to refer to them. **ID** is the PAN (personal area network) ID, used to broadcast messages to a certain network. The value for the PAN ID is 16-bit, so you can have any number in the range 0-0xFFFF. If you want to broadcast messages to all PANs, use 0xFFFF, otherwise the XBee will broadcast its messages only to the set PAN. We used a decimal value here (1234), but hexadecimal values are also valid (using the 0x prefix).

The MY command is the source address. Again, this is a 16-bit value in the range 0-0xFFFF. This is the address that the XBee will listen to. In our case, it's 1, but you could use anything in the 16-bit range, except from 0xFFFF. The DL command is the destination address (in fact, it's Destination Address Low). This is the address that the XBee will transmit to. This command reads the lower 32 bits of a 64-bit address, but we can use it with 16-bit values only, again in the range 0-0xFFFF, excluding 0xFFFF. **BD** is the baud rate command. Listing 3-5 shows the available baud rates.

Listing 3-5. Available Baud Rates for the XBee

```
0 = 1200 bps
1 = 2400
2 = 4800
3 = 9600
4 = 19200
5 = 38400
6 = 57600
7 = 115200
```

Here we have used the 57600 baud rate, since we provided number 6 to the command. Changing the baud rate of the XBee is not really necessary, but we do it to know how to, since it will very probably be necessary in some cases. The last command, WR, is the write command. If you exclude it, anything you've set will be discarded. With the WR command, all the data you've sent to the XBee will be stored even when the XBee is powered off, so you can use it in a plug-and-play mode.

Now that we've configured the first XBee, let's configure the second one too. The steps are identical up to writing the commands. On Linux and OS X, hit the Disconnect icon in CoolTerm first, and unplug the Explorer USB. On Windows, you can leave the PuTTY window open and unplug the USB. Mount the second XBee on the Explorer and plug it in. On Linux and OS X, click the Connect icon. On Windows, right-click the title bar of the PuTTY window and click **Restart Session**. We're now ready to configure the second XBee. The commands are essentially the same, only the MY and DL commands get reversed arguments. Listing 3-6 shows the commands for the second XBee.

Listing 3-6. Configuring the Second XBee, Using the Same Commands with the ATMY and ATDL Arguments Reversed

```
+++

ATID1234

ATMY2

ATDL1

ATBD6

ATWR
```

Again, don't hit Return when you type the first command (+++), and wait for the OK. After entering each of the rest of the commands, hit **Return** and again wait for the OK, to type the next command. When you receive the OK after the last command (WR), on Linux and OS X, click the Disconnect icon on top of the window. On Windows, just close the PuTTY window, and you're done.

The ID command takes the same argument so the two XBees are in the same network. As you see MY now got 2, and DL got 1, so the two XBees can talk to each other. The BD must definitely get the same argument, otherwise the XBees won't be able to communicate. Finally, WR will write the changes so they can take effect. The full command list for the XBee is available on the Internet, if you're interested in more configuration options. The ones here will be enough for us to use the Arduino wirelessly, however. Make sure you don't forget the baud rate of the XBee in case you want to reconfigure it in the future. Trying to log in the XBee with a 9600 baud rate won't work because it is now configured at a 57600 baud rate, and the computer won't be able to talk to the XBee.

Now it's time to test the two XBees. We can use the serial_write.ino Arduino sketch that from the GitHub page with the Pd abstractions. We'll use that because we've already configured the XBees to a 57600 baud rate, and the serial_write.ino sketch uses this baud rate. Apart from both XBees needing to have the same baud rate to communicate with each other, the Arduino must have the same baud rate as well, so that it can transmit its data via the XBee. Mount the XBee shield on top of the Arduino, an XBee on the XBee headers on the shield, and build the circuit with the three potentiometers and the three switches. To upload the sketch to your Arduino board you need to use a switch on the XBee shield, which is next to the XBee headers. This switch has two positions, DLINE and UART. To program the Arduino set it to DLINE and connect the Arduino to your computer via USB. Once you've programmed it, you can unplug it to power up it with a battery so you can clearly see the wireless connection is action (you can use the USB cable as well). Turn the XBee shield switch to the UART position, so you can use the XBee. Then connect the batter the same way you did with the Bluetooth module. Figure 3-28 shows the battery connections. Once you power it up you should see both the Arduino and the XBee turning on. Mount the other XBee on the XBee Explorer and connect it to your computer with a USB cable. The XBee is simultaneously a transmitter and a receiver, so it's not important which XBee will be connected to the Arduino and which will be connected to the computer.

Now open the help patch of the [serial_write] abstraction and click the "devices" message to see the available serial devices. Look for the serial port of the XBee, and type its number to the number atom connected to the "open $1" message. The serial port of the XBee will open and you should start seeing the values of the potentiometers and the switches in the Pd patch. Congratulations! You've gone wireless once again!

Figure 3-28. *Powering up the Arduino with a 9V battery*

Conclusion

This chapter covered the use of the Raspberry Pi (most of which applies to other embedded computers as well) and how to use the Arduino wirelessly. This concludes all introductions needed for this book. In some of the projects that will be developed here, we'll use either an embedded computer or a wireless connection to the Arduino, so this chapter is very important to realize these projects.

Apart from the interfaces built here, being able to use embedded computers and wireless connections, enables you to realize lots of very flexible and powerful projects. Embedded computers are very useful when building mobile projects where there's no need to look at a computer screen, so the computer can be hidden inside the project enclosure. You can build custom-made, very flexible interfaces that exactly fit your needs. Wireless communication between the computer and the Arduino can prove to be very helpful when cabling can make a setup problematic, and a cable-free setup seems more appropriate.

A subject we haven't covered here is the use of external sound cards with embedded computers. When using computer audio, it's generally advisable to use an external sound card, because the computer's sound card circuit is usually not very good. This is because computers tend to be rather compact, and since they are general purpose devices, they're not dedicated to audio, so their audio circuit is not of the highest priority, resulting in quality reduce. An external sound card is dedicated to audio, so the quality of the audio output is number one priority, making it more appropriate for making music. Also, an external sound card usually provides input channels, whereas the Raspberry Pi does not. Some sound cards provide more than two input/output channels, which might be desirable, plus most sound cards provide pre-amps for input and output, which helps the audio overall amplification. The Raspberry Pi has dedicated sound cards, because it's a very popular computer with a big community. These sound cards usually come with OS images including the necessary drivers so that the sound card can work.

In case you want to use another sound card, or in case you're using another embedded computer, you'll have to find another solution (if you want to use an external sound card). There is an issue here because most of the sound cards don't include drivers for Linux. In general, Linux is outside the commercial market, so sound card designers usually don't bother to build drivers for it. Sound card drivers for Linux depend on the Linux developers. If you're planning on buying a sound card for your embedded computer, do a thorough research first. Search the Internet for sound cards that are reported to work with your embedded computer, check forums, and ask around. A general rule is that a class-compliant USB sound card (which will most probably not need drivers for OS X, which, like Linux, is Unix-based) is very likely to work with Linux. If it doesn't work (or doesn't work properly) with the default settings of Pd, try to use it with the Jack audio server (we'll cover it in the first project that uses an embedded computer). There is a great possibility that Jack will work better with a sound card than another audio server.

Now that we have acquired all necessary background, we can start building musical projects. In the next chapter, we'll build some very simple interfaces, to start getting the hang of it. They won't be finalized interfaces and they will be rather limited. They will be the kickoff for building more complex, expressive tools to make music.

CHAPTER 4

■ ■ ■

Getting Started with Musical Applications

You have now been introduced to all the necessary tools, so we can start building some real interfaces. In this chapter, we're going to build a couple of rather simple interfaces on a breadboard. Since they're not going to be finished projects, we're not going to deal with soldering circuits on a perforated board, as the breadboard will serve just fine for our purposes. This chapter acts more like an introduction to using Pd and Arduino to create musical interfaces, and as a kickoff to making really interesting projects for your musical creations.

We're going to combine things learned in the first two chapters (we're not going to use any embedded computers or wireless communication) in a way that will create some more interesting sounds than the ones we've already made. We're going to build two interfaces—a phase modulation interface and a simple drum machine, and we're also going to combine them. The interfaces here will be fairly simple, but they cover the basis of a general approach to building finished projects.

Parts List

Table 4-1 lists the parts needed for each of the three interfaces. The third one is a combination of the first two, but it's listed here as well.

Table 4-1. *Parts List for the Interfaces of this Chapter*

Interface	Potentiometers	Switches	Push buttons	LEDs	Resistors
Phase modulation	4	2	0	1	$1 \times 220\Omega$
Drum machine	2	2	3	1	$1 \times 220\Omega$
Combination	5	3	4	1	$1 \times 220\Omega$

Phase Modulation Interface

The first project we'll realize is a phase modulation Pd patch, controlled by some potentiometers and some switches with the Arduino. Phase modulation is a technique similar to frequency modulation, but slightly different. I didn't cover it in the first chapter, but since we've already covered frequency modulation, it won't be very difficult to understand what's happening. It is a celebrated technique that can produce very interesting sound results with just a few elements. As in frequency modulation, in phase modulation there's a carrier oscillator that has its phase (instead of its frequency) modulated by another oscillator. Again, there are

141

three elements: the carrier oscillator, the modulator oscillator, and the index. To understand what is meant by *phase*, let's think of a sine wave oscillator. Figure 4-1 illustrates how the waveform of a sine wave occurs.

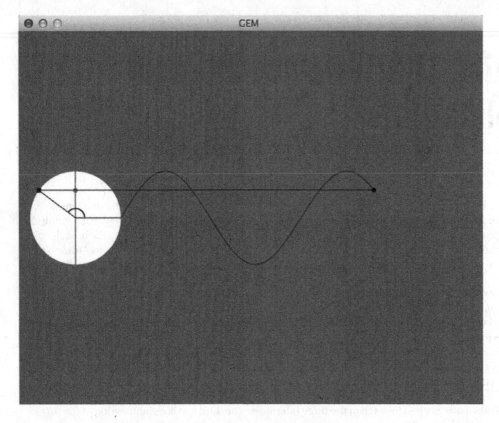

Figure 4-1. *Sinusoidal waveform produced by a rotating angle*

How Phase Modulation Works

The angle in Figure 4-1 (which is made in Pd, by the way) is rotating counter-clockwise, constantly. As it rotates, a straight line is drawn from the point where the rotating angle meets the circle circumference (if the sine wave has full amplitude) to the vertical axis of the circle. The end of the line that meets the vertical axis of the circle (indicated by a dot) is the sine of the angle. On the right side of the circle, the sine is translated in time, forming its waveform. When we modulate the phase of a sine wave oscillator, we're actually modulating the rotation of the angle. So, instead of going constantly counter-clockwise in a steady frequency, it will change both direction and frequency according to the wave form of the modulator oscillator. When these changes occur slowly, we hear a glissando that goes up and down, around the carrier frequency. When these changes occur fast, what we perceive is a tone rich in harmonics. You can see that this effect is very similar to frequency modulation. You can get similar results with the two techniques, if the right coefficients are provided. It might be a bit easier to use phase modulation to control timber richness (at least to my experience); whereas with frequency modulation, you can control frequency shifts easier.

Making the Pd Patch

Since I've started by explaining the technique, we'll use for the sound. Let's first look at the Pd patch for this project. Figure 4-2 illustrates the patch, which I'll explain in detail.

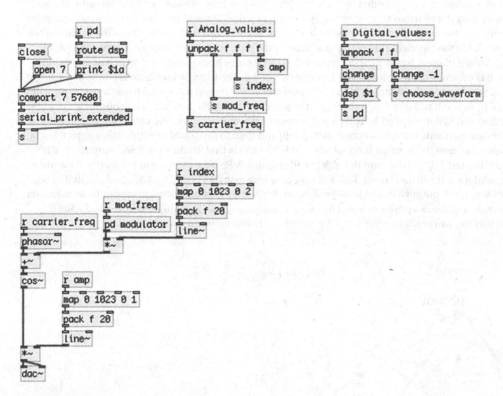

Figure 4-2. *Phase modulation Pd patch*

Receiving Values from the Arduino

On the top-left part of the patch we have [comport] with the serial port and baud rate arguments (the serial port argument is very likely to be different in your computer), along with some messages, which we'll look at in a while. On the top-right part of the patch, we have the [r Analog_values:] and [r Digital_values:] objects that receive the corresponding values from the Arduino, exactly like in the help patch of the [serial_print_extended] abstraction. This time we're using four analog values (four potentiometers) and two digital (two switches), which you'll see when I explain the Arduino code and circuit.

Implementing Phase Modulation in Pd

On the lower part of the patch, we have all the signal objects, which is where the phase modulation is happening. On the left side, you see a [phasor~] connected to a [+~] and then to a [cos~]. Connecting a [phasor~] straight to a [cos~] will produce a pure cosine oscillator. [cos~] takes a signal in its inlet and multiplies that by 2pi. Sending a [phasor~] to it, which is a rising ramp from 0 to 1, creates a rising ramp from 0 to 2pi. [cos~] will give the cosine of this, which results in a sinusoid oscillator. This makes the name of [phasor~] meaningful, as it is actually the phase of the oscillator. If we place a [+~] between [phasor~] and [cos~] and connect another oscillator to the right inlet of [+~], we are modulating the phase of [cos~].

143

Building the Modulator Oscillator

On the right side of [phasor~], there is a subpatch called *modulator*. Figure 4-3 shows its contents. Here we can choose between the two waveforms for the modulator oscillator, which are a sinusoid and a triangle oscillator. Both oscillators are controlled by the same [phasor~], so they will have exactly the same phase. Again, we're making the sinusoid by connecting [phasor~] to [cos~], instead of using [osc~], so we can share the phase between the two oscillators. On the right side we're creating a triangle oscillator the same way we did in Chapter 1. On the top-right side of the modulator subpatch, you see a [r choose_waveform], which takes a value from [s choose_waveform] in Figure 4-2, which comes from the second switch of the circuit. Since a switch is controlling this value, it will be either a 1 or a 0. The triangle oscillator is controlled straight by that value. So whenever the switch is in the OFF position, it will output a 0, so the triangle oscillator will be multiplied by 0, and it will give 0 output. When the switch is in the ON position, it will output a 1, and the triangle oscillator will be multiplied by it, so it will output its waveform intact. The value of the switch is also sent to the cosine oscillator, but it first connects to [== 0] and then to the signal multiplication object. [== 0] takes in a value and tests if that value is equal to 0. If it is, the test is true, so the object will output a 1. If it's not equal to 0, the test will be false, and the object will output a 0. So if we send a 0 to [== 0], it will output a 1, and if we send it a 1, it will output a 0. This is a very easy way to invert 1s and 0s. So, when the switch is in the OFF position, it will output a 0, and [cos~] will be multiplied by 1, therefore it will output its signal intact. When the switch is in the ON position, it will output a 1, and [cos~] will be multiplied by 0, producing 0 output. This way, we can easily use one ON/OFF switch to choose between two elements.

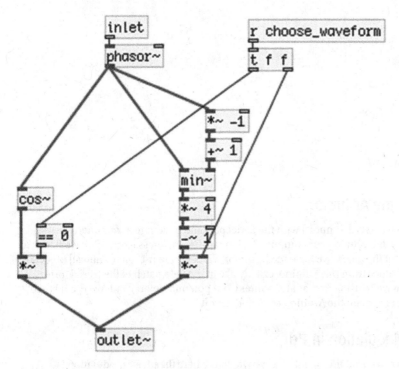

Figure 4-3. *Contents of the modulator subpatch*

Mapping the Index Values

On the right side of the modulator subpatch, you see a [r index] (in Figure 4-2), which takes its value from [s index], which comes from the third potentiometer of our circuit. [r index] is the connected to [map 0 1023 0 2]. This is an abstraction I've made that makes it very easy to map a range of values to another range. You can get it from my GitHub page at https://github.com/alexdrymonitis/miscellaneous_abstractions. This page contains various abstractions along with help patches, so it might be quite useful to you. Once unzipped, place all its contents to your "abstractions" folder, which is included in your Pd's search patch. [map] takes four arguments, the lowest value of the original range, the highest value of the original range, the lowest value of the desired range, and the highest value of the desired range. In this case, we're receiving an analog value from the Arduino, which has a 10-bit range.

This means that it will go from 0 to 2 to the 10th power minus 1, so from 0 to 1023. What we want is to get a range from 0 to 2, so the arguments we provide to [map] are 0, 1023, 0, 2. You can build a patch that does the same thing simply by dividing the input range by its highest value, and multiplying the result by the maximum desired value. In this case one would have to divide by 1023 (which is the maximum value of the analog pins of the Arduino) and multiply by 2. A division is more expensive as far as the CPU is concerned, and [map] avoids doing it. Open the abstraction if you want to see how it works. [map] then connects to [pack f 20]'s left (hot) inlet. [pack f 20] will pack the value coming in its left inlet along with the value 20, so it will output a list of these two values, which is sent to [line~]. To refresh your memory, [line~] takes a list of the target value and the ramp time in milliseconds, and it will make a ramp from its current value to the target value, which will last as long as the milliseconds provided by the second value of that list. It is used to combine the control domain with the signal domain, avoiding possible clicks that occur by this combination.

The output of [line~] goes to the right inlet of the signal multiplication object the modulator subpatch connects to. This way the value received by [r index] becomes the index of the phase modulation in our patch.[1]

At the bottom of the patch, we're multiplying the output of [cos~], which is the modulated oscillator, and what we actually hear, by the values received in [r amp], which comes from the fourth potentiometer. We're mapping that value to a range from 0 to 1, because that's the range used to control the signal amplitude sent to the speaker. Again, we're using [line~] to smooth out the value changes and avoid clicks, and we finally send that output to [dac~].

Handling the Values Received from the Arduino

Before we go on to the Arduino code, I must talk about the top-right part of the patch in Figure 4-2. [r Analog_values:] sends a list of four values (as many as the potentiometers we're using), which is unpacked by [unpack f f f f] and each value is sent to its destination by the corresponding [send] object. That is pretty straightforward. The digital values get a different kind of treatment to be used properly. [r Digital_values:] receives a list of two values (as many as the switches we're using) and unpacks it with [unpack f f]. Below [unpack], you see two [change] objects, where the right one has an argument, -1. [change] takes in a value and compares it to its argument (if no argument is provided, it compares it to 0). If the value is the same as the argument, [change] won't output it. If it is different, it will output it and it will also update its arguments with the value it has just output. When we send the values of a switch to [change], we'll get the switch value only when it becomes 1 (when the switch goes to the ON position), and [change] will update its internal value to 1. As long as the switch remains in the ON position, [change] won't output anything, because its internal value is now 1, and the switch keeps on sending the value 1. When we put the switch to the OFF position, it will output a 0, and [change] will output it, since it's different than 1, and it will again update its internal value. This way we can receive digital values only when they change, which is really necessary in some cases.

---

[1]It is said that "good" index values for phase modulation are between 0 and 1, but higher values can produce nice results.

The left [change] is connected to the message "dsp $1". "$1" will take the first value of an incoming list. Our list consists of one value only, so "$1" will take the value of the switch, when it goes through [change]. The message that will be constructed ("dsp 1" or "dsp 0", according to the position of the switch) is sent to [s pd]. We can send certain messages to Pd, which can control certain aspects. The "dsp" message controls the DSP, as you can imagine. So sending a "dsp 1" message will turn the DSP on, and sending a "dsp 0" message will turn the DSP off. This way we can control the DSP from the Arduino, without needing to our computer's keyboard or mouse.

The right [change] has an argument that is –1. This is there only to initialize this [change]. We do this because when we open the serial port, the data from the Arduino will start coming in, and if the corresponding switch is in its OFF position, it will send a 0. If we don't provide an argument to [change], it will compare its incoming value to 0, and if the incoming value is 0, it won't output it. We're using this switch to control the waveform of the modulator oscillator, where sending a 0 will give a sinusoid, and sending a 1 will give a triangle. If [change] has no arguments and the switch is in the OFF position, nothing will come out of [change], so no oscillator will be chosen. By giving –1 as an argument to [change], we make sure that any position of the switch will go through as soon as we open the serial port (whether OFF or ON, 0 or 1), and we'll immediately choose our modulator oscillator.

Sending Data from Pd to the Arduino

The last thing that I need to explain about this patch is the "print $1a" message in the top-right part of it. As we can send messages to Pd using [s pd], we can also receive messages from Pd, using [r pd]. In our case we connect [r pd] to [route dsp], so we filter out all messages sent by Pd, and receive only messages about the DSP. These will be either 1 or 0, according to the DSP state. So when the DSP is on, [route dsp] will output a 1, and when it's off, [route dsp] will output a 0. We send that value to the "print $1a" message, which goes to [comport]. The word "print" will convert the rest of the message to its ASCII values (this is a feature of [comport]). Using "a" is a convenient way to diffuse the data we send from Pd to Arduino, as you've already seen in Chapter 2. This message receives the DSP state and controls an LED in the Arduino circuit, which indicates whether the DSP is on or off. This will be really necessary when we'll be building interfaces with embedded computers, or when we'll be using a wireless connection between the Arduino and Pd. In general, we need some kind of visual feedback for the state of the DSP, whenever we won't have visual contact with the screen of the computer.

Arduino Code for Phase Modulation Patch

Now let's take a look at the Arduino sketch. This is shown in Listing 4-1.

Listing 4-1. Code for the Phase Modulation Patch

```
1.    // analog values array size, must be constant
2.    const int num_of_analog_pins = 4;
3.    // digital_values array size, must be constant
4.    const int num_of_digital_inputs = 2;
5.    // digital outputs, doesn't need to be a constant
6.    int num_of_digital_outputs = 1;
7.
8.    // create an array to store the values of the potentiometers
9.    int analog_values[num_of_analog_pins];
10.   // create an array to store the values of the push-buttons
11.   int digital_values[num_of_digital_inputs];
12.
```

```
13.   void setup() {
14.     // set digital input pin modes
15.     for(int i = 0; i < num_of_digital_inputs; i++)
16.       pinMode((i + 2), INPUT_PULLUP);
17.
18.     // set digital output pin modes
19.     for(int i = 0; i < num_of_digital_outputs; i++)
20.       pinMode((i + num_of_digital_inputs + 2), OUTPUT);
21.
22.     Serial.begin(57600);
23.   }
24.
25.   void loop() {
26.     if(Serial.available()){
27.       static int temp_val;
28.       int which_pin;
29.       byte in_byte = Serial.read();
30.       // check if in_byte is a number and assemble it
31.       if((in_byte >= '0') && (in_byte <= '9'))
32.         temp_val = temp_val * 10 + in_byte - '0';
33.       // check if in_byte is a letter and call digitalWrite
34.       else if((in_byte >= 'a') && (in_byte <= 'z')){
35.         which_pin = (in_byte - 'a');
36.         which_pin += (num_of_digital_inputs + 2);
37.         digitalWrite(which_pin, temp_val);
38.         temp_val = 0;
39.       }
40.     }
41.
42.     // read and store analog and digital pins
43.     for(int i = 0; i < num_of_analog_pins; i++) analog_values[i] = analogRead(i);
44.
45.     for(int i = 0; i < num_of_digital_inputs; i++) digital_values[i] = !digitalRead(i + 2);
46.
47.     // write the stored values to the serial line
48.     Serial.print("Analog_values: ");
49.     for(int i = 0; i < (num_of_analog_pins - 1); i++){
50.       Serial.print(analog_values[i]);
51.       Serial.print(" ");
52.     }
53.     Serial.println(analog_values[num_of_analog_pins - 1]);
54.
55.     Serial.print("Digital_values: ");
56.     for(int i = 0; i < (num_of_digital_inputs - 1); i++){
57.       Serial.print(digital_values[i]);
58.       Serial.print(" ");
59.     }
60.     Serial.println(digital_values[num_of_digital_inputs - 1]);
61.   }
```

This is a modified version of the serial_print.ino sketch that comes with the [serial_print_extended] Pd abstraction. Some things are the same, but there are some new things as well, which I am going to explain.

Defining Constants, Variables, and Pin Modes

Lines 2 and 4 set the number of analog and digital input pins we're using. We're using four potentiometers and two switches. Line 6 sets the number of digital output pins we're using, which is 1. This variable doesn't need to be a constant, as it doesn't define the size of an array, it will only be used in the test field of a `for` loop. In line 15 we run a `for` loop to set the pins where the switches are attached to as inputs with the pull-up resistors enabled. Notice that we add 2 to the i variable, because we use digital pins from 2 onward, as pins 0 and 1 are used for the serial communication between the Arduino and Pd. In line 19, we run another `for` loop to set the pin of the LED as output. Here we add the `num_of_digital_inputs` plus 2 to the i variable, because the variable starts counting from 0, but the pin we've attached the LED to is number 4.

Handling Input from the Serial Line

In lines 26 to 40, you see the same technique we used in the last sketch of Chapter 2. This is a technique that makes it easy to receive different kinds of information and diffuse it accordingly in the Arduino code. Line 27 defines a `static int` variable that will hold its value even after the function it belongs to has ended. Line 28 defines a variable that will hold the number of the pin we want to control, and line 29 stores the incoming bytes in a byte variable, one by one. Line 31 checks if the current byte is a numeric value, and if it is, it assembles all numeric bytes the original value sent. Here we'll be sending only 0s and 1s, so `temp_val * 10` could have been omitted, but if we want to send values with more than one digit, we'll need that line exactly as it is, so we're using it this way here as well. Line 34 checks if the current byte is an alphabetic character, and it if is, it stores that value to the `which_pin` variable. In line 35, we subtract `'a'` from in_byte, and in line 36 we add `num_of_digital_inputs + 2` to that, so we can obtain the first digital output with the letter a, the second digital output with the letter b, and so forth. If `in_byte` is `'a'`, then `in_byte - 'a'` is 0. Adding `num_of_digital_inputs + 2` makes it 4, which is the digital pin we've attached our LED to. Here, we're using only one digital output, but if we used more, we could use incrementing letters of the Latin alphabet this way, to map each value to the corresponding pin.

Line 37 writes the value stored in `temp_val` to the `which_pin` pin using the `digitalWrite` function. This function is called only when we receive input in the serial line, and only if that input is an alphabetic character. This way we can save some processing power, since we're not calling this function over and over again in each iteration of the `loop` function. After we call `digitalWrite`, we set `temp_val` to 0, so when we send a new value, it will be assembled anew.

Reading the Analog and Digital Pins and Writing the Values to the Serial Line

Lines 43 and 45 read and store the values read from the analog and digital pins respectively. In line 45 we use the exclamation mark to invert the readings of the digital pins because of the pull-up resistors, as if we didn't use it, when a switch would be off, we would read 1 instead of 0. Lines 48 to 53 and 55 to 60, write these values to the serial line, using the `print` and `println` functions, so we can retrieve these values with the [serial_print] abstraction in Pd.

Circuit for Arduino Code

Now that we've gone through the Arduino sketch too, let's see the circuit for this project. Figure 4-4 illustrates it.

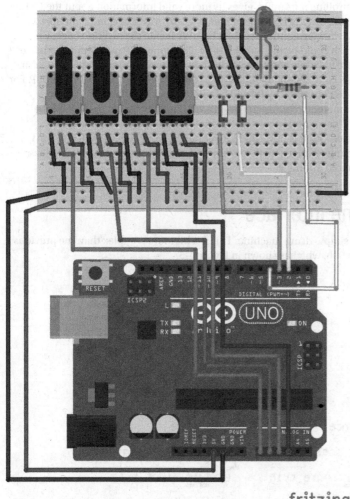

fritzing

Figure 4-4. *Phase modulation project circuit*

The potentiometers must use analog pins 0 and onward, and the switches must use the digital pins 2 and onward for the Arduino sketch to work properly. The LED must be attached to the first digital pin after the switches. We're not using resistors with the switches because we use the internal pull-up resistors of the Arduino processor, by using INPUT_PULLUP in the pinMode function. These resistors are different than the ones we need for LEDs, so we have to use a 220Ω external resistor with it. Notice that we use a jumper wire to connect the ground buses on bottom and top of the breadboard. This is because of the limited space on our breadboard. We've placed the LED on the upper part of the breadboard, and connected its ground pin to the upper ground bus. If we don't provide ground from the Arduino, the bus won't be grounded and the LED won't work. Since we connect Arduino's ground to the lower bus of the breadboard, we must connect the two buses with a jumper wire, so that the LED gets the ground too.

Once you've built the circuit, upload the sketch to your Arduino board, and open the Pd patch. Open the Arduino's port in [comport] (making sure you use the correct baud rate, here 57600) and put the first switch to the ON position. You should see the DSP indication in Pd's console turning green and being ticked, and the LED on your circuit lighting up. We could have controlled the LED straight from the Arduino code, without having Pd interfering. Controlling it from Pd, gives us more valid information about the DSP. If for any reason, we turn the DSP switch on, but the DSP doesn't really go on, the LED won't light up, so we'll know that the DSP is not running. If we were controlling the LED only inside the Arduino code, we could face a situation where the DSP switch could be on, the LED as well, but the DSP not running, and we wouldn't have an indication for it. Now we have eye contact with the computer screen, but in the case of an embedded computer, or a distant wireless connection, the LED would be our only indication for the DSP. For this reason, it's very important to use LEDs controlled by Pd when building such interfaces.

Now you can use the fourth potentiometer to control the overall amplitude of the Pd patch, and the first to third potentiometers to control the carrier frequency, the modulator frequency, and the index, respectively. This is a quite simple interface, but it is nevertheless a completed one. We've built it to start getting the hang of such processes. The projects that are yet to come will realize even more musical ideas. Congratulations, you've built your first audio interface!

A Simple Drum Machine Interface

For the second interface, we'll build is a simple drum machine. This is a bit more complex than the previous interface. Again, we'll start with the Pd patch, which is shown in Figure 4-5.

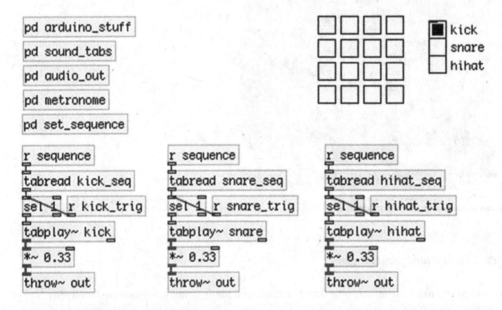

Figure 4-5. *Simple drum machine patch*

Building the Pd Patch

We're going to use drum sound samples, instead of procedural audio (drum sounds generated by oscillators). You can register at freesound.org to get loads of audio samples that you can use for free. Mind the license of each sample, depending on what you want to do with it. You can use some of the samples even for professional reasons and earn money, but not all of them. Read about the different licenses (they're three in total) if you want to do something else than personal use. Here we're just building a simple drum machine for fun, so all licenses allow such use.

The Parent Patch

Part of the whole interface is in the parent patch, and part of it is in subpatches. This helps to keep things clear. In the parent patch, you can see a matrix of toggles, a Vradio, the [tabread] objects that read the sequence for each drum sound, and the [tabplay~] objects that play the stored audio files. Since we have three sounds in total, we're multiplying their output by 0.33, so that their sum doesn't exceed 1, in case they are all triggered together.

Notice that the toggles don't have inlets or outlets. This is because we've set some things in their properties, which you'll see a bit further on. The same applies to the outlet of the Vradio. Once you've built the lower part of the patch, let's build the subpatches, one by one.

The arduino_stuff Subpatch

Figure 4-6 illustrates the contents of the *arduino_stuff* subpatch. This is where we have [comport] listening to the Arduino serial port, at a 57600 baud rate. Below it, we have the [serial_write] abstraction. We could have used [serial_print] again, but we'll use [serial_write] for variety, and to get the hang of using this one as well. In [serial_write] we first receive the digital values, which are five in total, and then the analog, which are two. You can see that in the order of the arguments of the abstraction. This is the way we're going to write the Arduino sketch too. If we were to store the analog values first in the Arduino sketch, we should invert the order of the arguments in [serial_write].

The left outlet of [serial_write] outputs a list of five values, which are the values of the five digital input pins we'll use in the Arduino. We're unpacking this list with [unpack f f f f f] and we send each value to its destination using [send]s. Before the values go to each [send], we send them through a [change], so that they output their values only when they are changed (we could do that in the Arduino sketch instead, but in Pd it's a bit easier, and since the Arduino has a rather limited processing power, we prefer to do it here).

The right outlet of [serial_write] outputs a list of two values, which are the values of the potentiometers. We unpack the list with [unpack] (no arguments to [unpack] is the same as [unpack f f]) and send the values to their destinations again using [send]s. You'll see further on where these values go.

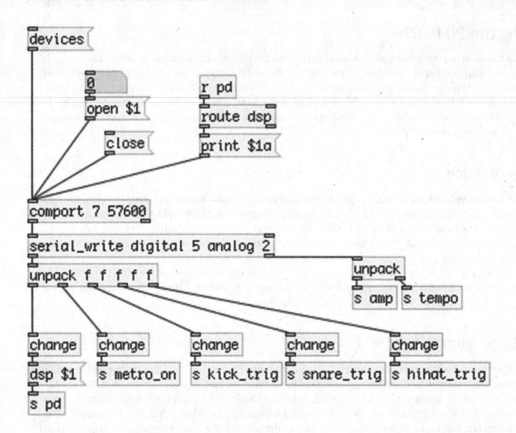

Figure 4-6. *Contents of the arduino_stuff subpatch*

The sound_tabs Subpatch

The next subpatch is *sound_tabs*, which is shown in Figure 4-7. In this subpatch, we're storing the audio samples for our drum machine. I have downloaded three samples from freesound.org, a kick-drum, a snare-drum, and a hi-hat sample, which I have named "kick," "snare," and "hihat," respectively. Inside the Pd patch folder, I've created a folder called sounds, and I put all three samples in there. All this is necessary so that Pd knows where to look for these audio samples, to load them to tables.

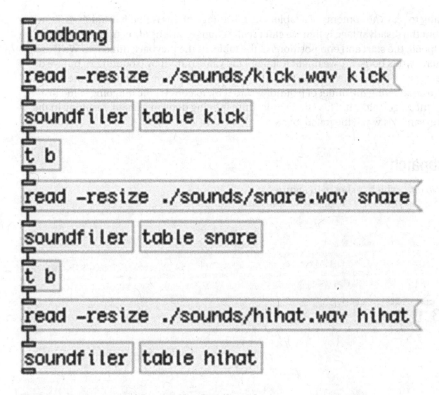

Figure 4-7. *Contents of the sound_tabs subpatch*

Since this is going to be a fixed interface, it's nice not having to load your audio samples manually. To do this, we're using [loadbang], which sends a band as soon as the patch is loaded, which goes to a message that reads "read -resize ./sounds/kick.wav kick". This messages is sent to [soundfiler], which will load the sample to the table it is told to. The message tells [soundfiler] to "read" the file called kick.wav (the file extension, .wav, must be included), which is located in the sounds directory, which directory is located in the current directory. This is the ./sounds/kick.wav part of the message. The dot means the current directory, and then we're being navigated to the sounds directory to retrieve the kick.wav file. This is Unix syntax, which you saw in Chapter 3, when we were navigating through the Linux system in the Raspberry Pi. This syntax will work in Windows too, even though it is not Unix. The message also contains the -resize flag, before the path to the audio file, which tells [soundfiler] to resize the table it will store the audio file to, to the size of the file. The last word in the message is the name of the table. To put it all together, when you want to load an audio file to a table, the message sent to [soundfiler] should contain the read command with the -resize flag, then the path to the audio sample, and lastly, the name of the table to load the file to.

Next to [soundfiler]. You can see the table that the audio file will be loaded to. Make sure that you give it the same name as the last word in the message. We could have used arrays, but we don't really need to look at the graph of the audio file, so we prefer to use [table] instead. If you click [table] (in a locked patch), you'll be able to see the graph of the file.

[soundfiler] outputs the number of samples the audio file consists of. Don't confuse the word sample with the way we use it for an audio file. In computer music, a sample is a discrete value representing the amplitude of sound. It is confusing when the same word is being used for a short audio file, like one kick of a kick drum. Since the drum machine we're building is based on audio files, we're using the word sample to refer to these files, as it is widely used this way, but the samples number [soundfiler] outputs, concern the discrete amplitude values. This number [soundfiler] outputs is necessary when we use [tabread~] to read the audio file from the table. In this patch, though, we're using [tabplay~], so we don't really need it.

[tabplay~] only takes a bang to play the contents of a table. The advantage of this is that it is much easier to set up a playback patch, but the disadvantage is that we can't really change various aspects of the playback of the audio file, like the speed, the start and end positions of the table, or the playback direction. We'll use [tabread~] in other chapters in this book. So, we might not need the value output by [soundfiler], but we do need to bang the next message to load the next audio file to the next table, once we're done with the current audio file. To do this, we're connecting the outlet of [soundfiler] to [t b], to convert the outgoing value to a bang, which will bang the message below it. This whole procedure is being done in a chain, resulting in the automatic load of all audio samples when the patch loads.

The audio_out Subpatch

The next subpatch is *audio_out*, which is shown in Figure 4-8.

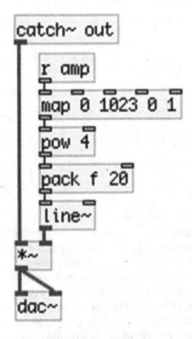

Figure 4-8. *Contents of the audio_out subpatch*

Here you see the first [receive], which takes input from [s amp], which comes from the first potentiometer, and it's located in the *arduino_stuff* subpatch (actually all [send]s are located there). We also have the [catch~], which takes audio from the three [throw~]s in the parent patch. We could have used [inlet~] instead and connect the output of all [tabplay~]s to it, but [throw~]/[catch~] makes the patch a bit more tidy. It might increase the CPU a bit, but the audio processing done in this patch is very small, so CPU is not a problem. Again, we're mapping the 10-bit range of the potentiometer to a range from 0 to 1, but this time we're sending the output to [pow 4]. This object raises the value that comes in its left inlet to a power set either via an argument, or via the right inlet. In this case, we're raising the value to the 4th power. We do this to make the audio fade in and out smoother, more like in audio equipment, like mixers for example. Figures 4-9 and 4-10 show a linear and an exponential curve respectively, the latter made by raising an incrementing value from 0 to 1, to the 4th power (you'll see [loop] in action later). Our perception of loudness is logarithmic, and using an exponential curve to control it, makes things sound more natural.

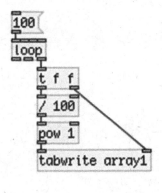

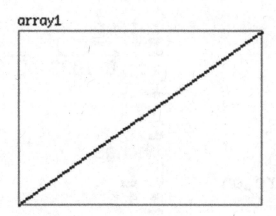

Figure 4-9. *Linear curve*

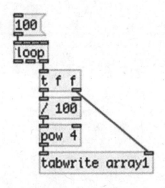

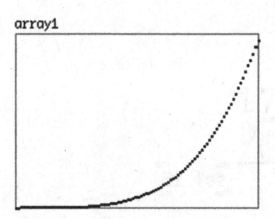

Figure 4-10. *Exponential curve by raising values to the 4th power*

The metronome Subpatch

The next subpatch is *metronome*, which is shown in Figure 4-11. This is the heart of the sequencer of our drum machine. [metro] is an object that outputs bangs at a specified time interval, in milliseconds. This interval can be set either via an argument, or via the right inlet. Any non-zero value in its left inlet will start it, and a zero will stop it. [r tempo] takes input from the second potentiometer, which we map to a range between 40 and 240. We send that to [i] to get the integral part of the value ("i" stands for *integer*) and we multiply it by 4. This will be our tempo in BPMs, but we're splitting each bit to four, because we want to bang 16-notes and not quarter notes, so we need four bangs per beat. Then we send that value to an abstraction called BPM2ms. This is quite similar to the [Hz2ms] abstraction we made in Chapter 1; it is shown in Figure 4-12.

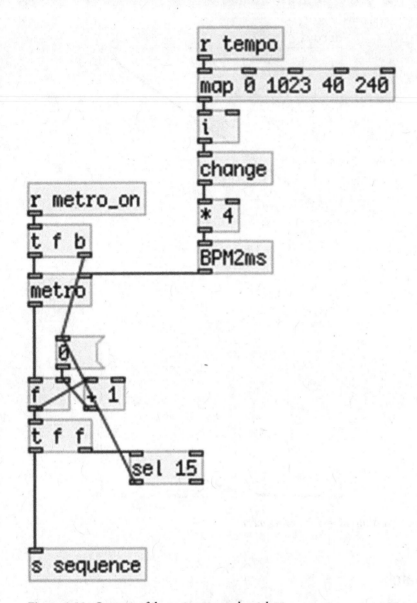

Figure 4-11. *Contents of the metronome subpatch*

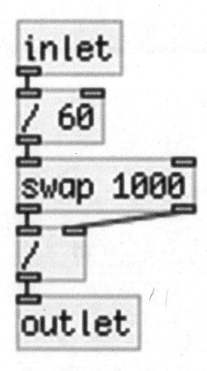

Figure 4-12. [BPM2ms] abstraction

I'm not going to explain how this abstraction works, as it is pretty simple, but try to understand it yourself (mind that [swap 1000] has both its outlets connect to both inlets of [/]). Since BPM is a more musical measure unit, it's better to use it in our interface. [metro] takes in milliseconds, so we need to do the conversion before we send the potentiometer value to the right inlet of [metro], otherwise, the tempo would rise as the potentiometer's value fell. In Figure 4-11, you see that before the BPM value goes to the multiplication object, it goes through a [change]. The analog pins of the Arduino tend to have a bit of noise. Since we're reducing the resolution of the potentiometer by mapping its value to a much smaller range, we also reduce the noise, making it much more stable. I have also mentioned that division in computers is quite CPU-intensive, so it's better to avoid it when you can. Since the potentiometer's output have become more steady, we can use [change] to make sure that the its value will be sent to [BPM2ms] only when it is changed, so we can avoid doing these two division shown in Figure 4-12 every time we receive input from [comport].

In Figure 4-11, below [metro], you can see a simple incrementing counter. This counter sends its values both to [sel 15] and to [s sequence]. [sel 15] will output a bang out its left outlet whenever it receives the value 15. This bang goes to the message "0", which goes to the right inlet of [f]. So, whenever our counter reaches 15, its internal value will be updated to 0, and it will start again (the right inlet of [f] is cold and will only store the value and won't output it, it will only be output when [f] receives the next bang). We're building a 16-step sequencer here, so we need to update our counter after 16 steps, and since the counter starts counting from 0, the last step will be step number 15. [s sequence] will send the counter's value to each [tabplay~] object in the parent patch. Finally, [metro] will be triggered when [r metro_on], the second switch in our circuit, sends a 1.

The set_sequence Subpatch

The next subpatch is *set_sequence*, which is shown in Figure 4-13. In this subpatch, we can set the sequence for each audio file.

Figure 4-13. Contents of the set_sequence subpatch

The diffuse_beats Subpatch

The first subpatch is [pd diffuse_beats] , which you can see in Figure 4-14. This subpatch takes input from the toggle matrix of the parent patch. To achieve this you have to go to the parent patch and change the properties of each toggle. Right-click the top-left toggle and select **Properties**. You should get a window like in Figure 4-15. Go to the **Send symbol:** field and type **tog0**, like in Figure 4-14, and in the **Receive symbol:** below, type **r-tog0**. Do this for every toggle, but increment the number, so the second from top-left toggle should get the symbols **tog1** and **r-tog1**, the third from top-left should get **tog2** and **r-tog2**, and so forth. You can use any name you like for these toggles, but using consistent names that don't need much explanation is a very good programming practice. When done with the top line of toggles, go to the line below and again start from left to right. You should end up with the bottom-right toggle, which should get the symbols **tog15** and **r-tog15**. This is a bit of a cumbersome procedure, but it will facilitate the use of the patch a lot. Also notice that we start numbering the toggles from 0, and we end with 15, instead of 16. This is because in programming, we usually start counting from 0, and since we're going to reference these toggles with some automation processes, it's easier if we do it this way, instead of numbering them from 1 to 16.

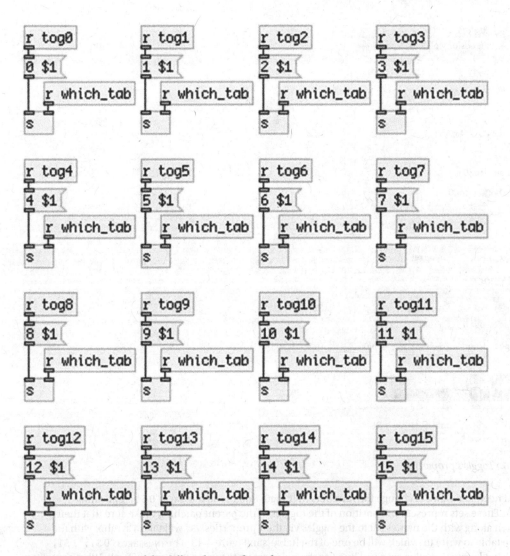

Figure 4-14. *Contents of the diffuse_beats subpatch*

Now let's go back to the [pd diffuse_beats] subpatch. Even though it looks like a lot of work to build this patch, it is actually quite easy if you use the duplicate feature. Go ahead and make the top-left set of objects, [r tog0], [r which_tab], [s], and the "0 $1" message, and connect them like in Figure 4-15. Then select them all and hit Ctrl/Cmd+D, for duplicate. Press **Shift+up arrow** to bring the duplicated objects to the same height with the original ones, and use **Shift+right arrow** to move them to the right. Select both sets of objects and again duplicate them. Again, use Shift and arrows to move them. Now you should have four sets of objects, all connected as they should. Now select all four and duplicate them to get eight sets. Lastly, select all eight sets and duplicate them, and you'll get sixteen sets.

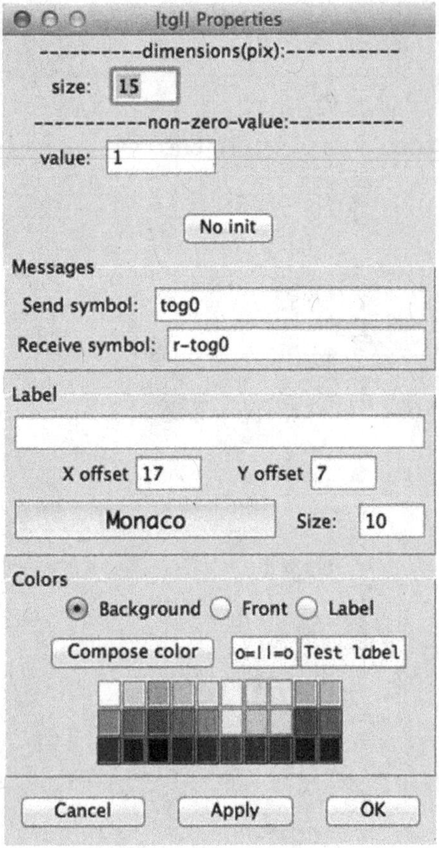

Figure 4-15. *Toggle's properties*

All you need to do now is change the numbers in [r tog0] and in the message "0 $1", replicating Figure 4-14. These sets represent the position of the toggles of the parent patch, so make sure that their names match along with the names set to the toggles via their properties. [s] will take a symbol with the name of the table to write to, which will be one of the [table]s in Figure 4-13. The messages "0 $1", "1 $1", and so forth. will write the value received by $1 to the index set by the value 0, 1, and so forth. What actually happens here is that [r tog0] will get a value from the corresponding toggle, and it will write it to the first index (index 0) of the table set by [r which_tab]. [r tog1] will get a value from the corresponding toggle and it will write it to the second index of the table, and so on. [r which_tab] receives a symbol from the *set_tab* subpatch you see in Figure 4-13, which is shown in Figure 4-16. The 1s and 0s that we'll write to the sequence tables will be read in the parent patch by the [tabread] objects, where a 1 will trigger the audio file read by [tabplay~]. This way we can use the counter to trigger each audio file at the location we want. This is how a sequencer works.

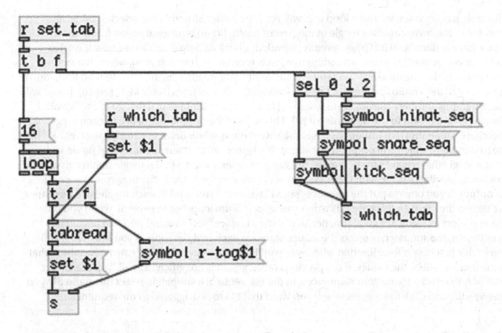

Figure 4-16. *Contents of the set_tab subpatch*

The set_tab Subpatch

This subpatch takes input from the Vradio shown of the parent patch. We've set the number of cells of the radio via its properties, in the **number:** field, which by default is 8, and we set it to 3 (right-click it and select **Properties**). Also, we've used comments next to each cell of the Vradio to specify what each cell will select (kick, snare, and hihat). In Figure 4-16, we receive that value and we send it to [sel 0 1 2], which will bang the corresponding symbol. In Pd, you can create a symbol from a message by prepending the word *symbol*. These symbols are the names of the tables in Figure 4-13, which will store the sequences for each audio file, and they are being sent to [s which_tab], which we receive with [r which_tab] in Figure 4-14. This way we can set the table to write our sequences in an easy-to-use way.

Once the table has been set, we run a loop that will read the values stored in the selected table, and will set each value to the corresponding toggle of the parent patch. [loop] is an abstraction found the in "miscellaneous_abstractions" GitHub page, already provided. Check its help patch to see how it works. [tabread] takes a message that is formed according to what is received in [r which_tab], which is a symbol with the name of the table. The message "set table_name" sent to [tabread], sets the table to read from. So, if we click the first cell of the Vradio in the parent patch, [tabread] will read from [table kick_seq 16]. [loop] will output 16 incrementing numbers starting from 0, ending in 15. These numbers will first go to the "symbol r-tog$1" message, which is sent to the right inlet of [s]. This will set the destination of [s], to each toggle of the parent patch, sequentially (remember, we've set their **Receive symbol:** fields to r-tog0, r-tog1, etc.). Then the values go to [tabread], which will output the value at that index. So, the value at index 0 of [table kick_seq 16] will go to the toggle that receives from "r-tog0", the value at index 1 will go to the toggle that receives from r-tog1, and so forth. Sending a value with the message "set", only sets the value to the toggle, and doesn't produce any output. If you omit to put the message "set $1", whatever you send to each toggle, will be output, and again written to the table. In this case this won't cause an infinite loop, but in general, when you only want a toggle to project its value, it's good practice to use the message "set" to avoid possible bugs. The mechanism of the set_tab subpatch projects the values stored at each table as soon as you select it. If you omit the loop in this patch, you'll realize that whenever you'll want to write a sequence to a new table, what you'll get in the toggles will be the values of the previous table, which is not efficient at all. You can try for yourself to see why that is. To create your sequences, in the set_sequence subpatch, select the audio file you want from the Vradio, and click the toggles where you want that file to be triggered in the sequence.

Concluding the Patch and Explaining the Data received from the Arduino

The patch we've built is rather complex, and yet it is a simple drum machine. This is a good demonstration of programming, since you can see that even the simplest things in computers need quite some work, when we program them from scratch. Still, the advantage of programming is that you can build unique interfaces and projects that don't exist in the market, as they are tailored to one person's needs. Now let's go back to Figures 4-5 and 4-6, the parent patch and the arduino_stuff subpatch. In the arduino_stuff subpatch, you see that the first switch controls the DSP state of Pd. The second goes to [s metro_on], which is sent to the metronome, inside the metronome abstraction. Whenever this switch is in the ON position, the drum machine will play the sequences stored in the set_sequence subpatch, with the tempo set by the second potentiometer. Whenever it is in the OFF, position, the sequence won't play, but you'll be able to trigger each audio file with the three push buttons of your circuit (we haven't built the circuit yet, but I'll explain it further on). In the arduino_stuff subpatch you can see three [send]s, [s kick_trig], [s snare_trig], and [s hihat_trig]. These take input from the three push buttons of the circuit, and they are sent to the parent patch. Each [receive] in the parent patch goes to [sel 1], so when you press a button, [sel 1] will output a bang, and [tabplay~] will play the audio file stored in the table it reads from. You can actually trigger the audio files manually, even when the sequence is playing. Finally, we receive input from Pd with [r pd], about the DSP state, which is sent to the Arduino, to control an LED that will indicate the DSP state.

Arduino Code for Drum Machine Patch

Let's now move on to the Arduino sketch, shown in Listing 4-2.

Listing 4-2. Drum Machine Code

```
1.    // analog values array size, must be constant
2.    const int num_of_analog_pins = 2;
3.    // digital values array size, must be constant
4.    const int num_of_digital_inputs = 5;
```

```
5.     // digital output pin
6.     int num_of_digital_outputs = 1;
7.
8.     // assemble number of bytes we need
9.     // analog values are being split in two, so their number times 2
10.    // and we need a unique byte to denote the beginning of the data stream
11.    const int num_of_bytes = (num_of_analog_pins * 2) + num_of_digital_inputs + 1;
12.
13.    // array to store all bytes
14.    byte transfer_array[num_of_bytes] = { 192 };
15.
16.    void setup() {
17.      for(int i = 0; i < num_of_digital_inputs; i++)
18.        pinMode((i + 2), INPUT_PULLUP);
19.
20.      for(int i = 0; i < num_of_digital_outputs; i++)
21.        pinMode((i + num_of_digital_inputs + 2), OUTPUT);
22.
23.      Serial.begin(57600);
24.    }
25.
26.    void loop() {
27.
28.      int index = 1; // index offset
29.
30.      if(Serial.available()){
31.        static int temp_val;
32.        int which_pin;
33.        byte in_byte = Serial.read();
34.        // check if in_byte is a number and assemble it
35.        if((in_byte >= '0') && (in_byte <= '9'))
36.          temp_val = temp_val * 10 + in_byte - '0';
37.        // check if in_byte is a letter and call digitalWrite
38.        else if((in_byte >= 'a') && (in_byte <= 'z')){
39.          which_pin = (in_byte - 'a');
40.          which_pin += (num_of_digital_inputs + 2);
41.          digitalWrite(which_pin, temp_val);
42.          temp_val = 0;
43.        }
44.      }
45.
46.      // store the digital values to the array
47.      for(int i = 0; i < num_of_digital_inputs; i++)
48.        transfer_array[index++] = !digitalRead(i + 2);
49.
50.      // store the analog values to the array
51.      for(int i = 0; i < num_of_analog_pins; i++){
52.        int analog_val = analogRead(i);
53.        // split analog values so they can retain their 10-bit resolution
```

```
54.        transfer_array[index++] = analog_val & 0x007f;
55.        transfer_array[index++] = analog_val >> 7;
56.      }
57.
58.    // transfer bytes over serial
59.    Serial.write(transfer_array, num_of_bytes);
60.  }
```

This sketch is a modified version of the serial_write.ino sketch that goes along with the help patch of the [serial_write] Pd abstraction. There's really not much to explain here. As with the previous Arduino sketch, in lines 2 and 4 we set constants that will set the size of the array we'll send to the serial line. This time we're using the write function instead of print, so we can send all the values in one function call. In line 6, we set the number of digital outputs, which is only one, and in line 11, we assemble the constants of lines 2 and 4, to set the size of the array we'll send to the serial line. Remember that write sends bytes, and the analog pins of the Arduino have a 10-bit resolution, therefore we need to split them to two, and reassemble them back in Pd. For this reason we need two bytes for each analog pin, and one for each digital (these will be either 1 or 0), plus one unique byte to denote the beginning of the data stream.

In the setup function, we set the modes of the digital pins and we begin the serial communication at a 57600 baud rate. In the loop function we first set the unique byte to the beginning of the array, and then we define an index offset for writing the rest of the values to the array. If you can't remember how this works and why we need to do it this way, go back to end of Chapter 2. Then in lines 29 to 45 we check whether we have input from Pd, to control the DSP LED. Afterward, we read and store the digital input values (remember in the Pd patch, we've set the digital pins first in the arguments of [serial_write]. The order of these arguments should be aligned with the order of the reading on pins in the Arduino sketch), and then the analog ones. Once we're done, we send the whole array to Pd.

Circuit for Arduino Code

Now let's check the circuit for this project, which is shown in Figure 4-17.

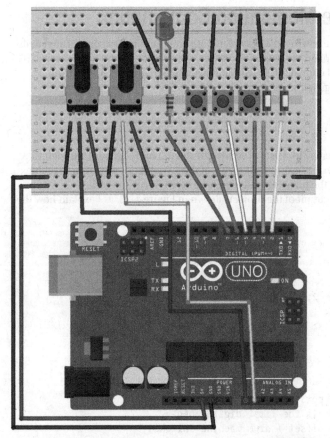

fritzing

Figure 4-17. *Simple drum machine circuit*

We follow the same philosophy concerning connections, so make sure that you connect the potentiometers from analog pin 0 onward, and the switches and push buttons from the digital pin2 onward. The LED should go to the first digital pin after the digital inputs. This helps automating some routines in the Arduino sketch, like setting the mode of the digital pins, and reading the correct pins; otherwise, we would need to set the pin numbers explicitly, and the for loops wouldn't be so simple.

Go ahead and store some sequences to the Pd patch, and start playing! One drawback of the Pd patch is that we cannot save the sequences we write, so every time we open the patch we need to create them anew. A simple way is to go to the toggle's properties and click the **No init** button, which will become **Init**. This way the patch will save the last value each toggle had before the patch was closed. If we want to store all sequences, we must save the contents of the [table]s to text files, but this adds some complexity, and since this is just an introductory project, we won't deal with it. If you want to check for yourselves, check [textfile]. An easy work around is to use arrays instead of [table]s, because in the array's properties we can set it to save its contents. Having the graphs of the arrays for the sequences is not so elegant, and we really don't need to look at them, as the toggles project the sequences in a much more user-friendly way. At the time of writing (September 2015), the latest Pd-vanilla has introduced the [array] object, which can save its contents, much like the array can. Unfortunately, this object doesn't exist in Pd-extended, so we must go for one of the other solutions mentioned.

Drum Machine and Phase Modulation Combination

Before we close this chapter, let's combine the two interfaces we built. It shouldn't be very complicated to do this, still there are some details I'll need to explain thoroughly. This time we'll start with the Arduino sketch. We're going to use the print function to send data to Pd, so we can retrieve it with objects like [r drum_machine_switches], which should make things self-explanatory.

Arduino Code

Listing 4-3 shows the Arduino sketch, which is the longest we've written so far. This is because we want to send groups of values separately, for example, one group will consist of the switches for the drum machine, another group will consist of the switches for the phase modulation, and so on. Write the code in Listing 4-3 to the Arduino IDE and check the circuit of the project, shown in Figure 4-18, as I explain how the code functions.

Listing 4-3. Combining the Two Projects

```
1.    const int drum_machine_pots = 2;
2.    const int phase_mod_pots = 3;
3.    const int drum_machine_switches = 2;
4.    const int phase_mod_switches = 1;
5.    const int drum_machine_buttons = 3;
6.    const int phase_mod_buttons = 1;
7.    int digital_outputs = 1;
8.
9.    // pin offsets
10.   int pm_pot_offset = drum_machine_pots;
11.   int dm_switch_offset = 2; // this is the first digital pin
12.   int pm_switch_offset = dm_switch_offset + drum_machine_switches;
13.   int dm_button_offset = pm_switch_offset + phase_mod_switches;
14.   int pm_button_offset = dm_button_offset + drum_machine_buttons;
15.   int led_offset = pm_button_offset + phase_mod_buttons;
16.
17.   // create arrays to store values from pins
18.   int dm_pots[drum_machine_pots];
19.   int pm_pots[phase_mod_pots];
20.   int dm_switches[drum_machine_switches];
21.   int pm_switches[phase_mod_switches];
22.   int dm_buttons[drum_machine_buttons];
23.   int pm_buttons[phase_mod_buttons];
24.
25.   void setup() {
26.     // set digital input pin modes
27.     for(int i = 0; i < drum_machine_switches; i++)
28.       pinMode((i + dm_switch_offset), INPUT_PULLUP);
29.     for(int i = 0; i < phase_mod_switches; i++)
30.       pinMode((i + pm_switch_offset), INPUT_PULLUP);
31.     for(int i = 0; i < drum_machine_buttons; i++)
32.       pinMode((i + dm_button_offset), INPUT_PULLUP);
```

```
33.     for(int i = 0; i < phase_mod_buttons; i++)
34.       pinMode((i + pm_button_offset), INPUT_PULLUP);
35.
36.     // set digital output pin modes
37.     for(int i = 0; i < digital_outputs; i++){
38.       pinMode((i + led_offset), OUTPUT);
39.     }
40.     Serial.begin(57600);
41.   }
42.
43.   void loop() {
44.     if(Serial.available()){
45.       static int temp_val;
46.       int which_pin;
47.       byte in_byte = Serial.read();
48.       // check if in_byte is a number and assemble it
49.       if((in_byte >= '0') && (in_byte <= '9'))
50.         temp_val = temp_val * 10 + in_byte - '0';
51.       // check if in_byte is a letter and call digitalWrite
52.       else if((in_byte >= 'a') && (in_byte <= 'z')){
53.         which_pin = (in_byte - 'a') + led_offset;
54.         digitalWrite(which_pin, temp_val);
55.         temp_val = 0;
56.       }
57.     }
58.
59.     // read and store analog pins
60.     for(int i = 0; i < drum_machine_pots; i++)
61.       dm_pots[i] = analogRead(i);
62.     for(int i=0; i < phase_mod_pots;i++)
63.       pm_pots[i]=analogRead(i + pm_pot_offset);
64.     // read and store digital pins
65.     for(int i = 0; i < drum_machine_switches; i++)
66.       dm_switches[i] = !digitalRead(i + dm_switch_offset);
67.     for(int i = 0; i < phase_mod_switches; i++)
68.       pm_switches[i] = !digitalRead(i + pm_switch_offset);
69.     for(int i = 0; i < drum_machine_buttons; i++)
70.       dm_buttons[i] = !digitalRead(i + dm_button_offset);
71.     for(int i = 0; i < phase_mod_buttons; i++)
72.       pm_buttons[i] = !digitalRead(i + pm_button_offset);
73.
74.     // write the stored values to the serial line
75.     Serial.print("drum_machine_pots ");
76.     for(int i = 0; i < (drum_machine_pots - 1); i++){
77.       Serial.print(dm_pots[i]);
78.       Serial.print(" ");
79.     }
80.     Serial.println(dm_pots[drum_machine_pots - 1]);
81.
```

```
82.     Serial.print("drum_machine_switches ");
83.     for(int i = 0; i < (drum_machine_switches - 1); i++){
84.        Serial.print(dm_switches[i]);
85.        Serial.print(" ");
86.     }
87.     Serial.println(dm_switches[drum_machine_switches - 1]);
88.
89.     Serial.print("drum_machine_buttons ");
90.     for(int i = 0; i < (drum_machine_buttons - 1); i++){
91.        Serial.print(dm_buttons[i]);
92.        Serial.print(" ");
93.     }
94.     Serial.println(dm_buttons[drum_machine_buttons - 1]);
95.
96.     Serial.print("phase_mod_pots ");
97.     for(int i = 0; i < (phase_mod_pots - 1); i++){
98.        Serial.print(pm_pots[i]);
99.        Serial.print(" ");
100.    }
101.    Serial.println(pm_pots[phase_mod_pots - 1]);
102.
103.    Serial.print("phase_mod_switches ");
104.    for(int i = 0; i < (phase_mod_switches - 1); i++){
105.       Serial.print(pm_switches[i]);
106.       Serial.print(" ");
107.    }
108.    Serial.println(pm_switches[phase_mod_switches - 1]);
109.
110.    Serial.print("phase_mod_buttons ");
101.    for(int i = 0; i < (phase_mod_buttons - 1); i++){
102.       Serial.print(pm_buttons[i]);
103.       Serial.print(" ");
104.    }
105.    Serial.println(pm_buttons[phase_mod_buttons - 1]);
106.  }
```

This sketch is long only because we have things being repeated. Lines 1 to 7 define some constants that hold the number of pins for each group. We're separating switches from push buttons and potentiometers into different groups, one of each for the drum machine and one for the phase modulation part of the patch. This makes a total of six groups, and lastly we define a variable for the LED, which will indicate the DSP state. Lines 10 to 15 define offsets for the pins of each group, according to the number of pins used by each group. The drum machine potentiometer group, doesn't need an offset, because it starts from analog pin 0.

The drum machine switch group needs an offset, because even though it's the first digital pin group, we start using digital pins from 2 onward. Combine this part of the sketch with the circuit in Figure 4-18. In the circuit, switches, buttons, and potentiometers, are being grouped by being placed close to each other. Check which pin each switch, button, or potentiometer of each group uses, and try to understand how these offsets will be helpful to use. Lines 18 to 23 define arrays for each value group.

In the setup function, we set the mode of all digital pins used. Here is the first part where we use the offsets we've set in lines 10 to 15. We need the variable i of the for loop to start from 0, so the loop will run as many times as we need, using the constant that stores the number of pins of each group (drum_machine_ switches for example, in line 27). Since i will start from 0, we need the pin offsets so that the pinMode function will use the correct pin. In the loop function, the first thing we do is check if we have input in the serial line, and use that to control the DSP LED. This happens in lines 44 to 57. Afterward, in lines 60 to 72, we read and store the values read from each group, starting with the potentiometers, then reading the switches, and finally the push buttons of each group, alternately. Finally, in lines 75 to 115, we print the values of each group to the serial line, each time using an appropriate tag. Notice that we don't need to print the groups in some specific order, since all values have already been grouped and stored in different arrays. It might have been easier to use the write function instead of print, and save ourselves from quite some code writing. If we did that, we would be receiving all values together and we would have to split them in the Pd patch. Now we'll be receiving each group according to its tag, using [receive], which will make things easier. I can't say that one technique is superior to the other; they're just different approaches. I'm presenting both approaches in this chapter so that you can see which one works best for you.

Arduino Circuit

As you can imagine, the circuit is a combination of the circuits of the two previous interfaces. It is shown in Figure 4-18.

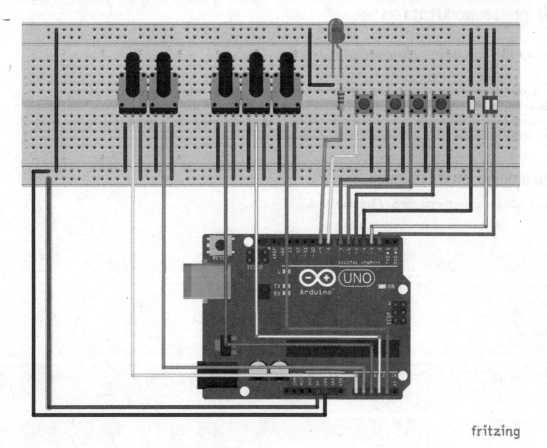

fritzing

Figure 4-18. *Drum machine-phase modulation circuit*

169

Pd Patch for Drum Machine-Phase Modulation Interface

Now let's check the Pd patch. It is obviously a combination of the two previous patches, but it has some enhancements. Figure 4-19 illustrates it.

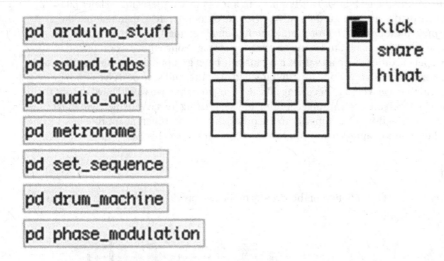

Figure 4-19. *Drum machine-phase modulation patch*

Like with the previous interface, we've put most of the stuff in subpatches to make things clear. The subpatches sound_tabs, audio_out, and set_sequence, are the same with the previous project of this chapter, so I'm not going to explain them here. The arduino_stuff and metronome have changed a bit, and we now have two new ones, drum_machine and phase_modulation.

The arduino_stuff Subpatch

Figure 4-20 illustrates the contents of arduino_stuff.

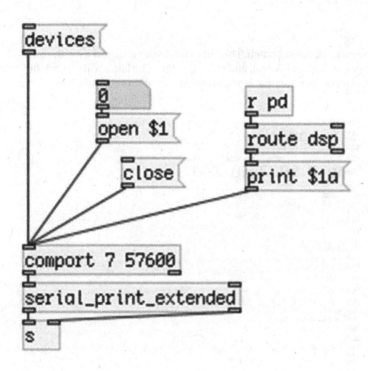

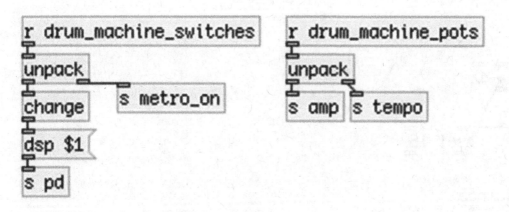

Figure 4-20. *Contents of the arduino_stuff subpatch*

Here we send the bytes [comport] outputs to [serial_print_extended], which sends them to [s]. Below we receive two of the value groups, drum_machine_switches and drum_machine_pots, which unpack the values of their groups and diffuse them using [send]s. The first switch controls the DSP, and the second controls [metro], which is in the metronome subpatch. The first potentiometer controls the overall amplitude, and the second the BPMs of the metronome. The first switch and the first potentiometer don't really belong to the drum machine group, but we've left them there, from the previous interface.

The metronome Subpatch

Figure 4-21 illustrates the contents of the metronome subpatch. This subpatch hasn't changed a lot, the only difference is that we're sending the BPM value both to [metro], and to [s env_dur]. The latter will control the duration of an amplitude envelope for the phase modulation, in the phase_modulation subpatch.

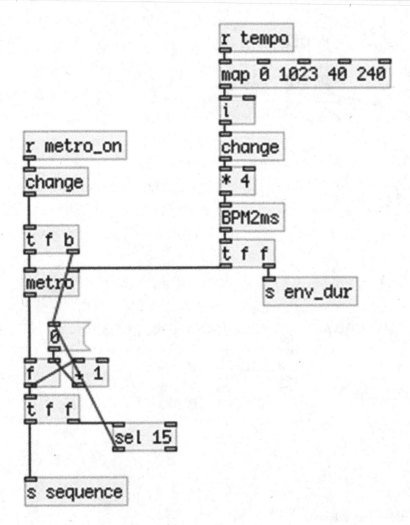

Figure 4-21. Contents of the metronome subpatch

The drum_machine Subpatch

Figure 4-22 illustrates the contents of the drum_machine subpatch. This is the main part of the parent patch of the previous interface. Only two things have changed here. One is the amplitude attenuation of each audio file. Now every [tabplay~] is being multiplied by 0.25, instead of 0.33, because we have the phase modulation part as well, so we have four different audio sources in total. Multiplying each one by 0.25, makes sure that they'll never exceed 1, even if they are all triggered together. The other thing that has changed is that we have removed the [r kick_trig], [r snare_trig], and [r hihat_trig] objects, because we're now receiving the drum machine push-button values straight in this subpatch. After unpacking them, we send each to a [change], so that their values will be output only when they're changed.

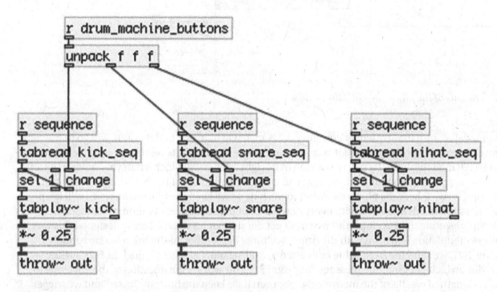

Figure 4-22. *Contents of the drum_machine subpatch*

The phase_modulation Subpatch

Finally, Figure 4-23 illustrates the contents of the *phase_modulation* subpatch. The bottom-left part of the patch is copied from the patch of the first interface we built in this chapter. This time we're using an amplitude envelope made with the [ggee/envgen] external object (the GUI in Figure 4-23). In the left part of the patch, you see a [r phase_mod_pots], which, as its arguments states, receives the values from the potentiometers for the phase modulation. We're unpacking the list of values and sending them to the carrier frequency, the modulator frequency, and the index. We map the value for the index to the range from 0 to 2, using the [map] abstraction, like we did before. What I need to explain in detail is the right part of the subpatch, where we have the envelope for the amplitude. In Figure 4-23 I have made some sort of an ADSR envelope (the sustain part is also decaying here). Above [ggee/envgen] there is a [r sequence] that receives the sequencer counter, but converts the counter values to two bangs with [t b b]. The first bang (from the right outlet), goes to [f]. [f] stores the value from the fourth push button, which is received via [r phase_mod_buttons].

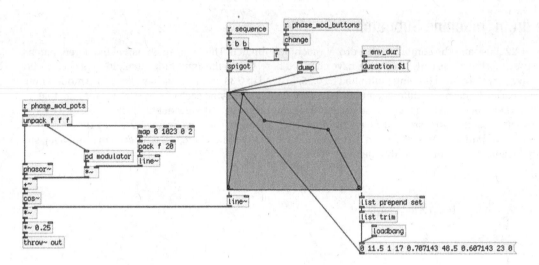

Figure 4-23. *Contents of the phase_modulation subpatch*

When a new value from the sequencer counter comes in, it will bang [f], and its value will go to the right inlet of [spigot]. [spigot] works like a gate. If it receives any non-zero value in its right inlet, it will let anything that comes in its left inlet through. If it receives a zero in its right inlet, it will block whatever comes in its left inlet. So, when we keep the button pressed, [spigot] will let the second bang of [t b b] through, which will bang the envelope, so we'll hear the phase modulation. As long as we keep the button pressed, we'll keep on triggering the sound of this subpatch with every new count of the sequencer. As soon as we release the button, we'll stop triggering the envelope and we won't get the subpatch's sound. This is designed this way so that the phase modulation is synced with the drum machine. Remember, in the metronome subpatch shown in Figure 4-21, we send the BPM value converted to milliseconds to a [s env_dur]. In Figure 4-23, you see the [r env_dur], which goes into the message "duration $1". This will set the duration of the envelope to exactly the same length of one bit of the metronome. So, even if we keep the button pressed and we trigger the phase modulation sound on every beat, [ggee/envgen] will have enough time to complete its envelope.

On the bottom-right part in Figure 4-23, we connect the second outlet of [ggee/envgen] to [list prepend set], and then to [list trim] and to a message. Don't type the values you see in the message of the figure, just leave it blank. Above [ggee/envgen] there is a message "dump." If you click it, [ggee/envgen] will output the values of its graph out its right outlet. To save our envelope, we store these values to a message that we bang with [loadbang] and send it to [ggee/envgen]'s inlet, so it can create the same envelope by itself. [list prepend set] takes in a list and prepends the last argument, in this case "set". If we send the message "set something" to an empty message, the word "something" will be printed in the empty message, try it. [list trim] trims the "list" selector off from a list, which has been added by [list prepend set]. If we omit [list trim], what will come out from [list prepend set] will be "list set" along with the values from [ggee/envgen], and the empty message won't understand it. Trimming out "list" from the list will result in the message "set" along with the values of [ggee/envgen], and this way the values will be saved in the message. Go ahead and create an envelope (make any kind of graph you like in [ggee/envgen]) and click [dump]. You'll immediately see some values in the empty message. Save your patch, and the next time you'll open it, you'll see your envelope still there.

The modulator Subpatch

The last thing to mention about this patch is the modulator subpatch of the phase_modulation subpatch, which is shown in Figure 4-24. This is essentially the same subpatch with the first interface of this chapter, only instead of a [r choose_waveform] in the top-right part of it, there's a [r phase_mod_switches], which receives the switch value for the phase modulation straight from the [serial_print_extended] abstraction. Again, we're sending the value to [change –1] so it will give output even if we start the patch with the switch to the OFF position, as we need to select the type of the modulator oscillator right away.

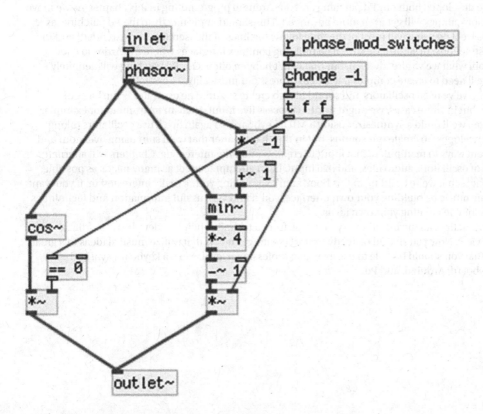

Figure 4-24. *Contents of the modulator subpatch*

This concludes the description of this interface. We extensively used the print function along with the [serial_print_extended] abstraction to make the Pd patch a bit easier to make and understand. It's up to you what kind of approach you'll take for your own projects. Go ahead and play with this interface, it should be fun!

Conclusion

In this chapter, we went through the process of creating a musical interface. The three interfaces made here might not be appealing to everyone, but they server as a generic platform on top of which you can build many different kinds of interfaces. We focused on the two different techniques for communication between the Arduino and Pd, which we have been introduced to in Chapter 2, showing the advantages and disadvantages of each. You should pick the technique that fits your needs best, there's no definite rule as to which technique is better over the other.

We have also developed both our Pd patching and our Arduino programming in this chapter. As we move further into this book, things will get even more advanced. This should apply mostly to the Pd patching, as the Arduino code we'll be using will most of the time follow the lines of the [serial_print_extended] and the [serial_write] abstractions. In some cases, we'll be building complex interfaces where the Arduino code will also differ from what we've already seen, but that won't happen often. Our Pd patching will definitely grow larger, as we'll need to create complex patches to express our musicality.

Up until now, we've used oscillators and sound files to create sound. The oscillator is the basis of electronic music, but in some cases, we might want to process live input, from an instrument for example. In the next chapter, we'll build a synthesizer using a MIDI keyboard, so again, we'll use oscillators (along with filters and envelopes) to create our sounds, but in the chapter after that we'll start using live input and we'll learn different ways to manipulate that input to create something interesting. Chapters will alternate between the use of oscillators, audio files, and live input, in an attempt to meet as many needs as possible. Even if the interfaces that we're building in this book aren't something you're really interested in, if you want to make electronic music by building your own interface, you'll find very useful information and techniques that will prove helpful in creating your own ideas.

This chapter finalizes all the introductory material, from a technical point of view, but also from a creative point of view. Now you should be ready to start creating projects that utilize musical ideas to a great extent; projects that you should be able to use even at a professional level. Next, a keyboard synthesizer using a MIDI keyboard, Arduino, and Pd.

CHAPTER 5

■ ■ ■

A Simple Synthesizer Using a MIDI Keyboard and Arduino

In this chapter, we'll start building interfaces that can really be used for expressing musical ideas. Our first project will be a clavier synthesizer. For this interface, we'll use a MIDI keyboard to control the pitch of the synthesizer, and the Arduino for various other types of control parameters. If you have a MIDI keyboard that provides potentiometers, pitch-bend wheels, and push buttons, you might want to avoid using the Arduino altogether. Still, the Arduino has a better resolution in its analog pins, so you might find that it provides more room for expression when it comes to sound quality control parameters. In this chapter, we're going to cover different types of MIDI keyboards, from the new USB MIDI to the old-style 5-pin DIN cable MIDI. In the case of the latter, using an Arduino will be necessary if you want to avoid buying a MIDI-to-USB converter, which might be rather expensive.

The idea behind this project is to revive any old equipment that you might have lying around. I found a 30-year-old keyboard synthesizer in my house, which hasn't been used for decades. Even though it's a very old instrument, it does have MIDI, although with the old 5-pin DIN cable. This is where the Arduino came in handy—with a simple circuit and very cheap components, I managed to receive input from the synthesizer and send it over to my computer. This is not a real MIDI-to-USB converter, as my computer doesn't see the Arduino as a MIDI device. Still, it's possible to send the messages received from the keyboard over to the computer "MIDI style," so you can use the keyboard to control you oscillators in your Pd patch. If you have a USB MIDI keyboard that you'd like to use to hack and create a brand-new synthesizer, use that instead, as it will save you from some circuit building and some coding too.

In this project, we'll end up with an interface with an embedded computer (a Raspberry Pi) that will be enclosed in the synthesizer. For this reason, I chose to use an Arduino Nano instead of the Uno, as it is much smaller and destined to be enclosed in projects. It is a bit more expensive than the Uno, but it has the same capabilities and fits in very small enclosures. Also, it provides eight analog inputs, which seems quite necessary for this project. You are welcome to use whichever Arduino you like best. If you use an Uno, which has fewer analog pins, you'll have to compensate by reducing your control parameters.

I'll provide some suggestions as to how this project can be implemented with fewer control pins as we build the Pd patch. You are not obliged to embed a computer to your own implementation of this project. If you want, you can do all the coding and patching, but keep your devices outside enclosures. This may be because you have a MIDI keyboard that you use for other reasons too, and you don't want to hack it and transform it to something else. You might also want to use your laptop instead of a Raspberry Pi, because you already have an external sound card. The way this project is built is only a suggestion. You are welcome to make your own version of this or any other project in this book.

Another thing that we'll cover to some extent is external sound cards. We'll go through a few different options for the Raspberry Pi, and how to use them with Pd. Not using one won't affect the project very much, as we're not going to receive audio input. The difference will be that the audio without an external sound card will be a bit lower in volume, and it will have some noise. If you don't have one, you can always build this interface without it, and if you want, you can get one in the future and come back to this chapter to get information as to how to use it with Pd. Now let's get started.

You may find some of the patching or coding in this project a bit complex, so you might want to go through the chapter more than once. In any case, stay focused during the whole process, and if there is something you don't grasp during your first read, go ahead and read it again. It's all a matter of programming with the two languages—Pd and Arduino, so eventually it should all come together and make sense.

Parts List

Table 5-1 lists the parts we'll need to build this project.

Table 5-1. *Parts for this Project*

Part	Quantity
Arduino	1 (preferably Nano)
MIDI keyboard	1 (USB or 5-pin DIN)
Optocoupler	1 × 4N35 or 4N28
Potentiometers	8 × 10 kiloohm
Push buttons	4 (to mount on panel)
Switches	5 (to mount on panel)
LED	1
Resistors	1 × 220Ω
	1 × 3K3Ω
	1 × 100KΩ
Diodes	1 × 1N4248 (diode type)

What Is MIDI?

Before we start using MIDI with Pd, I should talk a bit about MIDI itself. MIDI stands for Musical Instrument Digital Interface, which is a communication protocol, digital interface, and connector type (although the standard MIDI cable has mostly been replaced by the USB type one), that allows for communication between different devices, for control of musical applications. There are quite some different kinds of MIDI messages a MIDI device and deliver or accept, and we're going to use some of the here. MIDI was initiated in 1983 and it's still a prevailing interface in digital (sometimes interfering with analog too) electronic music. Even though it's so old, it still hasn't been replaced by another protocol. The OSC (Open Sound Control) protocol started off as a replacement for MIDI, but as it developed, it took another route, and ended up in a more network oriented communication protocol.

MIDI devices enable us to physically control various aspects of our musical applications. Different types of control interfaces, like keyboards, potentiometers, pads, foot pedals, push buttons, and so forth provide a wide variety of controllers, which we can use to create various musical interfaces. All this might sound a bit similar to the use of the Arduino we make in this book. Although MIDI and Arduino are two different things, they both aim at control and communication (the Arduino has a wider range of capabilities, like motors and solenoids, for example). At least the way we use the Arduino here is very similar to the way MIDI controllers are used. Why not use MIDI devices and avoid Arduino programming in the first place? One drawback of MIDI (at least in my point of view) is that first of all, MIDI controllers are rather expensive compared to their quality, and secondly, one depends on the products of the market. With Arduino you can build really unique control interfaces that cannot be found on the market (you can only find the components of such an interface on the market, but not the interface assembled and programmed), and the overall cost will be much smaller. In addition to that, the high resolution of the Arduino's analog pins provide much more flexibility in controlling the sound aspects.

In this chapter, we're using MIDI mostly because of the keyboard interface you can easily find in the market. Building a keyboard is a difficult task that can cause a lot of frustration to someone who is not experienced in electronics. Since a MIDI keyboard is something very common, and many people using a computer for music (in any way) do have one lying around, it's a good idea to start with something like this, as the first finished project of this book. I'll assume you have a MIDI keyboard. If it has a MIDI cable, I'll assume you have a MIDI-to-USB converter or a sound card that has a MIDI connector, so that you can connect the keyboard to your computer. If you have a MIDI cable controller, later in this chapter, you'll see how to get its input using the Arduino.

Pd and MIDI

Using MIDI in Pd is quite simple. This project is based on a keyboard, so we'll first cover the use of MIDI keyboards in Pd. To be able to use a MIDI device with Pd, you need to plug the device in your computer before launching Pd. Once you launch Pd, set your keyboard as a MIDI input device, as shown in Chapter 1. The object that receives data from MIDI keyboards is [notein]. Figure 5-1 shows [notein] receiving input from a MIDI keyboard.

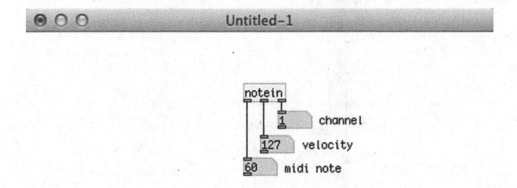

Figure 5-1. *The [notein] object*

How [notein] Works

[notein] has three outlets. The leftmost outlet outputs the MIDI note number. In Figure 5-1 we see MIDI note 60, which is the middle C. The middle outlet outputs the velocity, which means how hard the key is pressed (hence, the dynamic of the note, in more musical terms). The rightmost outlet outputs the MIDI channel the device is talking to. The channel can be set via an argument, in which case [notein] will have only two outlets, one for the MIDI note number, and one for the velocity. MIDI notes, as well as velocity, counts from 0 to 127, with 127 velocity being full amplitude. The channel is used to avoid clashes with other MIDI controllers used simultaneously. Check its help patch for more information.

To use [notein] in a meaningful way, we also an object to convert the MIDI note numbers to frequencies, so we can control oscillators this way. Pd has a built-in object for that, [mtof], which stands for *MIDI to frequency*. Figure 5-2 shows how to use these two objects to control the pitch and amplitude of an oscillator. In this figure, we see that the incoming note is MIDI note number 69 (which by the way is the tuning A) and the velocity is 116. The MIDI note goes through [mtof] which converts it to the corresponding frequency (440 in this case), and then sent to [osc~]. The velocity is first divided by 127, because 127 is the maximum velocity we get from MIDI keyboards, and the maximum amplitude value we should use is 1. The result of the division is sent to [line~] to smooth the possible clicks of the control and signal domain combination, and is then multiplied with the oscillator's output to control its amplitude. The number atoms used in the figure are not necessary, and it's better to avoid them when not needed, as they require a CPU, but there're there to demonstrate the conversions from one type to the other. Go ahead and build it to see how MIDI keyboards are used in Pd.

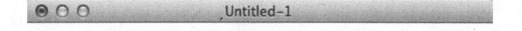

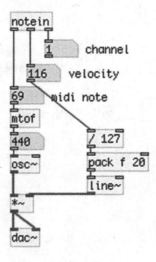

Figure 5-2. *Using [notein] and [mtof] to control an oscillator*

Using a Different Tuning and an Amplitude Envelope with [notein]

The patch in Figure 5-2 demonstrates the simplest use of [notein], and we can see that two issues arise from this kind of use. First, we have a fixed tuning at 440 Hz, which is not the case with most acoustic instruments, as modern tuning is usually at 442 Hz or higher, and other kinds of music might use different tunings, like early music for example. The other issue is that this patch doesn't make use of an amplitude envelope, making the audio output sound rather mechanic. A solution to the first issue is provided by the [mtof_tune] abstraction, which is included in the miscellaneous abstractions GitHub page already provided. This abstraction lets you set the tuning A pitch to any frequency you want, and will then convert MIDI-notes according to that pitch. The second issue can be solved by using [ggee/envgen], which helps us create amplitude envelopes in a user-friendly way. Figure 5-3 illustrates the use of the [mtof_tune] abstraction and the [ggee/envgen] external.

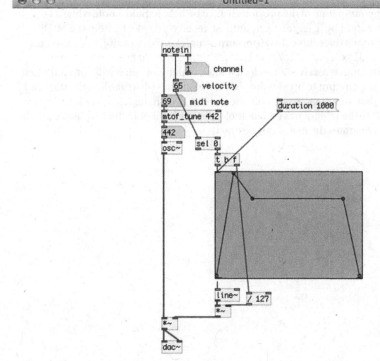

Figure 5-3. *Using [mtof_tune] and [ggee/envgen] to set tuning and amplitude envelope*

Our patch has now become a bit more complex. The easy part is the left side of the patch, where we use [mtof_tune] instead of [mtof], with a 442 argument, and we can see that the tuning A produces a 442 Hz frequency, which is very likely to be desired if a synthesizer is combined with acoustic instruments. The right side of the patch applies an amplitude envelope with [ggee/envgen]. We first connect the velocity value to [sel 0], and connect the right outlet of [sel 0] to [t b f]. [sel 0] will output a bang out its left outlet when it receives a 0, and anything else will be output out its right outlet intact. We do this to strip the 0 velocity from the message, as we only care for the maximum velocity of each key press since the amplitude envelope we use has a fixed length and it doesn't need a 0 velocity to end. [t b f] will output the maximum velocity of the key press out its right outlet which goes to [/ 127] below [ggee/envgen], and a bang out its left outlet which goes into [ggee/envgen], triggering the envelope. We multiply the output of [line~], which is triggered by the envelope, with the normalized velocity of the key press, because if we used the envelope as is, no matter how hard or soft we hit the keys, we would always get the same amplitude envelope. This way we map the envelope to the velocity of the key press, making it responsive to our playing. [ggee/envgen] has a very short duration by default, but we can set the duration we desire with the "duration" message, as shown in Figure 5-3. Mind that if you hit a key before an envelope has finished, you'll get a click. This is solved later on with the use of an abstraction.

Holding the Sustain Part of the Envelope

We have solved the frequency and envelope issue, but the [ggee/envgen] external has introduced another issue. It's very common among synthesizers to hold a note as long as the corresponding key is pressed. This is usually done by maintaining the sustain part of an ADSR envelope while a key is pressed. [ggee/envgen] has a fixed duration, and doesn't provide this feature. What this object does is output lists for [line~], where each list is delay by the amount of the ramp time of the previous list. To be able to hold a note while a key is pressed, we must hold the last list output by [ggee/envgen] until we receive a 0 velocity from the MIDI keyboard. I have made an abstraction that does this called (not surprisingly) [hold_sustain], which you can find on GitHub https://github.com/alexdrymonitis/envelopes. This abstraction takes in lists from [ggee/env] and holds the very last list, until it receives a 0 in its left inlet, which is when it will output the last list. It takes one argument, which is the amount of breakpoints our envelope has (disregarding the start and end breakpoints), and according to that argument, it will hold the last list. Even if the release ramp takes a lot of time, and we send a new note before the release has ended, [hold_sustain] will hold the new lists, until the release ramp has ended. Figure 5-4 illustrates the use of this abstraction.

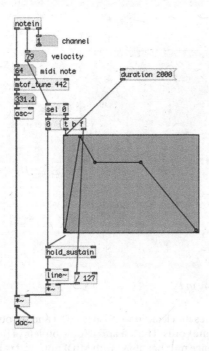

Figure 5-4. *Using the [hold_sustain] abstraction to hold the sustain part of an ADSR envelope while a key is pressed*

The patch in Figure 5-4 is almost identical to the one in Figure 5-3, only this time when we receive a 0 velocity we bang a 0 message and send it to [hold_sustain] (remember, [sel 0] outputs a bang from its left outlet when it receives a 0, it doesn't output 0). The envelope now has a longer release ramp, so you can test the delay applied to a new note when the envelope is retriggered before the release ramp has ended.

Polyphony with MIDI in Pd

The preceding examples utilized MIDI to control one oscillator. This setup is monophonic, as there's no way to have two distinct notes at the same time with only one oscillator. To have polyphony, we'll need to use more oscillators, and a way to allocate the MIDI notes to each voice. Luckily, Pd has a built-in object for voice allocation, [poly]. [poly] takes in MIDI notes and velocities and outputs a list consisting of a voice number along with the MIDI note and the velocity for that voice. When a new note comes in, while the previous is still playing, [poly] will send the new note to another voice, keeping the previous voice reserved until it is freed, which will happen when [poly] receives the MIDI note number of that voice with a 0 velocity. The number of voices is set via an argument to [poly]. Sending the MIDI note and velocity to the voice specified by [poly] is achieved using [route]. Figure 5-5 illustrates the use of [poly].

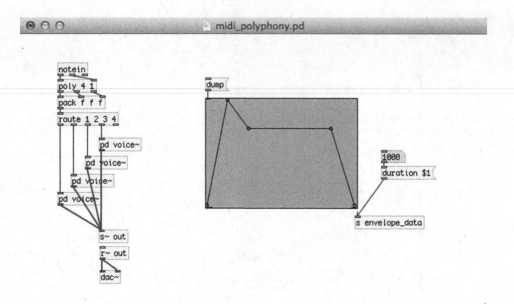

Figure 5-5. *Using [poly] for MIDI polyphony in Pd*

We have placed the voices in subpatches for cleanliness. Also [s~ out] and [r~ out] is there for patch cleanliness, as it saves us from patching some cords. The left and middle outlets of [notein] go to the two inlets of [poly], which outputs the lists of voice number along with MIDI note and velocity. These values are packed and sent to [route], which will route the MIDI note number and the velocity, according to the voice number. Figure 5-6 illustrates the contents of the [pd voice~] subpatch.

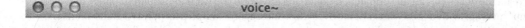

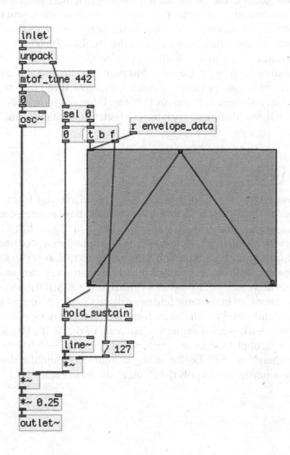

Figure 5-6. *Contents of the voice~ subpatch*

At the top of the patch, we're receiving the list with the MIDI note and the velocity, which is unpacked and sent to its destinations. There's a [r envelope_data] connecting to the inlet of [ggee/envgen], which will receive various data concerning the envelope. This is because it is cumbersome to make the envelope for each voice by hand, plus there's no way that the envelopes will all be the same, if done manually. Also, if we want to set the duration for the envelopes, it's easier if there is a global "duration" message that will affect all envelopes. Even for this patch that has only four voices, doing all this by hand can be tiring, let alone if we want to create more voices, like 10 or 20!

To create the envelopes for all subpatches, create the envelope you want in the parent patch, and click the "dump" message. When [ggee/envgen] receives these messages, it outputs its points' values out its right outlet. This list is sent to [s envelope_data] going to the inlet of each [ggee/envgen] in all subpatches, which will create exactly the same envelope. The same applies to the "duration" message, which is sent to all subpatches too. Now if you play with your keyboard, you'll see that you can have four voices simultaneously.

In Figure 5-5, we see that [poly] has two arguments. The first one is the number of voices it will create. The second one enables or disables voice stealing (which is disabled by default). *Voice stealing* is the technique where if one note too many arrives, [poly] will zero the oldest note in its list and will allocate the new note to that voice. The same goes for even more notes that arrive. For example, if you play an arpeggio of a 9th chord, you'll first send C, then E, then G, then B, and lastly D. When D arrives, [poly] will zero C and will send D with its velocity to the voice that was playing C. Try it to get a better understanding of how voice stealing works.

MIDI Control Messages in Pd

Let's now look at another type of MIDI message, the control message. This type of message is sent from MIDI devices with potentiometers and foot pedals. You MIDI keyboard is very likely to have potentiometers as well. In Pd, you can retrieve these messages with [ctlin], which stands for *control in*. If your MIDI keyboard has potentiometers on it, use that, otherwise plug in a MIDI controller with potentiometers or pedals (keep your keyboard plugged in too). If you plug in a new controller, you'll have to restart Pd, so it can see it. Go to **Media** ä **MIDI Settings...** and click **Use multiple devices** to use both MIDI controllers. On Linux, you'll need to set the **In Ports:** to 2 and then use aconnect -lio on your terminal to see all MIDI devices and applications and make the appropriate connections in the same line as we did in Chapter 1 (using ALSA-MIDI). Choose your keyboard as **Input device 1:**, and your potentiometer controller as **Input device 2:**, click **Apply** and then **OK**. Now open the polyphony patch we created earlier so that we can work on it a bit more. Place a [ctlin] object and check its help patch. This object has three outlets as well: [notein], which output the controller value, the controller number (there's one number for each potentiometer), and the channel the controller is talking at. Open one of the subpatches and apply the changes shown in Figure 5-7.

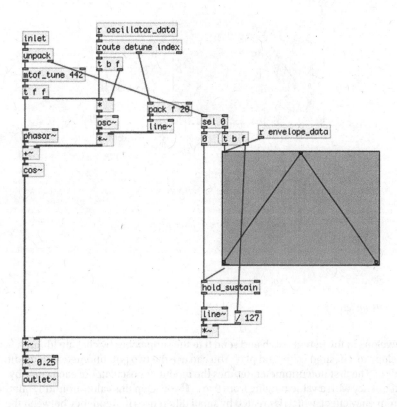

Figure 5-7. *Contents of the voice~ subpatch, receiving data from [ctlin]*

In this subpatch we're applying phase modulation, like we did in the first project in Chapter 4. [r oscillator_data] on the top part of the subpatch receives input from the parent patch, and we're using [route] to diffuse that input to where we want. We're using two potentiometers (or sliders) to control the detune of the modulator oscillator and the index of the modulation. The detune value goes to [t b f], because we're sending it to the cold inlet of [*]. Using [t b f] we're banging the hot inlet of [*] whenever we send a new value to the cold inlet, thus producing output. If we press and hold a key, we can use this value to instantly change the modulator frequency while the voice is still sounding. If we didn't use [t b f], we would have to wait until we hit a new key to hear the new frequency, and that is not very nice. Try the subpatch with and without [t b f] to hear the difference (without [t b f], the left outlet of [route] goes to the right inlet of [*]).

Figure 5-8 shows the parent patch. The only change in the parent patch is on the bottom part of it, where we have two [ctlin] objects, one listening to controller number 14, and another one listening to controller number 15. They both map their incoming range from 0 to 127, which is the MIDI range of 0 to 2, and send that to [s oscillator_data]—the first one as *detune*, and the second as *index*, so they can be routed in the subpatches using [route].

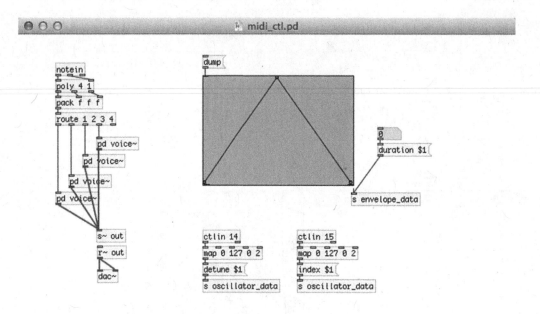

Figure 5-8. *Parent patch using [ctlin]*

Again, create an envelope in the parent patch and send it to the subpatches by clicking [dump]. Set the duration of the envelope to a desired value and play. You can use the two potentiometers to modify the sound of the oscillators. The first potentiometer detunes the modulator oscillator of each voice by multiplying the voice frequency with a value ranging from 0 to 2. If you keep this value around 1, but not 1, you'll hear the beat frequency effect, which is created by small differences in frequency between the carrier and the modulator, resulting in an alternating boosting and attenuating of the amplitude (due to phase alignment and offset between the two oscillators). The second potentiometer controls the index of modulation, which we have already seen a few times in this book. By using [t b f] in the voices subpatches for the modulator detune, we can use the detune potentiometer to change the sound of the voice while the voice is playing.

Pitch Bend Messages in Pd

Many MIDI keyboards have a pitch bend wheel, which sends pitch bend MIDI messages. This is different from control messages in the sense that it has a much wider range, from 0 to 168383, where the value of the middle position (these wheels return to their center position) is 8192. In Pd, we can retrieve these values with [bendin]. This object has two outlets, one for the value of the pitch bend wheel, and one for its channel. The channel can be set via an argument, in which case the object will have only one outlet and will listen to the set channel only.

Modify the "voice~ subpatches, as shown in Figure 5-9. The only addition is that [route] now takes a "bend" message and sends its value to [t b f] (for the same reason as the "detune" value) and then multiplies it with the frequency of the voice. Figure 5-10 shows the parent patch. The only addition to that is the [bendin] message, which maps its values from its own range (0 to 16383) to a range from 0.5 to 2. Again, create an envelope and set its duration, but this time you'll have to give some input from the pitch-bend wheel before you start playing, because the frequency of each voice is multiplied by it, and if you give no input, all frequencies will be multiplied by 0, and you'll get no sound. So, spin the wheel once and then start playing. Now the pitch of all voices is controlled by the pitch-bend wheel, where when the wheel is all the way down, the oscillators will go one octave down (their frequencies will be multiplied by 0.5, so they will be halved, which is one octave down), and when the wheel is all the way up, they will go one octave up (their frequencies will be multiplied by 2, so they will be doubled, which is one octave up).

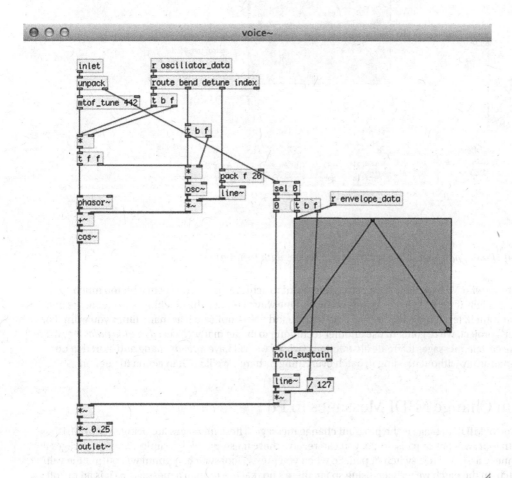

Figure 5-9. *The voice~ subpatch receiving pitch bend messages*

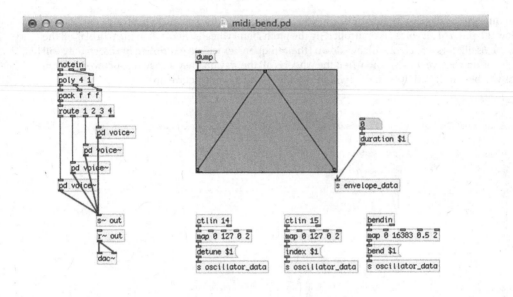

Figure 5-10. *Parent patch receiving pitch bend messages with [bendin]*

Polyphony with MIDI in Pd is rather straightforward to achieve, but might seem a bit too much patching, especially if we want to create many voices. One way to reduce the patching is to create the first voice as you want it, test it until you are satisfied with it, and then duplicate it as many times you want. For this chapter's project, we're going to use another technique to create many voices: *dynamic patching*. We're going to use certain messages to create abstractions (which we will have already made and stored in our system) dynamically without needing to patch everything by hand. We'll see it in action further on.

Program Change MIDI Messages in Pd

Another type of MIDI message is the program change message. These messages are being sent by MIDI devices with foot switches or pads. In Pd, you can receive these messages with [pgmin]. These messages are single numbers, assigned to a switch or pad. So when you press a foot switch, [pgmin] will output the value of that switch. In the patch we've been using so far, there's no real use of such a message, as it is an on/off kind of message. Later on, when we build the entire project, I'll mention where this can be helpful, replacing switches used with the Arduino.

This concludes our introduction to MIDI in Pd. We can see that with a MIDI keyboard that includes potentiometers and a pitch-bend wheel, you can create a rather complete synthesizer with Pd. If you want to make it a stand-alone instrument, then the Raspberry Pi (or any other embedded computer) comes in, which we will cover later on in this chapter. Also, not all MIDI keyboards have potentiometers and pitch-bend wheels, and that is where the Arduino comes in, especially if the keyboard has an old-style 5-pin DIN cable. Next is the Arduino code in different versions to include all cases (no potentiometers, 5-pin DIN cable) so you can build a stand-alone synthesizer with any kind of MIDI keyboard.

Arduino Code

Now that we've been introduced to MIDI in Pd, let's see the possible version of the Arduino code that will complement this project. We have three different cases, one where we don't need Arduino at all, because the MIDI USB keyboard we're using has potentiometers and a pitch-bend wheel, plus some pads. The second case is a MIDI USB keyboard that's only that and doesn't have any potentiometers or anything else apart from the keys. The third case is a MIDI keyboard with a 5-pin DIN cable with no potentiometers; we're going to use the Arduino for attaching potentiometers and switches, and also for receiving MIDI messages from the keyboard and transferring them to Pd. We're not going to deal with the first case here, because it doesn't include an Arduino. I'll talk about it when we build the Pd patch. So we'll start with the second case, where we'll be using the Arduino only with potentiometers and switches. Listing 5-1 shows the sketch. If you are in the third case, go through this section, as including MIDI messages in Arduino will only be complementary to the code in Listing 5-1.

Listing 5-1. Code Without Receiving MIDI Messages in the Arduino

```
1.   // potentiometer variables and constants
2.   const int num_synth_pots = 4;
3.   int synth_pots[num_synth_pots];
4.   const int num_filter_pots = 3;
5.   int filter_pots[num_filter_pots];
6.   int amp_pot;
7.   // potentiometer pin offsets
8.   int filter_pot_offset = num_synth_pots;
9.   int amp_pot_offset = num_filter_pots + num_synth_pots;
10.
11.  // digital pins arrays, variables, and pin numbers
12.  const int oscillator_types = 4;
13.  int osc_type_buttons[oscillator_types];
14.  int old_osc_type_buttons[oscillator_types];
15.  int filter_switch, old_filter_switch;
16.  int tune_switch, old_tune_switch;
17.  int sustain_switch, old_sustain_switch;
18.  int dsp_switch, old_dsp_switch;
19.  int carrier_mod_switch_pin = 6;
20.  int filter_switch_pin = 7;
21.  int tune_switch_pin = 8;
22.  int sustain_switch_pin = 9;
23.  int dsp_switch_pin = 10;
24.  // use pin 13 so you don't need to use an external resistor
25.  int dsp_led_pin = 13;
26.
27.  void setup() {
28.    for(int i = 2; i < 11; i++) pinMode(i, INPUT_PULLUP);
29.    pinMode(dsp_led_pin, OUTPUT);
30.
31.    // Arduino's baud rate
32.    Serial.begin(38400);
33.  }
34.
```

```
35.   void loop () {
36.     if(Serial.available()){
37.       byte in_byte = Serial.read();
38.       digitalWrite(dsp_led_pin, in_byte);
39.     }
40.
41.     // read the potentiometers
42.     for(int i = 0; i < num_synth_pots; i++)
43.       synth_pots[i] = analogRead(i);
44.     for(int i = 0; i < num_filter_pots; i++)
45.       filter_pots[i] = analogRead(i + filter_pot_offset);
46.     amp_pot = analogRead(amp_pot_offset);
47.
48.     // print their values
49.     Serial.print("synth_pots ");
50.     for(int i = 0; i < (num_synth_pots - 1); i++){
51.       Serial.print(synth_pots[i]);
52.       Serial.print(" ");
53.     }
54.     Serial.println(synth_pots[num_synth_pots - 1]);
55.
56.     Serial.print("filter_pots ");
57.     for(int i = 0; i < (num_filter_pots - 1); i++){
58.       Serial.print(filter_pots[i]);
59.       Serial.print(" ");
60.     }
61.     Serial.println(filter_pots[num_filter_pots - 1]);
62.
63.     Serial.print("amp ");
64.     Serial.println(amp_pot);
65.
66.     // read the switches
67.     int carrier_mod_switch = digitalRead(carrier_mod_switch_pin);
68.     for(int i = 0; i < oscillator_types; i++){
69.       osc_type_buttons[i] = digitalRead(i + 2);
70.       if(osc_type_buttons[i] != old_osc_type_buttons[i]){
71.         if(!osc_type_buttons[i]){ // if button is pressed
72.           if(carrier_mod_switch){
73.             Serial.print("carrier_type ");
74.             Serial.println(i);
75.           }
76.           else{
77.             Serial.print("modulator_type ");
78.             Serial.println(i);
79.           }
80.         }
81.         old_osc_type_buttons[i] = osc_type_buttons[i]; // update old value
82.       }
83.     }
84.
```

```
85.    filter_switch = digitalRead(filter_switch_pin);
86.    if(filter_switch != old_filter_switch){
87.      Serial.print("filter_switch ");
88.      Serial.println(!filter_switch);
89.      old_filter_switch = filter_switch;
90.    }
91.
92.    tune_switch = digitalRead(tune_switch_pin);
93.    if(tune_switch != old_tune_switch){
94.      Serial.print("tune_switch ");
95.      Serial.println(!tune_switch);
96.      old_tune_switch = tune_switch;
97.    }
98.
99.    sustain_switch = digitalRead(sustain_switch_pin);
100.   if(sustain_switch != old_sustain_switch){
101.     Serial.print("sustain_switch ");
102.     Serial.println(!sustain_switch);
103.     old_sustain_switch = sustain_switch;
104.   }
105.
106.   dsp_switch = digitalRead(dsp_switch_pin);
107.   if(dsp_switch != old_dsp_switch){
108.     Serial.print("dsp_switch ");
109.     Serial.println(!dsp_switch);
110.     old_dsp_switch = dsp_switch;
111.   }
112. }
```

Before I explain the code, we should take a look at the circuit, which is illustrated in Figure 5-11. For now just look at the parts of the circuit and which component connects to which pin on the Arduino. It will help to understand the approach to writing the code in Listing 5-1.

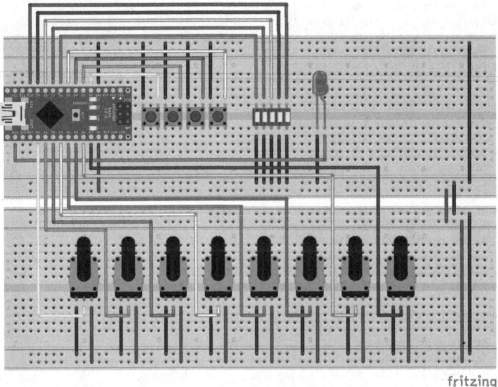

Figure 5-11. *Test circuit for synthesizer project*

First of all, we can see that now we're using an Arduino Nano, which we have placed on the breadboard. We have four push buttons patched to the first four digital pins of the Arduino (starting from digital pin 2, as always). Then we have five switches patched to the next five digital pins, and then we have an LED connected to the digital pin 13. We use this pin for the LED so we can avoid an external resistor, which is rather useful when building circuits on perforated boards. The circuit shown in Figure 5-11 is only a test circuit. It's always a good idea to test your circuit on a breadboard before you attempt to solder it on a perforated board.

On our synthesizer, we'll have four potentiometers for the two oscillators of each voice, three potentiometers for a filter we'll apply to the synth, and one potentiometer for the amplitude. The four push buttons will select the waveform for both the carrier and the modulator. The first switch will set whether the push buttons will set the waveform for the carrier, or the modulator, the second switch will control whether we'll use the filter or not, the third switch will control whether we want to tune our synthesizer, using the [mtof_tune] abstraction, and the fifth switch will control the DSP. The LED will indicate whether the DSP is on or off.

Explaining the Code

Now that we know what each component of the circuit does, let's go through the code. We're using the print function to send data to Pd instead of write. This is mostly because in the next version of the code, where we'll be receiving data from the MIDI keyboard, we prefer to use this function, because it makes it easy to separate data before they arrive to Pd. The first 25 lines define variables, constants and arrays to read and store data. The comments should make the code rather clear. Notice the offset variable in lines 8 and 9, which will be used along with a for loop, so we can read the correct pins, as we've already done. In line 13, we define an array called osc_type_buttons, which will store the values of the push buttons. We group them all in one array, since they are used for the same purpose: to choose a waveform for an oscillator. In line 14, we define another array, called old_osc_type_buttons. This array will hold the previous value of each button, which will be used for comparison when we read these pins, so we can print their values to Pd only when they are changed (actually, when we press a button). You'll see how this works further on in the code. Lines 15 to 18 define pairs of variables for the pins of the switches in the same manner as with the push buttons arrays. We use two variables—one to hold the current value of each pin and one to hold the previous value—so we can compare them and see if the values have changed. In these lines, you can see that we define two variables in each line, separated by a comma, like this:

```
int filter_switch, old_filter_switch;
```

This syntax is perfectly legal in C++, and we use it to make our code a bit more compact and easier to read, since the two variables cooperate. Lines 19 to 25 set pin numbers to variables, which we'll use to read the switches and write to the LED pin. We could have used the numbers straight in the functions that read and write to pins, but using self-explanatory names makes the code more readable.

The setup and loop Functions

Our setup function is nothing special, we just set the mode of the digital pins we're using and start the serial communication. This time we're using a 38400 baud rate, because while testing (including receiving data from the MIDI keyboard, although this should take long) it proved that a higher baud rate stressed the Arduino. In this baud rate, everything works fine. In the loop function, the first thing we do is check for data coming in the serial line, and use that to control the LED of the circuit. This time we're reading data in a much simpler way than we did before, because the only thing we want to control is one LED, so we don't need the generic technique we used till now. All we do is create a byte variable to store the incoming data, and write that byte to the pin of the LED. Then we read through the analog pins, which we separate in groups, and afterward we write these values to the serial line. We don't bother to test whether a value has changed or not, because the analog pins have a little bit of noise and they kind of constantly change, so we print them to Pd anyway.

Enter the New Technique

Now the new part comes in. In line 67, we read the first switch, which sets whether the push buttons will set the waveform for the carrier or the modulator. In line 68, we run a for loop and read through the push buttons. Line 70 checks with an if control structure whether the newly read value is different from the old one, and if it is, its code will be executed. This test will be true whenever we press or release a button. If the value has indeed changed, line 71 will check whether we are pressing the button. We're using the exclamation mark here, since when we press the button, we'll read 0 because of the pull-up resistors. Using if this way, its test will be true when the value is indeed zero, since we're reversing the value by placing the exclamation mark, so when we press the switch, this test will be true. If the test is true, line 72 will test whether the first switch is in the on or off position. This test will be true when the switch is in the off position, because here we're not using an exclamation mark. If the test is true, we're printing the i variable to the serial line with the carrier_type tag, and if it's false, with the modulator_type tag. The i variable will actually hold the number of the switch, since it's the variable used to control the for loop and goes from 0 to 3. So, if we press the third switch for example, the tests in lines 70 and 71 will be true, and the value we'll print to Pd will be 2 (we start counting from 0). This way, not only we print a value to Pd only when it's changed, and more precisely, when a button is pressed, but we also don't need to print all four switches, but use them all as one entity. Finally, in line 81 we update the element of the old_osc_type_buttons array. If we fail to do this, the test in line 70 will hold true only when we press a switch. This might sound ideal, as this is what we want. What will actually happen is that, while we keep a button pressed, this test will keep on being true, and its code will keep on being executed, so we'll be printing data to Pd constantly. By updating the variable that holds the old value, we make sure that the test in line 70 will hold true only when the value has indeed changed.

Lines 85 to 90 read the second switch, which is used to define whether we'll use the filter or not. Again, we're testing if the value has changed, in line 86, and if it has, we're printing it with the appropriate tag. We're using the exclamation mark to invert the value because of the pull-up resistor. We repeat this process for the rest of the switches, each time reading the appropriate pin and using the appropriate tag. And this brings us to the end of the code. It is a rather long sketch, but we're using the same principle for all digital pins. This technique is very helpful because it facilitates things in Pd to a great extent. Next is the Pd patch, where we'll see the code and circuit in action, and things will become even clearer.

Pd Patch

Figure 5-12 shows the Pd patch. Most of the stuff is again hidden in subpatches for clarity's sake. The lower part of the patch should be rather clear. We're using [notein] with [poly] to receive and diffuse MIDI notes from the keyboard. [poly_synth~] is an abstraction which creates voices according to its argument (in this case 10). We use it this way both for keeping the patch clean, but also because with an abstraction we can automate the process of voice creation so we don't need to patch each voice separately. We'll see how this is achieved further on. [pd filter~] contains the filter we're using in our synth, and after that we have the [dac~] controlled by a potentiometer from Arduino. On the top part of the patch we see [declare -path ./polysynth]. This objects sets a search path exclusively for this patch. If we're using abstractions that are project specific, and we don't want to place them in Pd's standard search paths, we can put them anywhere in our system and use [declare] to set their path for this patch. The -path flag means that what follows is a path. With [declare] we can also import libraries, where we use the -lib flag. The path for our abstractions is located in a directory called polysynth, which is in the directory of the patch, hence ./polysynth. Again, we're using Unix syntax to specify the path.

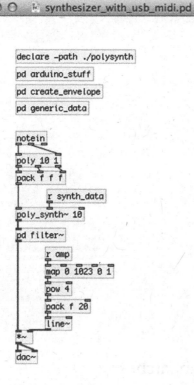

Figure 5-12. *Synthesizer Pd patch*

The arduino_stuff Subpatch

Now let's start looking at the subpatches one by one, starting with the arduino_stuff subpatch, which is shown in Figure 5-13. The only different thing here is that we're sending the DSP value to [comport] as is, without the "print" message, because we're not assembling the value in the Arduino code, but use the incoming byte as is, so we don't need its ASCII form here.

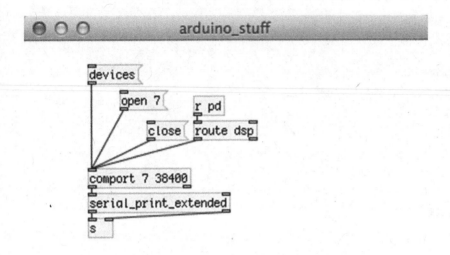

Figure 5-13. *Contents of the arduino_stuff subpatch*

The create_envelope and generic_data Subpatches

Figure 5-14 illustrates the contents of the create_envelope subpatch. Nothing new here as well. Only create the envelope you want and click [dump]. The next time you'll open the patch, it will be banged on load and will be sent to all voice abstractions.

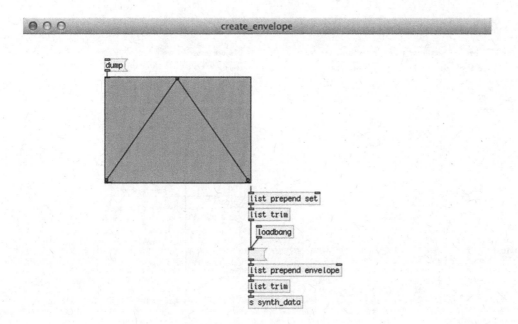

Figure 5-14. *Contents of the "create_envelope" subpatch*

Figure 5-15 shows the contents of the generic_data subpatch. Here we receive most of the data from the Arduino. At the top-left part you can see [r carrier_type] and [r modulator_type], which receive a number from 0 to 3 from the Arduino, depending on which of the osc_type_buttons is pressed, and according to the position of the carrier_mod_switch. All this happens in lines 67 to 83 in the Arduino sketch. We can see here that we have done some processing in the Arduino code, and sent the resulting data to Pd instead of only reading the pins, and printed them as is. [r sustain_switch] receives a value straight from the switch that controls whether we'll hold the sustain of our amplitude envelope or not. On the bottom-left part we receive the value from the DSP switch and control the DSP state accordingly. On the right side of the patch we receive the values from the oscillator potentiometers, which we map and then send to their destinations. [r tune_switch] is there to either let the frequency tune potentiometer through, or not. Since the analog pins have some noise, it's good practice to not let that value through all the time. We can use the switch that controls this to tune our synthesizer, and when we're done tuning, we can put the switch in the OFF position, and our tuning won't be affected any more. Each of these values is being sent to our [poly_synth~] abstraction via [s synth_data].

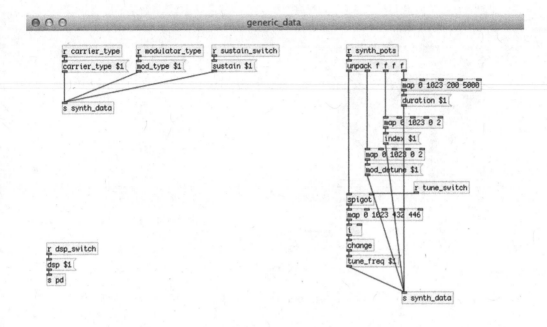

Figure 5-15. *Contents of the "generic_data" subpatch*

The [poly_synth~] Abstraction

Now let's look at the [poly_synth~] abstraction, which is shown in Figure 5-16. There are quite some new things here, which I'll explain in detail. This is a rather simple abstraction, but it facilitates polyphony a great deal. It has three control inlets and one signal outlet. The control inlets send their data to various destinations using [send]. Notice the $0 used in each [send], and in the name of the two subpatches. This is a special sign in Pd. When we use it in abstractions it generates a unique number, usually starting from something over 1000, which is used to create unique names for objects like [send]/[receive], or for arrays and [table]s. If we want to create more than one instance of this abstraction, and we didn't use $0, whatever would arrive in an inlet of any abstraction, would be sent to the corresponding [receive]s of all instances of the abstraction. Using $0 we make sure that whatever arrives in an inlet of an instance of the abstraction, will be diffused only locally to that specific inlet. For example, if we created two [poly_synth~] abstractions, $0 of the first would be something like 1009, and of the second something like 1013. When data arrived in the first inlet of the first abstraction, it would be sent to [s 1009-voice-list], and would be received by all [r 1009-voice-list], which would exist in the first abstraction only, so this data wouldn't be received in the second abstraction as well. If we sent data to the first inlet of the second abstraction, it would be sent to [s 1013-voice-list] and would be received by [r 1013-voice-list], without interfering with [r 1009-voice-list] at all.

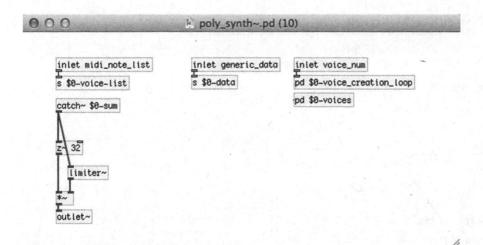

Figure 5-16. *[poly_synth~] abstraction*

The [pd $0-voice_creation_loop] subpatch

We cannot see this unique this unique number printed, but wherever in the abstraction you see $0, it is actually replaced by this unique number. [pd $0-voice_creation_loop] demonstrates its use clearer. Its contents are illustrated in Figure 5-17.

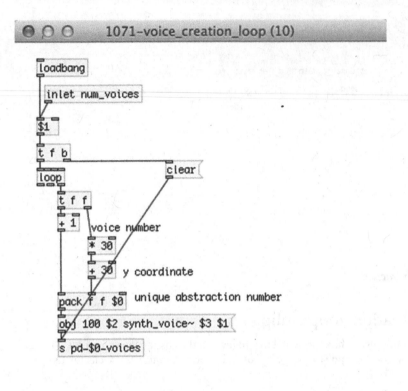

Figure 5-17. Contents of the "$0-voice_creation_loop" subpatch

This subpatch is banged on load with [loadbang], which bangs [$1]. [$1] takes the value of the first argument of the abstraction. In Figure 5-12, which illustrates the main patch for this project, we can see that we have initiated [poly_synth~ 10] with the argument 10. This is the value [$1] will take, which is the number of voices we want to create. When this value is banged, it first bangs the message "clear", which is sent to [s pd-$0-voices]. This [send] sends its data to the "$0-voices" subpatch you see in Figure 5-16. By using the name of a subpatch with "pd-" prepended to it, we can send data directly to the subpatch. The message "clear" will erase everything in the subpatch. Then a loop will be initiated with 10 iterations (due to the argument provided to [poly_synth~]) using the [loop] abstraction. At each iteration of the loop, the incremented value coming out the rightmost outlet of [loop] will first be multiplied by 30 and added to 30. It will be stored to the middle inlet of [pack f f $0], and then added to 1 and stored to the leftmost inlet of [pack f f $0]. This will cause [pack] to output its values, because the leftmost inlet is hot. The last value of [pack], which is $0, will be the unique value generated by $0, as already mentioned.

The list output by [pack] goes to the message "obj 100 $2 synth_voice~ $3 $1". This message is sent to [s pd-$0-voices], hence to the "$0-voices" subpatch. Sending a message of this type to a patch (or subpatch) will create an object, much like putting an object from the Put menu, or by hitting Ctrl/Cmd+1. This specific message will create the object [synth_voice~] with two arguments, the third and first values output by [pack] ($3 and $1 in the message), positioned at 100 pixels to the right and $2 (the second value of [pack]) pixels down. As the loop iterates, $2 will take the following values: 30, 60, 90, and so forth (remember, [loop] starts counting from 0). So the [synth_voice~] objects that will be created, will be placed at the following coordinates in the "$0-voices" subpatch: 100 30, 100 60, 100 90, and so forth. This means that the objects will be placed one below the other, all aligned at vertically. The two arguments each [synth_voice~] will take, are the unique number of [poy_synth~], and the number of each voice, starting from 1. So the objects that will

202

be created inside [pd $0-voices] will be [synth_voice~ 1009 1], [synth_voice~ 1009 2], [synth_voice~ 1009 3], and so forth, given that the unique number generated by $0 is 1009.

How Dollar Signs Work in Pd

The dollar-sign numbers in Pd are very useful and it's important that you understand them. In objects, like [$1], dollar-sign numbers starting from 1 take the value of the corresponding argument in an abstraction. So [$1] will take the value of the first argument, [$2] will take the value of the second argument, and so forth $0, as already mentioned, generates a unique value for abstractions (for subpatches too, but it behaves a bit differently there), which is used to avoid name clashes with [send]/[receive] pairs and [table]s and arrays. In messages, dollar-sign numbers take the corresponding value from an incoming list. So, when the list "6 13 27 9" comes in the message "$1 $2 $3 $4", $1 will take the value 6, $2 will be 13, $3 will be 27, and $4 will be 9. Messages cannot take values directly from arguments, but only from lists that arrive in their inlet. Also $0 won't work with messages. That's why in Figure 5-17, we store $0 in [pack] and send it to the message, which is retrieved by the message's $3. Dollar-sign numbers might be a bit confusing in the beginning, but once you get the grasp of it, you'll see how powerful this feature is.

The [synth_voice~] Abstraction

In [poly_synth~], create a subpatch called "$0-voices", like in Figure 5-16, and leave it empty. That's where all the [synth_voice~] objects will be created. Of course [synth_voice~] is no native Pd object, but another abstraction we'll create ourselves. You can see the abstraction in Figure 5-18.

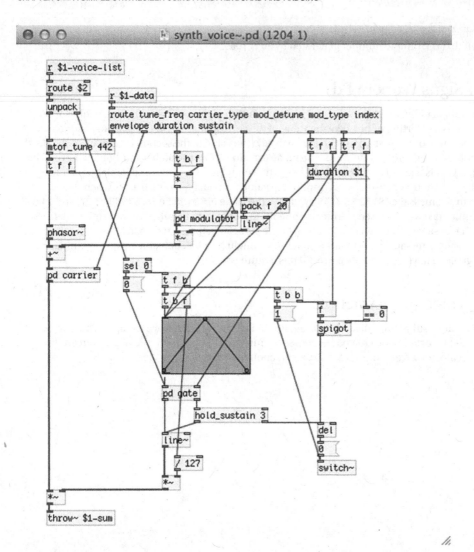

Figure 5-18. *The [synth_voice~] abstraction*

At the top part of the patch we have a [r $1-voice-list], which will receive values from [s $0-voice-list] of [poly_synth~]. Since we'll be creating the [synth_voice~] as described earlier, the first argument of this abstraction will be the unique number generated by $0 of the [poly_synth~] abstraction. We cannot use $0 straight in [synth_voice~] because it will generate a new unique number which won't be the same with that of [poly_synth~], plus every instance of [synth_voice~] will get its own unique number, and now we don't want that, since we want to send the same data to all voices. All $1 values in Figure 5-18, will be the same as $0 in Figures 5-16 and 5-17.

On the right part of the abstraction we receive data with [r $1-data], which is diffused with [route]. The leftmost outlet of [route] will output the tuning frequency for our synth, the second outlet will output the oscillator type for the carrier oscillator, and so on. The arguments of [route] should make clear what each value does (don't hit Return after the "index" argument, just type all the arguments with spaces between them. When they become too many to stay in one line, Pd will start printing them one line below by itself).

204

[r $1-voice_list] sends its list to [route $2], which will give its output according to the second argument of [synth_voice~], which is the voice number. When [poly] in the parent patch (Figure 5-12) outputs the list "1 69 127", [route $2] of the first [synth_voice~] will output the list "69 127". [route $2] of the rest of the [synth_voice~] abstractions won't output anything, because the first element of the list "1 69 127" won't match their argument.

The list coming from [route $2] in unpacked and the note number is sent to [mtof_tune 442], which will convert the note number to its frequency, tuned at 442 Hz. The velocity (the second element of the list output by [route $2]) is sent to [sel 0], like with the MIDI examples earlier in this chapter. The main structure of this abstraction is very similar to the subpatches of the MIDI examples, where we apply phase modulation and we use an amplitude envelope. This time we can choose between different waveforms for both the carrier and the modulator. The "modulator" subpatch in Figure 5-18 is shown in Figure 5-19. And the contents of each subpatch in [pd modulator] is shown in Figures 5-20 to 5-23.

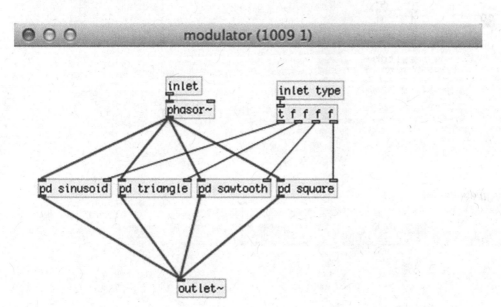

Figure 5-19. *Contents of the "modulator" subpatch of the [synth_voice] abstraction*

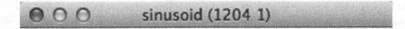

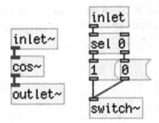

Figure 5-20. [pd sinusoid]

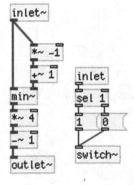

Figure 5-21. [pd triangle]

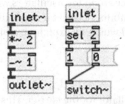

Figure 5-22. *[pd sawtooth]*

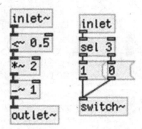

Figure 5-23. *[pd square]*

The modulator Subpatch and Its Subpatches

When [synth_voice~] receives the message "mod_type $1", it sends this value to [pd modulator], and from there it goes to the subpatches in Figures 5-20 to 5-23. In these figures each [sel] will control the [switch~] of each subpatch. [switch~] enables turning the DSP of a subpatch or abstraction on or off, independently of the DSP state of Pd. This helps us both to choose the oscillator type we want, but also to save some CPU since we have four oscillators for the modulator and four for the carrier of each voice, eight in total, for ten voices, which makes them 80! If we want to use more voices then the oscillators we'll be using will start becoming a bit too many. Using [switch~] reduces the amount of CPU Pd needs drastically. We've also used [switch~] in the parent patch of [synth_voice~], the use of which I'll explain in a bit.

The carrier Subpatch

[synth_voice~] also has a [pd carrier], which sets the oscillator for the carrier. This subpatch is almost identical to [pd modulator], only [phasor~] is outside the subpatch so we can modulate its phase. Figure 5-24 shows the subpatch. This subpatch has a signal inlet for the oscillators, which first connects to [wrap~]. This object wraps the value in its input between 0 and 1. So if it receives 1.1, it will wrap it back to 0.1. If it receives –0.1, it will wrap it back to 0.9. Check its help patch for more information. We've put it there because we add the modulator oscillator to the phase of the carrier. The phase must always be between 0 and 1, whether it is modulated or not. When we built the MIDI examples earlier in this chapter we didn't use [wrap~] because we used a cosine oscillator with [cos~], and [cos~] wraps the values it receives internally, so we didn't need to do this explicitly. This time, though, we use more oscillators which consist of objects that don't wrap their input internally, so we must use [wrap~] to avoid strange results. The subpatches of [pd carrier] are identical to those of [pd modulator], as shown in Figures 5-20 to 5-23.

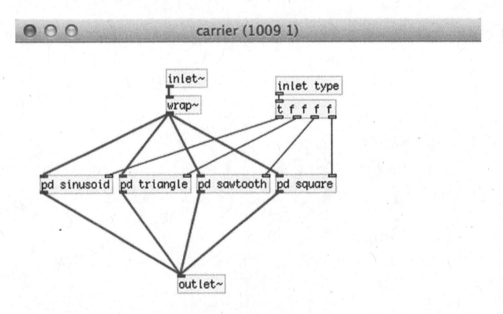

Figure 5-24. *Contents of [pd carrier]*

The Amplitude Envelope

Now I'll explain the bottom-left part of [synth_voice~] where we have the amplitude envelope plus some other things. In the middle part of Figure 5-18, we see a [sel 0] that receives the velocity of the MIDI note, and connects to [t f b], which first outputs a bang out its right outlet, which goes to [t b b]. This last [trigger] first bangs [f], which takes its value from "duration" from the [route] above it. This value is the duration of the envelope, which also goes to the message "duration $1" and is sent to [ggee/envgen] to set this duration to the envelope. The output of [f] goes through [spigot], and if [spigot] is open, it goes to the left inlet of [del]. "del" stands for *delay*, and it's a delay object for the control domain. It is somewhat different from the signal domain delay. It receives a value and it outputs a bang delayed by the value it received, in milliseconds. Check its help patch for information. When [del] receives the duration amount in its inlet, it will send a bang out its outlet after the set delay, which will send 0 to [switch~]. The left outlet of [t b b] sends 1 to [switch~], which means that the abstraction will turn its DSP on. After all this is done, [t f b] will send the velocity out its left outlet to a [t b f], which will first output the velocity from its right outlet and it will go to [/ 127] at the bottom part of the patch. Then [t b f] will output a bang from its left outlet, which will trigger the envelope.

The gate Subpatch, How It Works, and How We Manage the Audio of [synth_voice~]

Below [ggee/envgen] there's [pd gate], which we can see in Figure 5-25. This subpatch takes the output of [ggee/envgen] in its left inlet, and the "sustain" value from [route] in its right inlet. This value is either a 1 or a 0, controlled by the sustain switch in the Arduino circuit. It sets whether we'll hold the sustain part of the envelope as long as we keep the key pressed, or not. If the "sustain" switch value is 0, then the output of [ggee/envgen] will connect straight to [line~]. If it's 1, it will connect to [hold_sustain] and then to [line~]. The "sustain" switch value also controls the [spigot] after [f] in the parent patch of the abstraction, which holds the duration of the envelope. If the "sustain" switch is off, then the [spigot] will be open (because of [== 0]), and if it's on, [spigot] will be closed. If we keep the "sustain" switch off, then [switch~] will receive 1, the envelope of [ggee/envgen] will be triggered, when after the duration of the envelope has ended, [switch~] will receive 0 banged by [del], since [spigot] will be open. If the "sustain" switch if on, then [switch~] will receive 1, the envelope will be triggered, but [spigot] will be closed, so [del] won't receive the duration value from [f], but [hold_sustain] will output the appropriate delay time out its right outlet when it receives 0, and [switch~] will be closed when it should. Check the help patch of [hold_sustain] to see how this works.

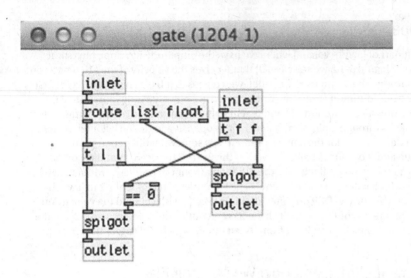

Figure 5-25. *Contents of the "gate" subpatch*

Each [synth_voice~] sends its output to [throw~ $1-sum], which all go to [catch~ $0-sum] in [poly_synth~], in Figure 5-16. The output of [catch~] is controlled by [limiter~]an external of the zexy library— to make sure we'll never exceed an amplitude of 1. We could have multiplied the output of each [synth_voice~] by the reciprocal of the number of voices, but using [limiter~] gives a better result, as if we play only one voice, it will have full amplitude, instead of one-tenth (in the case of 10 voices) of it. Check the help patch of [limiter~] to see how it works.

The filter~ Subpatch

Now that we've discussed and built [poly_synth~] and [synth_voice~], let's go back to the parent patch of our project, the one shown in Figure 5-12. There's one subpatch left to show and discuss: filter~. It is shown in Figure 5-26. Compared to all we've already done, this should be fairly easy. It receives the output of [poly_synth~] and sends it both to its [outlet~], intact, but also to [omniFilter_abs~ lowshelf]. [omniFilter_abs~] is an abstraction for many types of filters. The filter type is sent via an argument (or from the rightmost inlet; check its help patch). [r filter_pots] receives the filter potentiometer values from Arduino, and [r filter_switch] receives the value of the filter switch of the Arduino circuit. If the switch is in the OFF position, the input of [pd filter~] will be output intact. If it the switch is in the ON position, the input will go through [omniFilter_abs~]. The output of [pd filter~] is being sent to [dac~] having its amplitude controlled by the last potentiometer of the Arduino circuit.

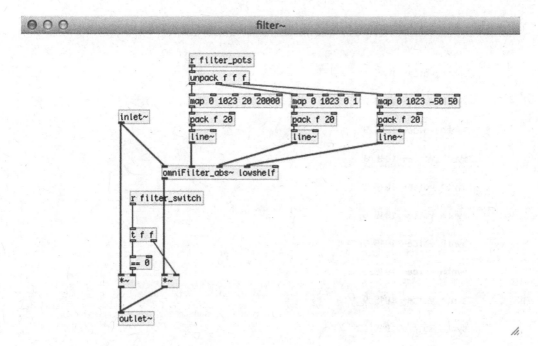

Figure 5-26. *[pd filter~]*

Done with Patching, Time to Test

We have now completed the Pd patch for this project. If your MIDI keyboard doesn't have a USB cable, the modifications to the patch will be minor, and we'll go through them in a bit. Now close the patch and re-open it, click [poly_synth~] and check [pd $0-voices]. This subpatch should look like Figure 5-27.

211

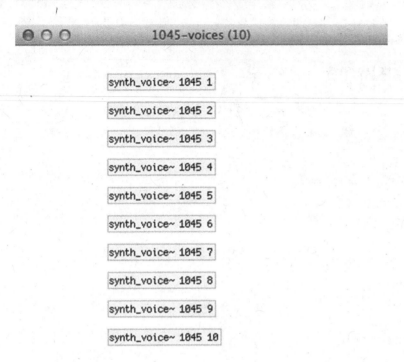

Figure 5-27. *Contents of [pd $0-voices]*

In the title bar of the window, you should see something like **1045-voices**. 1045 is the unique number created by $0 (yours is very likely to be different), so this title is the result of "$0-voices". The number in the parenthesis is the argument passed to [poly_synth~], which appears on the title bar of all subpatches of the abstraction. Using $0 in [pack f f $0], in [pd $0-voice_creation_loop] results in creating all [synth_voice~] abstractions with this unique number as the first argument. It doesn't really matter what your unique number is. What is really important is that the number on the title bar of [pd $0-voices] should be the same as the first argument of all [synth_voice~] abstractions; otherwise, things are not going to work properly. The second argument of [synth_voice~] is the voice number, and it should be an incrementing number starting from 1.

Building this project with all the subpatches, and especially with the abstractions, might be a bit complicated, but if you build the abstraction carefully, and go through the explanations (maybe more than once) you should be able to understand how everything works. Pd, being a visual programming language, projects the data flow very clearly, so it's easy for you to follow the data from beginning to end, or the other way round, to understand how a patch works. You should first have a good understanding of the Arduino circuit and code, and what each component is supposed to do, and then apply this to the Pd patch and abstractions, so you can fully understand their behavior.

Arduino Receiving MIDI Messages from the MIDI Keyboard

In case you have an old MIDI keyboard that has a 5-pin DIN cable, you'll need to do some tweaking to the code and circuit to be able to use it with the Raspberry Pi. If you intend to build this project with a laptop (or a desktop), and you have a MIDI-to-USB converter, or you have an external sound card that takes MIDI cables, then you don't need to go through this section, except if you're just curious.

Figure 5-28 shows a female MIDI connector. If you connect one end of a MIDI cable to your keyboard and the other end to such a connector, then you'll have easy access to it pins. You might need a multimeter to make sure you connect to the right pins. This device is very useful when building electronics, so it's advisable to get one. Learning how to use it is rather easy, and there are a lot of resources online. To build the circuit you'll also need an optocoupler. This is a very small, inexpensive chip that will help the circuit work. I have used the 4N35 optocoupler, but the 4N28, or probable any from these two families would do. Instead of building the circuit yourself, you could get an Arduino MIDI shield that exists on the market, but building such a circuit will cost very little money (probably less than one dollar), and can also be fun to do.

Figure 5-28. *Female MIDI connector*

Arduino Circuit Additions

Figure 5-29 shows the additional circuit we'll need to receive MIDI messages from our keyboard. The previous circuit still holds, the one in Figure 5-29 is supposed to be added on top of the previous circuit. There could be other ways to achieve this, but I found that this one was rather simple and used components I had lying around, so it took me very little time to build it. You are encouraged to search yourself to find other solutions for this, but the circuit in Figure 5-29 should do the trick.

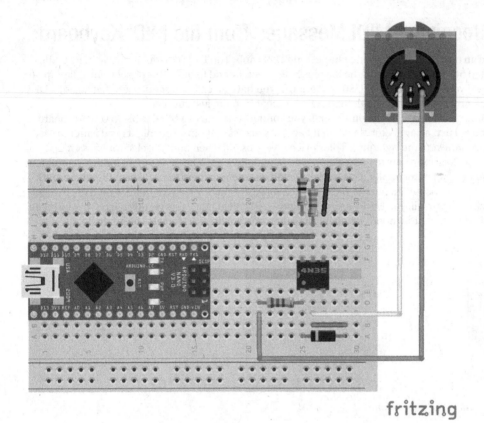

Figure 5-29. *Connecting a MIDI cable to Arduino*

In this circuit, we're connecting MIDI pin 4 (the second from right) to the first pin of the optocoupler, via a 220Ω resistor. The first pin of the optocoupler is marked with a dot (some chips might only have a notch on one side, where having that side on top, the top pin on the left side is the first pin of the chip), and the rest are numbered sequentially. Figure 5-30 shows the pin numbers of 4N35. All chips follow the same numbering order. The second pin of the optocoupler connects to MIDI pin 5 (the second from right, MIDI pins are not numbered sequentially), and also to the first pin of itself via a 1N4148 diode, with the ring of the diode on the side of the first pin of the chip. Pin 4 of the optocoupler connects to ground, pin 5 connects to 5V via a 3K3 resistor (3K3 notation means 3.3KΩ) and to Arduino's digital pin 11. Finally, pin 6 of the optocoupler connects to ground via a 100K resistor.

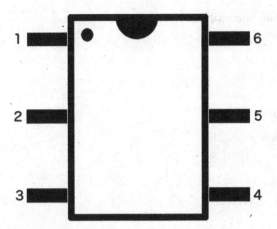

Figure 5-30. *Pin numbering of 4N35 optocoupler*

Arduino Code Additions

The code in Listing 5-1 stays, and we're just going to add a few things to it, so it can receive MIDI messages and transfer them to Pd. First of all, at the beginning of the code type the contents in Listing 5-2.

Listing 5-2. Create a Software Serial Line and Variables for the MIDI Messages

```
1.   #include <SoftwareSerial.h>
2.
3.   // create a SoftwareSerial object with RX and TX pins 11 and 12
4.   SoftwareSerial MIDI_serial (11, 12);
5.
6.   byte MIDI_note;
7.   byte MIDI_velocity;
```

As with the Bluetooth module from Chapter 3, where we needed an extra serial line to configure it, we're going to need another serial line now to receive data from the MIDI keyboard. To do this, we need to import the SoftwareSerial library and create an object with two pins, an RX (for receiving data) and a TX (for transferring data). We're going to use only the RX pin, since we won't be sending any data to the MIDI keyboard, but when we initialize an object of the SoftwareSerial class we need to include both pins. Then we'll need to define two variables, one for the MIDI note number and one for the velocity.

Then in the setup function we must start the serial communication of the new, SoftwareSerial port. Listing 5-3 shows what you should type in the setup function.

Listing 5-3. Begin the Serial Communication in the Software Serial Line

```
1.   // MIDI baud rate
2.   MIDI_serial.begin(31250);
```

The baud rate used here is the baud rate of the MIDI protocol. We need to use that to be able to communicate with the MIDI keyboard. The baud rate we're using for the communication between the Arduino and Pd is (Serial.begin(38400);), it doesn't need to be the same (actually [comport] doesn't support this baud rate), so you can still use 38400. Lastly, Listing 5-4 shows what you should write in your loop function.

Listing 5-4. Check for Incoming MIDI Messages in the Software Serial Line

```
1.   if(MIDI_serial.available()) {
2.     byte cmd_byte = MIDI_serial.read();
3.     if((cmd_byte >= 144) && (cmd_byte <= 160)){
4.       MIDI_note = MIDI_serial.read();
5.       MIDI_velocity = MIDI_serial.read();
6.       // print the MIDI note to the serial port
7.       Serial.print("MIDI_note ");
8.       Serial.print(MIDI_note);
9.       Serial.print(" ");
10.      Serial.println(MIDI_velocity);
11.    }
12.    else if((cmd_byte >= 128) && (cmd_byte <= 143)){
13.      MIDI_note = MIDI_serial.read();
14.      MIDI_velocity = MIDI_serial.read();
15.      // print the MIDI note to the serial port
16.      Serial.print("MIDI_note ");
17.      Serial.print(MIDI_note);
18.      Serial.print(" ");
19.      Serial.println(MIDI_velocity);
20.    }
21.  }
```

This can go anywhere in your code, I put it right after reading data from the serial line, in the beginning of the loop function. What happens here is that we're checking if there are data in the serial port of the MIDI, and if there are, we read them. In the MIDI protocol, the first byte that is sent is called a *status byte*, which tells the receiving device what kind of command it is receiving, and the rest of the bytes are called *data bytes*, which hold the data of the command. What we care about is the note-on and note-off messages. These messages are 144 for note-on on channel 1, 145 for note-on on channel 2 and so on, for sixteen channels, so from 144 to 160. 128 is note-off for channel 1, 129 is note-off for channel 2 and so forth, for sixteen channels, so from 128 to 143. So line 3 in Listing 5-4 checks if we're receiving a note-on message (we don't really care about the channel). If it's true it will assign the next two bytes to the MIDI_note and MIDI_velocity variables, respectively, because this is the order these data bytes are sent. Then we print these two values to the serial line Pd is listening to with the MIDI_note tag, including a space in between values so they can be properly interpreted by the [serial_print] abstraction.

Line 12 checks if we're receiving a note-off message, and if it's true, it will execute the same code with the note-on message. Again, what we care about is the note number and the velocity. Of course, the note-off velocity is zero, so this is kind of trivial. Many MIDI keyboards don't even send note-off messages, but instead they send a note-on message with a 0 velocity. Even if that's the case with your keyboard, this code should work. Again, we print the data bytes to Pd's serial line with the MIDI_note tag. You can imagine that in Pd we'll be receiving this data with [r MIDI_note].

Mind that all this code is added to the code we've already written for the Arduino, nothing should be replaced by anything (including Serial.begin(38400);).

Your keyboard might also send control messages, which are indicated by the status bytes—176 up to 190, again for 16 channels. The data bytes that follow are the controller number and the controller value. The keyboard I have doesn't send control messages, so I can't test it. Applying the values of control messages to the code in Listing 5-4 should be fairly easy, and printing these values to Pd, are receiving them there should also be rather easy. If your keyboard sends control messages, you're encouraged to modify your code to receive them in Pd.

The basic status bytes, along with their values, can be found at http://www.midi.org/techspecs/midimessages.php.

You can test if your keyboard sends any of the status bytes in Table 5-2. All the status bytes in the table are followed by two data bytes, except from program change and channel pressure, which are followed by one data byte only. Make sure you read all data bytes in your Arduino code.

The pitch bend change status byte is followed by the 7 bits of the least significant byte, and the 7 bits of the most significant byte. To assemble these two bytes in Pd, you'll need to multiply the first byte (the least significant byte) by 128, and add it to the second byte (the same technique we apply with the [serial_write] abstraction to reassemble the analog pin values from the Arduino). This will yield a maximum value of 16383, which is the 14-bit value of the Pitch Bend wheel on MIDI keyboards.

To determine what data your MIDI keyboard is sending, [print] will come in handy in Pd. This object prints whatever it receives in its inlet to Pd's console. Replace the code in Listing 5-4 with that in Listing 5-5 to determine what kind of data your keyboard is sending.

Listing 5-5. Checking for Other Types of MIDI Messages

```
1.   if(MIDI_serial.available()) {
2.     byte cmd_byte = MIDI_serial.read();
3.     Serial.print("MIDI_data ");
4.     Serial.println(cmd_byte);
5.   }
```

In a Pd patch connect [r MIDI_data] to [print] and use all potentiometers and wheels and other controllers on your keyboard. [print] will print to the console anything it receives. Use Table 5-2 to determine what kind of data you're receiving, whether it's Note On/Note Off events, Control Change, or Program Change messages, and so forth. For example, if you're receiving a Control Change message on channel 1, from controller number 8, sending the value 100, in Pd's console you should get the bytes 176, 8, 100 one after the other. It's up to you how you use this information.

Pd Patch Receiving MIDI Data from Arduino

Once you've built the circuit and modified the Arduino code, you should make a very small modification to the Pd patch. The patch is shown in Figure 5-31. The only actual change is that [notein] has been replaced by [r MIDI_note], which sends its list to the left inlet of [poly], which is the same as sending the velocity to the right inlet and the note number to the left, like we did with [notein].

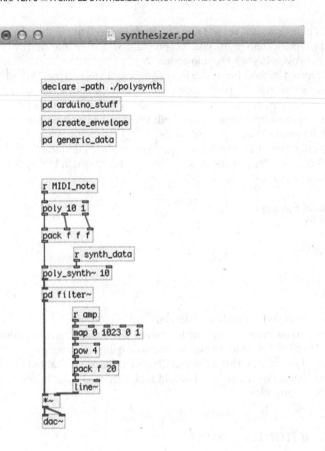

Figure 5-31. Pd patch receiving MIDI data from the Arduino

Pd Patch Receiving a Constant Velocity

Some MIDI keyboards may be so old that they won't really send a velocity value, because their keys might not be force sensitive. This is the case with my MIDI keyboard, which instead of a velocity, on a key press it sends the value 64 as velocity (which I found out by connecting [r MIDI_note] to [print]), and on a key release it sends the value 0, both with a Note On event. In this case unfortunately you won't have amplitude control with the key presses, but you can still use your keyboard. What I've done is send a 127 with a note-on, and let the 0 through. The patch is shown in Figure 5-32. I use [sel 64], so that whenever I press a key and get 64, I send 127. When [sel 64] receives 0 it will let it through intact out its right outlet. This minor modification to the patch in Figure 5-31 seems to fix the constant velocity problem.

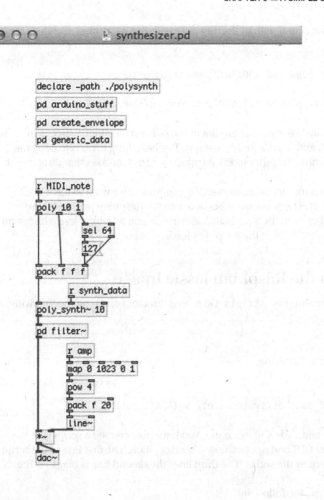

Figure 5-32. *Pd patch receiving a fake velocity*

This concludes our programming for this project, now we're going to use the Raspberry Pi to embed it to our keyboard to make it a stand-alone instrument.

Running the Pd patch on Raspberry Pi

If you want to embed a computer to your keyboard, so it's a stand-alone instrument, you'll need to use a Raspberry Pi, or any other embedded computer you like. Most of the process should be clear by now, there are a few things that I should explain. First of all, copy the Pd patch and the Arduino sketch to your embedded computer using SCP, like we did in Chapter 3. Since the Pd patch contains a folder with the abstractions, you'll need to copy this whole directory (the one that contains the patch and the polysynth folder) to your embedded computer. Also, the Arduino sketches must be inside a directory with the same name as the sketch, so again you'll need to copy a whole directory. Use scp -r /path/to/directory to copy the whole directory. The commands for copying to Pi's pd_patches and sketchbook directories are shown next. Replace /path/to/... with the actual path of the directory with the Pd patches and the Arduino sketch.

```
scp -r /path/to/pd_patch/directory/pi@192.168.100.10:/home/pi/pd_patches
```

```
scp -r /path/to/arduino_code/directory/pi@192.168.100.10:/home/pi/sketchbook
```

You'll also need to copy any abstractions we used that are not in the polysynth folder, which is in the folder of this project, like [loop], [map], and maybe [hold_sustain]. The best thing to do is to copy the contents of pd_abstractions folder from your computer to Pi's pd_abstractions, and set that directory to Pd's search path.

Since we're going to use the Pi headless (or any other embedded computer), we want to force it to launch Pd and the synthesizer patch on boot. The process differs between the Raspbian Jessie and the Raspbian Wheezy images, and very likely between the Raspbian Jessie image and a Debian Jessie image on another type of embedded computer. First, we'll cover the Raspbian Jessie procedure.

Launching Pd on Boot with the Raspbian Jessie Image

In your home directory, create a directory named my_scripts. Go to that directory and type the following:

```
nano launch_pd.sh
```

The nano editor will open. In it, type the following:

```
#!/bin/sh
/usr/bin/pd -nogui -open /home/pi/pd_patches/synthesizer/synthesizer.pd &
```

Hit Ctrl+O and Return to save the file and Ctrl+X to exit nano. You have now created a script that launches Pd in the background without the GUI and opens the synthesizer patch. The first line of the script tells the shell which program to use to interpret the script. The third line (the second line is blank) is the command that launches Pd.

Now, being in your home directory, type the following:

```
crontab -e
```

This will open the crontab file of the user pi. If it's the first time you run the crontab editor, you'll be asked which editor you want to use to edit crontab, while being prompted to use nano, as it's the simplest one. The choices will be numbered, and nano will probably be the second choice, so type 2, and hit **Return**. The crontab file will open in nano and you'll be able to edit it.

If you haven't edited this file it should only have a few lines with comments, telling you what this file does and how to use it (well, giving some minimal instructions). After all the comments add the following line:

```
@reboot sleep 20 ; sh /home/pi/my_scripts/launch_pd.sh
```

This tells the computer that when it reboots (or boots) it should first do nothing for 20 seconds, and then it should run the script we've just written. Take care to place the semi-colon between the sleep 20 command and the sh command. You might want to experiment with the number 20, and try a smaller number. Give the Pi enough time to fully boot, but 20 might be too much. Try a few different values and see which is the lowest that works.

Launching Pd on Boot with the Raspbian Wheezy Image or with Another Embedded Computer Running Debian Jessie

Linux systems include a script that runs on boot (this holds true for the Raspbian Jessie image, but it won't work), so we can use that to tell the computer to do what we want it to do. This script is in the /etc directory and it's called rc.local. In your home directory (or wherever), type the following:

```
sudo nano /etc/rc.local
```

This sketch by default looks like Listing 5-6.

Listing 5-6. The /etc/rc.local script of the Raspberry Pi

```
#!/bin/sh -e
#
# rc.local
#
# This script is executed at the end of each multiuser runlevel.
# Make sure that the script will "exit 0" on success or any other
# value on error.
#
# In order to enable or disable this script just change the execution
# bits.
#
# By default this script does nothing.
# Print the IP address
_IP=$(hostname -I) || true
if [ "$_IP" ]; then
  printf "My IP address is %s\n" "$_IP"
fi
exit 0
```

This is what it looks like in a Raspberry Pi. Maybe it will be a little bit different on some other embedded computers, but essentially it does the same thing, run on boot. We shouldn't care about what these lines do, it's not of our interest. Before exit 0 we're going to write a command to launch Pd along with the patch we want. The command you need to place there is shown here:

```
su -c '/usr/bin/pd -nogui -open /home/user/pd_patches/synthesizer/synthesizer.pd &' - user &
```

This command tells the computer to launch Pd without the GUI, and open the synthesizer patch (here we suppose that the patch is called synthesizer.pd that is in the directory called synthesizer in the pd_patches directory in the computer's home). There seems to be an issue with [comport], which doesn't like to be used by root. su is the name of the root, and since we're using a script in the / directory, we must have root privileges. The -c flag tells the computer to run the command inside the single quotes as the user set after the single quotes close. This means that you must replace the word *user* with the name of the user, where in the Raspberry Pi it's *pi*, in the Odroid it's *odroid*, in Udoo's Debian images it's probably *debian*, and so forth. To find the name of the user, type the following:

```
$ whoami
```

The user name will be printed in the terminal monitor. The command in the /etc/rc.local script will result in Pd being launched by the "user" user, and not by root (again "user" refers to the name of the user). This way the patch will open and [comport] will run just fine. We're launching Pd in the background, and the command that tells the computer to do this as user user, by using the ampersand (&) in both commands. Use this command as is if you want things to run properly. After you type the command in the script, hit **Ctrl+O** and **Return** to save it, and **Ctrl+X** to exit nano.

Mind that it's not definite that editing this script will work. If it doesn't, try the procedure of the Raspbian Jessie image explained earlier.

Using a USB MIDI Keyboard with the Pi

If you have a USB MIDI keyboard that you want to use with the Pi there are a few more steps you'll need to take. First, you'll need to launch Pd with the -alsamidi flag with the GUI. Once launched open a new patch and put a [r pd] connected to a [print]. Then go to **Media** ä **MIDI Settings...** and make sure you have one **In Port:**. If you changed the number of ports, click **Apply**. Whether you changed the ports number or not, click **OK**. In Pd's console you'll get a message of the type "midi-dialog 0 0 0 0 0 0 0 0 1 1" (not necessarily the same). Copy this message and paste it in a message box in your synthesizer patch. Connect a [loadbang] to the message and the message to [s pd], like in Figure 5-33.

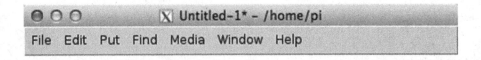

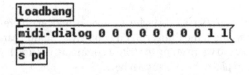

Figure 5-33. Sending the MIDI dialog message to Pd on launch

Now, on Pi's terminal, type the following and check the output:

```
aconnect -lio
```

Your USB MIDI device and Pd should be in that list. In the script that launches Pd, after launching it write the following:

```
sleep 3
aconnect yourUSBmidiDevice:0 'Pure Data':0
```

Change yourUSBmidiDevice with the name of your MIDI keyboard. The two 0s are the output port number of the MIDI keyboard and the input port number of Pd. The output list of aconnect should print the correct values. Check Chapter 1 if you don't remember the exact process of getting MIDI input in Linux.

Shutting Down the Pi (or Any Embedded Computer) While Running Headless

If we're going to use Pd without the GUI, we'll need to be able to shut the Pi down somehow. We have used all pins of the Arduino, so we can't put another push button that will enable shutting down the Pi. We have to use the existing components in some way that will help us achieve what we want. The most logical thing is that when we'll want to shut the Pi down, we'll have turned the DSP off. We can use the DSP switch to determine whether we can actually shut the Pi down or not. In the Arduino code add the code in Listing 5-7 to the end.

Listing 5-7. Sending the DSP Switch Data

```
1.   if(dsp_switch){
2.     int shut_down = digitalRead(2);
3.     Serial.print("shut_down ");
4.     Serial.println(!shut_down);
5.   }
```

This code will run only if the DSP switch is in the OFF position (because of the pull-up resistors). If the switch is indeed off, we'll read digital pin 2, which is the first push button, and we'll print that to Pd with the shut_down tag. In the Pd patch, make a subpatch and fill it in with the contents in Figure 5-34.

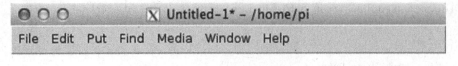

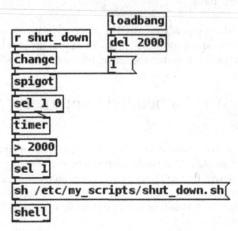

Figure 5-34. Contents of the subpatch that shuts the Pi down

In this subpatch we're receiving the value of the first push button, only when the DSP switch is in the OFF position. We're printing the value of the button with the exclamation mark prepended to it (line 4 in Listing 5-7), so that when we press it we'll receive a one, and when it's released we'll receive a 0. We're using [timer] in this subpatch, which measures logical time. When this object receives a bang in its left inlet, it is reset. When it receives a bang in its right inlet, it outputs the amount of milliseconds passed since it was reset. We're sending the output of [timer] to [> 2000] to check if we held the button pressed for at least 2000 milliseconds (2 seconds). Think about it, when we press the button (with the DSP switch in the OFF position), [timer] will be reset, and when we release it, we'll get the amount of milliseconds we kept the button pressed. If this value is above 2000, [> 2000] will output 1, and [sel 1] will bang the message, which will tell [shell] to run the script to shut the Pi down. This is the same script we wrote in Chapter 3, so if you skipped that chapter, go back and check it. [change] connects to [spigot], which opens two seconds after the patch has loaded, to prevent a shut down as soon as the patch opens, as the code in 5-7 will be printing the value of the first button constantly, as long as the DSP switch is in the OFF position.

A Small Caveat for the Raspbian and Debian Wheezy Images

If you're using the Raspbian Wheezy image, or another embedded computer with a Debian Wheezy image, you should have noticed that you can't make [serial_print_extended] work. That is because it's missing the [bytes2any] and [any2bytes] externals from the moocow library. The Raspbian repositories don't have this library so we haven't installed it. Go to GitHub and search for **pdstring**. Once you find it, download it as a .zip file. In your Pi, make a directory called pd_externals and move the pdstring-libdir-master directory in there (the directory you unzipped from GitHub). Now navigate to pdstring-libdir-master and hit **ls** to see its contents. There are three README files, which include instructions for installing the new externals. The instructions say that you must type **./configure**, but there's no configure file there. All you need to do is this:

```
make
```

and

```
sudo make install
```

First, run the make command, which will create binary files (executable files) out of the .c files, and when it's done (you'll get back to the $ prompt) run the second command. This will put the new binary files to /usr/local/lib/pd-extenals/pdstring. Now launch Pd and go to **Media ä Preferences ä Path** and choose this path so Pd can look up for these objects. Click **Apply** and **OK** and quit Pd. Re-launch it and open a new window and put [serial_print_extended]. You'll get error messages saying that Pd couldn't create [moocow/bytes2any] and [moocow/any2bytes]. Open the [serial_print_extended] abstraction and go to [pd $0-strings]. In there, change [moocow/bytes2any] to [bytes2any]. Then go to [pd $0-set_argument] in the parent patch of the abstraction, and from there to [pd $0-split_symbols]. In there, change [moocow/bytes2any] to [bytes2any], and [moocow/any2bytes] to [any2bytes]. Close the subpatches and save the abstraction. The next time you open it, it should work. I should mention that these are not my externals. I have only forked the repository in GitHub, so you can easily find these files.

Using an External Sound Card with the Pi

If you have an external sound card you want to use, you might need to take some steps to use it. I'll suppose that it either doesn't require drivers, or that you already have installed all necessary drivers for it. In Pd, you can go to **Media ä Audio Settings...** and choose your sound card. That might not work though, and you might either get audio dropouts (they produce clicks in the sound), or perhaps no audio at all. When using external sound cards in Linux, many time we prefer to use the Jack audio server. If the Pi doesn't have Jack installed, you can install it with apt-get. In Pi's terminal type **jackd**, and if you get a "command not found" error, then type **sudo apt-get install jackd**.

Using QJackctl offers an easy way to find out which is your sound card on Jack's list. This is a user-friendly GUI version of Jack. In Pi's terminal, type the following:

```
/usr/bin/qjackctl &
```

QJackctl will open, as shown in Figure 5-35. Make sure that you use the ampersand (&), because we'll need to use the terminal while QJackctl is running.

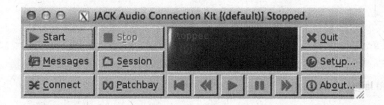

Figure 5-35. QJackctl interface

In QJackctl click **Setup** and you'll get the window shown in Figure 5-36. In the Setup, click the **Input Device:** menu and select your sound card. Figure 5-37 shows some choices provided by QJackctl, where a Focusrite Scarlett 2i4 sound card is included. What we care about is its *hw* number (for hardware), which is 1. Go ahead and choose your sound card both as an input and as an output device, and then click **OK** (leave **Driver:** on top right at "alsa"). This will take you back to the main window of QJackctl, as shown in Figure 5-35. Click **Start** and wait till the QJackctl screen shows that it has started.

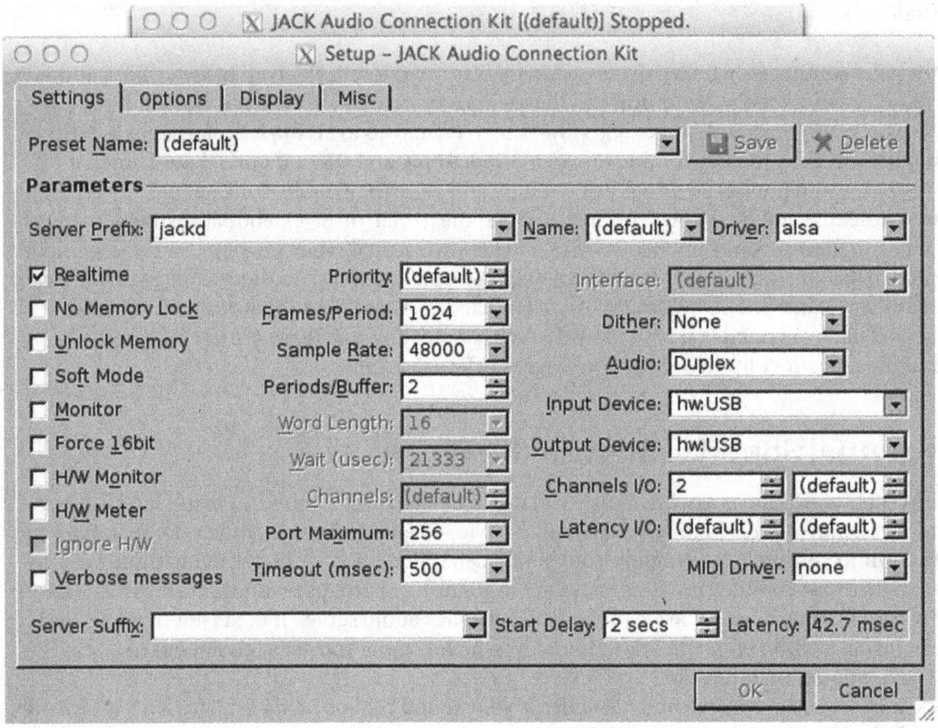

Figure 5-36. *Qjackctl's setup*

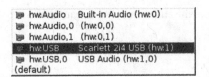

Figure 5-37. *QJackctl sound card choices*

Now go back to Pi's terminal and launch Pd like this:

```
/usr/bin/pd -jack &
```

This command will launch Pd in the background with Jack as its audio server. Go to **Media** ä **Test Audio and MIDI...** and see if your sound card is working properly. If it is, quit Pd, and click **Stop** on Qjackctl's window, and then **Quit**.

Since we'll be using the Pi headless, we can't use Jack's GUI, so we need to launch it without it. In Pi's terminal, type the following:

```
/usr/bin/jackd -d alsa -d hw:1 &
```

This will launch Jack, using Alsa as a driver and a hardware 1 sound card. Now launch Pd like before and again test if your sound card is working properly. If all goes well, you'll need to modify the scripts that launches Pd a bit. Open ~/my_scripts/launch_pd.sh and before the command that launches Pd, type the following:

```
/usr/bin/jackd -d alsa -d hw:1 &
sleep 3
```

This will launch Jack with the necessary driver and sound card, and then the Pi will wait for three seconds, before it launches Pd. The command to launch Pd stays the same. If you're using another embedded computer or the Raspbian Wheezy image, you'll need to add these two lines in the /etc/rc.local script, before the command that launches Pd. No need to run this command as a simple user, so the command will do just the way it's written earlier. Now edit the script that quits Pd and shuts the Pi down. Type the following to edit the script:

```
sudo nano /etc/my_scripts/shut_down.sh
```

We want to first quit Pd, then Jack, and then shut the Pi down. We already have this in our script:

```
sudo killall pd
sleep 3
sudo halt
```

Change it to this:

```
sudo killall pd
sleep 3
sudo killall jackd
sleep 3
sudo halt
```

So now the Pi will first quit Pd, then wait for three seconds, then it will quit Jack, again it will wait for three seconds, and finally it will shut down. Now your Pi is ready to run your patch using your external sound card.

Editing the Pd Patch When Pd is Running on Boot

Once you have configured your Pi with the settings we've been through, whenever your power it up, it will launch Pd and open the synthesizer patch. If you want to make changes to the patch, or use the Pi for something else, as soon as you log in, quit Pd like this:

```
sudo killall pd
```

so you there's no Pd instance running while you make your changes. If you need to quit Jack as well, do the same for "jackd". If you make changes to the already used patch, you won't need to do anything else. If you want to use another patch, or do something completely different, you'll have to edit at least the rc.local script, so that the Pi won't launch Pd, or won't open the old patch on boot.

Enclosing the Pi in the Keyboard

To enclose the Pi in the keyboard so that it's a stand-alone instrument is a process of trial and error. Since all keyboards are different in size and shape, I can't give definite guidance through this process. There are some "global" rules that apply to such processes. Bear in mind that most electronic devices have very limited space inside, so it might be very difficult or even impossible to enclose the circuit, the Arduino and the Pi inside your keyboard. In this case, you might want to enclose all this stuff in a separate box and use that along with your keyboard.

Choosing the Perforated Board

Whether you enclose the Pi inside the keyboard enclosure or you make a separate box, you'll need to choose the perforated board you'll solder your circuit on. Breadboards won't do here because they're not stable. A perforated board is preferred because you solder the parts of your circuit on it, making it much more stable than a breadboard. There are two kinds of perforated boards: boards with solder lines and boards with solder points. Figure 5-38 shows both boards.

Figure 5-38. *A perforated board with solder lines and a perforated board with solder points*

The advantage of the board with lines is that whatever you solder on a line, will be connected with all other components soldered on the same line, which saves us from some wiring. The drawback of such a board is that you'll very likely need to cut the copper at certain points to avoid shorts. For example, if you horizontally solder an Arduino Nano on it , you'll need to break the copper so as not to have pins on opposite sides of the Arduino connected to each other. You can do that with a drill. Figure 5-39 shows a perforated board with lines, and with the copper cut, exactly for this reason. You'll have to make sure that your circuit doesn't short anywhere, otherwise you might damage your Arduino.

Figure 5-39. A perforated board with lines and with the copper cut to avoid shorts

The board with points, on the other hand, offers some more flexibility, as you won't need to cut the copper anywhere, but you'll need more wires to make your connections. It's up to you what you find more suitable for your projects.

If you're using an Arduino Uno, you can use the Proto Shield, which is a perforated board with points that mount on top of the Arduino, giving you easy access to its pins and breaking out the 5V and GND pins to many points, so you have access to power and ground for many components.

Reducing the Wires of Your Circuit Board

Another tip is that it's maybe preferable to daisy chain all grounding points of all components, and have only one of them connect to the circuit board. If you're using a lot of potentiometers and switches, it's better to connect the potentiometer legs that connect to ground between them, and connect them to the ground pins of the switches too, so that only one potentiometer connects to the ground of the circuit board, and it then passes the ground to the other components with wires. The same applies to voltage. So, potentiometers should have their ground legs all daisy chained, their voltage legs all daisy chained, and their middle legs should all go to the circuit board, each to an analog pin of the Arduino. Figure 5-40 shows an example of this.

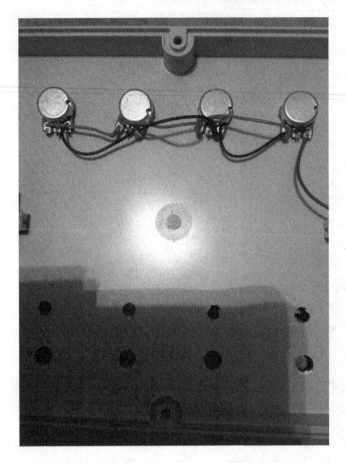

Figure 5-40. *Wiring the voltage and ground pins of the potentiometers*

Choosing the Right Power Supply and Getting Access to the Power Pins

You might want to use only one power supply. If your keyboard is powered by the USB cable, then you're good to go, but in case it needs an external power supply, you'll have to use two, one for the keyboard, and one for the Raspberry Pi. It is very likely that your keyboard operates at 9V, or maybe 12V. You can use a power supply at this voltage with enough Ambers for both the keyboard and the Pi (1 Amber is more than enough for the Pi), and use a voltage regulator to drop down the voltage for the Pi. If you open your keyboard case, you'll get access to its power jack. You can solder wires on the bottom side of its circuit board and connect it this way to the voltage regulator, which will power the Pi. Consult your local electronics store as to which voltage regulator is best for your setup, and which capacitors you'll need to make it work.

Accessing the MIDI Pins

In a similar way you can access the MIDI or USB cable pins. When enclosing a device in a box, usually we don't want to have cables floating around, so we solder everything inside the box. Since the USB or MIDI connectors will be facing the outer part of the enclosure, we have to access them from their connections on the back side of their circuit boards. Figure 5-41 demonstrates accessing the pins of the MIDI connector of a keyboard. All connectors, (USB, power supply) can be accessed this way. The proper way is to remove the connector you want to have access to, and in its pins holes, connect your wires. Removing circuit components can be quite difficult. Still, you are encouraged to try it. You'll need a solder pump and perhaps some solder wick to remove the solder from a circuit board, and then remove the component. Sometimes you might need to be a bit violent, as the solder will probably not be totally removed.

Figure 5-41. *Accessing the MIDI pins of a keyboard from inside the enclosure*

Whichever way you choose to make your connections, make sure you connect the correct pins, as making a mistake in the circuit can lead to damage. A multimeter can become a very good friend, since you can use its continuity to determine connections in a circuit.

Bringing the Power Input and Audio Output out of the Enclosure

Generally, it would be best if you visit your local electronics store or an online store to see what's available for your project. The powering cable for the Pi can be an issue, for example. It might be a good idea to get a power jack to a micro USB or a USB type B–to–micro USB to power up the Pi. A roll-up USB cable is also a something to consider, as it keeps things tidy inside limited enclosure spaces. Also, the way you have access to the audio jack in the Pi is an important aspect of your project, which you should bear in mind. Figure 5-42 shows a suggestion of how you can get the Pi power input and audio output out of the enclosure.

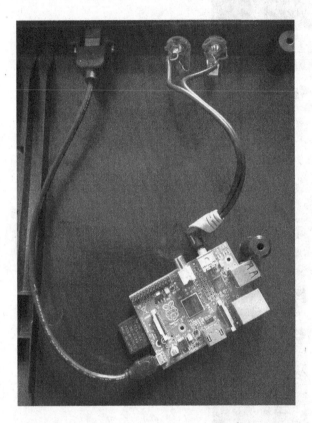

Figure 5-42. *Getting the power input and audio output of the Pi outside the enclosure*

Conclusion

We have now finished our first finalized project. Even if you used your computer with a USB keyboard, and you didn't deal with enclosing your project at all, you still went through the software process, which is a fundamental process for digital electronics. If you went all the way through the enclosure process, you have been introduced to some serious project building, which enables you to realize independent, very flexible, and powerful projects. We'll be using the Pi in other projects in this book, but the process will be very similar (if not identical) to the one we went through in this chapter.

CHAPTER 6

■ ■ ■

An Interactive Bow

In this chapter, we're going to build a sensor circuit to use with a bow to manipulate the sound of a live instrument, with input given by the gestures of the performer, who will control the instrument's audio input. This circuit can actually fit any instrument that involves some movement, such as a trombone, for example. The sensor we'll use is an accelerometer, which in this particular case was chosen for its tilt-sensing capability. *Tilt sensing* means that the sensor is able to orientate itself with respect to the Earth's surface. This sensor gives values for three axes, which we can use independently. In this project, we'll give ourselves as much freedom as possible so that we can use any of the three axes at any time (including all three together). We can even receive no input from the sensor, so that the performer can move freely in case he/she doesn't want to affect the instrument's sound at certain moments.

We'll also use the XBee transceivers to make our project wireless and give the performer more freedom of movement. Of course, since we'll be manipulating the instrument's audio input, we must use a microphone—whether a contact microphone, a condenser, or a dynamic microphone. Still, having a microphone cable hanging out of the instrument is not as limiting as having a cable (or more than one cables) hanging out of the performer's arm. For this reason, we're going to use a wireless connection between the Arduino and the computer. We'll also be using an Arduino Pro Mini, which is a very small and inexpensive Arduino that suits wireless projects very well. You're free to use any Arduino type that you want, but the circuit building will be your own responsibility.

Parts List

Table 6-1 lists all the parts that you'll need to build this project.

Table 6-1. Interactive Bow Parts List

Arduino	XBee	XBee Explorer	XBee USB Explorer	Accelerometer	Push buttons	Battery
1 × Pro Mini	2	1	1	1 (ADXL335 accelerometer breakout used in this chapter)	4	1 × 9V

233

Writing Custom Functions

Let's start building this project by writing the code first. In this sketch, we'll write a custom function to make things clearer. Writing our own functions makes the code more modular and easier to read and modify. This is also where I'll explain what void means. Before we write the code for this project, let's first write a simple sketch that defines a custom function. Listing 6-1 shows the code.

Listing 6-1. Simple Custom Function

```
1.  byte my_val;
2.
3.  void setup() {
4.    Serial.begin(9600);
5.  }
6.
7.  void loop() {
8.    if(Serial.available()){
9.      my_val = Serial.read();
10.
11.     byte val_times_2 = double_val(my_val);
12.     Serial.write(val_times_2);
13.   }
14. }
15.
16. // simple function to double value
17. byte double_val(byte val){
18.   val *= 2;
19.
20.   return val;
21. }
```

In this simple sketch, we've defined a function that receives a value and multiplies it by 2; we've named it double_val. Of course, this function could have been included in the main loop function. Since we want to get the hang of creating our own functions, this sketch serves as a first step toward that. Also, it will help us understand the concept of the void data type, which we have been using since the beginning of this book, but I haven't explained it yet. In line 8, we check whether we've received a byte in the serial line, and if so, we store that byte to the my_val variable. Then we create a new variable of the type byte, and assign to it the value returned by our new function double_val. Let's jump to line 17, which reads as follows:

```
byte double_val(byte val){}
```

This is the way we define a function in C++. The first thing we need to do is define the data type this function will return. In this case, we'll be returning a byte. Then we write the name of the function (this is anything we want, apart from standard keywords of the C++ and the Arduino language), and lastly we open a parenthesis and place the types of data we're expecting to receive, which are the arguments of the function. This function receives one argument only, which is of the byte type. Inside the curly brackets of the function we write the code that will be executed whenever this function is called (in the loop function, it is called in line 11). This function takes its argument (the val variable) and multiplies it by 2 (line 18). Before it exits, it returns the result of the multiplication by using the keyword return along with the value we want to return, in this case val. Since val is a byte, that's the data type we're returning, and that's the data type our function needs to be, which is defined in line 17. This line uses the keyword byte twice, and that may be a bit confusing. The first time we use it, outside the parentheses, we're defining the type of the data that this function will be returning. The second time we use it, inside the parentheses, we define the type of data that this function is expecting as an argument. These two types don't necessarily need to be the same. It just happens here because the argument the function receives comes from the serial line, which accepts bytes, and the value it returns will be written to the serial line, which writes bytes. In line 11, we read the following:

```
byte val_times_2 = double_val(my_val);
```

On the right side of this line, we have the function we wrote with the argument in the parentheses. my_val is a byte, and we can see that in line 1. The double_val function returns a byte, and that's why we define the val_times_2 variable as such. If we defined this variable as another data type, for example, int, the Arduino IDE would throw an error and the code wouldn't be compiled. By "returning" a value, I mean that we can assign the result of the function to a variable, which is what we do in line 11. If a function returns no value, it must be of the type void, which is the type of the two main functions in Arduino: setup and loop. These two functions don't include the keyword return at the end, so they're returning nothing, that's why they're of the data type void. This concept may be a bit confusing to grasp, but Listing 6-2 shows some code that will help clarify it.

Figure 6-1 shows a Pd patch that works with the code in Listing 6-1. It is a very simple patch where all we need to do is send a raw value to [comport] and it will output the double of this value. Instead of a number atom, we're connecting [comport] to [print] to see that it outputs a value only when it receives one. This is because all the code of the loop function in the Arduino sketch is included in the curly brackets of the if control structure. If you place lines 11 and 12, or just line 12, outside the curly brackets of if, you'll see that [comport] will be outputting values constantly. By placing the Serial.write function inside the code of if, we make sure that it will be called only when there is data coming in the serial line of the Arduino.

Figure 6-1. Simple patch that works with the double_val function Arduino sketch

Type some numbers in the number atom and check Pd's console. It should print the double of the value you've entered. Mind that the serial communication uses bytes, and since we're calling the Serial.write function and not Serial.print, we'll be receiving raw bytes and not their ASCII values. If you send a value higher than 127, its double will be wrapped back to 0. If, for example, you send 128, [print] will print 0; if you send 129, it will print 2; and so forth.

This Arduino sketch and Pd patch are not really useful for any other reason than clarifying the concept of custom functions and the void data type, which is explained even further in the code in Listing 6-2.

A Function of the Data Type void

Now let's write another simple function, but this time of the type void. What we'll do is send a value in Arduino's serial line, and use that value to read from the corresponding analog pin. The function we'll build will read the analog pin set by the incoming value and will print the reading to the serial line. Check the code in Listing 6-2.

Listing 6-2. A Void Custom Function

```
1.   void setup() {
2.      Serial.begin(9600);
3.   }
4.
5.   void loop() {
6.      if(Serial.available()){
7.         byte pin = Serial.read();
8.
9.         print_analog_pin(pin);
10.     }
11.  }
12.
13.  void print_analog_pin(byte pin){
14.     int val = analogRead(pin);
15.
16.     Serial.print("value ");
17.     Serial.println(val);
18.  }
```

In line 9, we're calling our custom function print_analog_pin, but this time we're not assigning its output to any variable, like we did with the custom function in Listing 6-1. This is because print_analog_pin is of the type void, and it doesn't return anything. If we try to assign print_analog_pin to a variable, the Arduino compiler will throw an error and won't compile the code. In line 13, we're defining the print_analog_pin function like this:

```
void print_analog_pin(byte pin){}
```

This function is of the type void but its argument is of the type byte. As stated in the example in Listing 6-1, having one type for the function and a different one for its argument(s) is perfectly legal (you can actually have more than one arguments to a function, each of a different type). In this function we're creating a variable and we store in it the reading of the analog pin set by the argument of the function. Afterward, we print this value to the serial line (this time using Serial.print) and the function ends. Since the printing of the value is being done inside the function, there's no need for the function to return any value, that's why it's of the type void. We could instead return the value of the analog pin, and in the main loop function, assign that value to a variable, and print it from there. This code is made like this to clarify the concept of the data type void.

Figure 6-2 shows the Pd patch that works with this Arduino sketch. Provide a number to [comport] and it will print the value of that analog pin. If you provide a value beyond the number of the analog pins of the Arduino, you'll get random readings. Try this sketch with a few potentiometers to see how it really works.

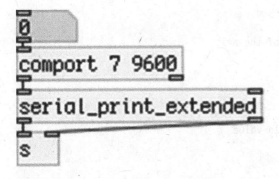

Figure 6-2. *Pd patch that works with print_analog_pin function*

Actual Code for the Project

As I've already mentioned, in this project we'll use an accelerometer, which has three analog outputs, one for each axis, so we'll need to read three analog pins from the Arduino. This sounds pretty simple, but what we essentially want to do is choose whether an axis is active or not. To do this we're going to use three push buttons. Using switches is easier to use in the code, since a switch has two states, on or off, and it can stay at any of the two, whereas the push button is at its on state only while kept pressed, so we'll need to alternate between states at each button press. This sensor circuit is aimed at being used with a violin, viola, cello, or double bass bow, and must be as small as possible, making the use of switches not very well suited. Listing 6-3 shows the code.

Listing 6-3. Accelerometer Code

```
1.   bool activity[3];
2.   int old_val[3];
3.
4.   void setup() {
5.     for(int i = 0; i < 3; i++){
6.       pinMode((i + 2), INPUT_PULLUP);
7.       activity[i] = false;
8.       old_val[i] = 1;
9.     }
10.
11.    Serial.begin(57600);
12.  }
13.
14.  void loop() {
15.    for(int i = 0; i < 3; i++){
16.      int button_val = digitalRead(i + 2);
17.      if(button_val != old_val[i]){
```

237

```
18.          if(!button_val){ // if button is pressed
19.          activity[i] = !activity[i];
20.            // show on patch the on/off state of the axis
21.            Serial.print("axis");
22.            Serial.print(i);
23.            Serial.print("\t");
24.            Serial.println(activity[i]);
25.          }
26.          old_val[i] = button_val; // update old value
27.        }
28.    }
29.
30.    // call axes according to the button presses
31.    for(int i = 0; i < 3; i++) if(activity[i]) axes(i);
32.  }
33.
34.  void axes(int which){
35.    int value;
36.    switch(which){
37.      case 0:
38.        value = analogRead(which);
39.        Serial.print("x ");
40.        Serial.println(value);
41.        break;
42.
43.      case 1:
44.        value = analogRead(which);
45.        Serial.print("y ");
46.        Serial.println(value);
47.        break;
48.
49.      case 2:
50.        value = analogRead(which);
51.        Serial.print("z ");
52.        Serial.println(value);
53.        break;
54.    }
55.  }
```

In the first two lines, we define two arrays, one of the type bool and one of the type int. These will be used for the output of the accelerometer, and since the sensor gives three values, they must have a size of 3. In the setup function we set the mode of the first three digital pins to INPUT_PULLUP, so we can use the integrated resistors of the Arduino and save some space in our circuit. We also initialize the two arrays with values, the activity array with all its elements set to false, and the old_val array with all its elements set to 1.

Detecting Button Pressed and Acting Accordingly

In the loop function we read through these digital pins one by one, and compare them to their previous state. Since we're using the integrated pull-up resistors of the Arduino, when the push buttons are not pressed, their digital pins will read HIGH, which is equal to 1. That's why we've initialized the old_val array with all its elements to 1. Line 17 checks if the current reading has changed, by comparing it to the corresponding element of the old_val array, and if it has changed, in line 18 we're checking if the push button is being pressed, by checking if the reading is 0 (using the exclamation mark). If the push button is being pressed, in line 19, we're reversing the corresponding element of the activity array. This line reads as follows:

```
activity[i] = !activity[i];
```

This way we're assigning to the current element of the array its reversed value. So if the current element was false, it will now be assigned to true.

Lines 21 to 24 print to the serial line the axis that we're activating or deactivating. We're achieving that by printing the string "axis", then the i variable, which holds the number of the push button being pressed (starting from 0, not 2), and then a horizontal tab, "\t". We'll use the [serial_print_extended] abstraction in Pd, which uses the horizontal tab, the white space, and the comma as delimiters between the tag string and the rest of the values. This way the string "axis" and the value of the i variable will be concatenated, resulting in one of the following strings: "axis0", "axis1", or "axis2". After we print the horizontal tab, we print the value of the current element of the activity array. This way we can receive these values in Pd with the objects [r axis0], [r axis1], and [r axis2] each corresponding to a push button of our circuit. We'll use these values in Pd to visualize which axis is active and which in not.

Line 25 closes the curly bracket of the if(!button_val) test, and in line 26, we update the value of the current element of the old_val array. By updating this value we make sure that while a push button is being pressed, the if(button_val != old_val[i]) test in line 17 will not be true, until we release the push button. When we release the push button, this test will be true, but the if(!button_val) test in line 18 will not be true, since releasing the push button will result in the value 1 (because of the pull-up resistors used), so the code in this test will not be executed. With this technique we can use the push buttons like switches. Every time we press a button the corresponding value of the activity array will be reversed, which is what happens when we change the position of a switch. Using the value of each element of the activity array in Pd to visualize the activity state of each push button gives the appropriate visual feedback we need.

Line 27 closes the curly bracket of the if(button_val != old_val[i]) test, and line 28 closes the curly bracket of the for loop. In line 31, we run another for loop. In this loop we're checking the value of each element of the activity array. If it's true, we're calling the axes function with the index of the array passed as an argument to that function.

The Custom Function of the Project

Now to our custom-made function, axes. In line 34, we define this function like this:

```
void axes(int which){}
```

This function is of the type void and takes an int as an argument, which argument is the number of the push button that has been activated. The first thing we do in this function is define a variable of the type int, value. This variable will be assigned the value read by the analog pin set by the argument of the function, which will be one of the three axis of the sensor readings. In line 36, we begin a switch control structure, which takes the function's argument as its argument. Mind that in the definition of a function we must define the type of the argument(s), but in the switch control structure used here we should not define this type, since it has already been defined in the definition of the function itself. We have named this argument which because it tells to the program which of the three axis of the accelerometer has been activated. This value will be either 0, 1, or 2. In the case tests of the switch control structure, we're using these three values to test which part of the code we'll execute. If the which variable holds 0, the first case will be met, and its code will be executed.

The code in all three cases is very similar; the only thing that changes is the string tag. If the which variable holds 0, the value variable will be assigned the value read by the analog pin 0, and then we'll print it to the serial line, preceded by the string "x " (the white space in the string is used as a delimiter in the [serial_print_extended] abstraction in Pd). Once we print the value to the serial line, we call break to exit switch. If we don't call this, switch will go on and test all three cases, which will result in unexpected behavior. If the which variable holds 1, the second case will met and the value variable will be assigned the value read by the analog pin 1, and printed to the serial line preceded by the string "y ". If which holds 2, value will be assigned the value read by the analog pin 2, and will be printed to the serial line preceded by the string "z ". This way we can retrieve the value of each axis in Pd, using [r x], [r y], and [r z].

Once a case has been met and its code has been executed, the axes function ends and we'll go back to the main loop function, at the point where we called the axes function, which is the for loop, in line 31. This enables us to read any of the three axis in any combination, including all three at the same time. If we have activated all three axis, the for loop will go on and test the value of all the elements of the activity array. If all are true, axes will be called at each iteration of the for loop. The first time it will be called with the value 0 passed as its argument, so it will print the value of the analog pin 0 with the "x " string tag. Once this value has been printed, we'll go back to the for loop, which will test the second element of the activity array, and axes will be called with the value 1 passed as its arguments, and it will print the value of the analog pin 1 with the "y " string tag. After that, we'll again go back to the for loop and this time we'll call axes with the value 2 passed as its argument, so it will print the value of the analog pin 2 with the "z " string tag. This technique makes our code very flexible since we have full control on which axis of the accelerometer will be read and printed to the serial line.

One thing to mention is that in the axes function we could have avoided defining the value variable altogether, and in the parentheses of each Serial.println, we could have included analogRead(which) right away (in which case lines 38, 44, and 50 should be removed). This would result in exactly the same behavior, and we would also save the space the value variable takes. Since this variable is created only when axes is called, and when the function ends, the variable is being destroyed, it's not so much space loss, plus the code is quite small in size and we have a lot of space in memory to use. It is up to the coding style of each programmer as to which of the two techniques he/she will use.

The Test Circuit of the Accelerometer

Now let's take a look at the circuit for this code. Figure 6-3 illustrates it.

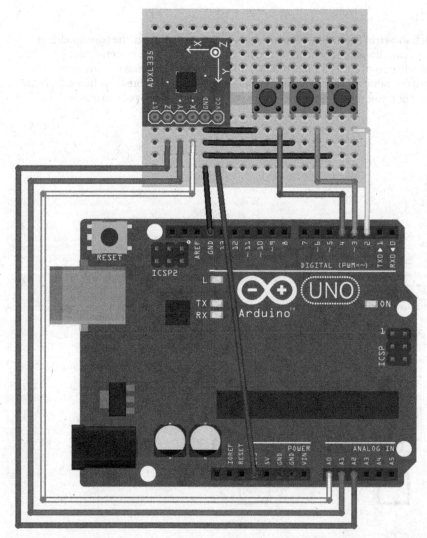

Figure 6-3. *Accelerometer circuit*

This is a test circuit, that's why it's built on a breadboard using an Arduino Uno instead of a Pro Mini, as mentioned earlier. In this circuit, we're using the ADXL335 accelerometer breakout board made by SparkFun, an electronics company (https://www.sparkfun.com). Its pins are being labeled with VCC for power, GND for ground, and X, Y, and Z for the three axis. There's also an ST pin, but we don't need to mind about that. This sensor is powered with 3.3V, and we can see that in Figure 6-3. The X, Y, and Z pins of the sensor breakout are being wired to the first three analog pins of the Arduino, with X connected to 0, Y to 1, and Z to 2. The three push buttons are wired to the first three digital pins, starting from pin 2, as pins 0 and 1 are being reserved for the serial communication. As in all circuits, all components share the same ground. The breadboard used here is a mini, so that you can easily hold it with your hand and shake it to test if the sensor is working as expected. Make sure that you use long jumper wires so that you can easily test the sensor.

The Test Pd Patch

To test it, build the Pd patch shown in Figure 6-4. Whenever you press a push button, the corresponding axis should be activated and you should be receiving its activity and values with the corresponding [receive]s in Pd. [r axis0] and so forth will output either a 1 or a 0, depending on whether the axis is active or not. [r x] and so forth will output the values of each axis from the sensor. On the sensor breakout board there are indications as to which direction you should tilt the sensor so you can get the values of that axis.

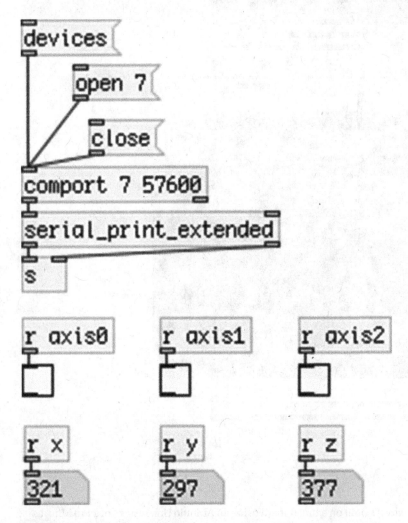

Figure 6-4. Accelerometer test patch

To get the values of the X axis and the Y axis, hold the sensor horizontally to the ground and tilt it 180 degrees (toward the board pins for the Y axis and toward the sensor name label for the X axis). This should give you the full range of these two axes. For the Z axis, hold the breakout with the sensor chip facing upward, and tilt it 180 degrees till it faces downward. These values are flickering and that's because the sensor is very sensitive. Also, its range should be from around 260 to around 400 (at least this is the approximate range of my sensor—yours might differ a bit). These values may seem a bit difficult to use, so it's better if we map them to a specific range that seems more usable. To make this whole procedure a bit easier we'll build a subpatch for each axis and we'll make it in such a way that it will be easy to map the values and save them for later use.

Building the Pd Patch

In a new Pd window, create a subpatch and name it **x_axis** (put a new object and write **pd x_axis**). What you need to do is put all the objects shown in Figure 6-5. The red rectangle in the figure is what is called a *Graph-On-Parent* in Pd (also referred to as *GOP*). Right-click in the empty subpatch and select **Properties**. This will open the subpatch's properties window, which is shown in Figure 6-6.

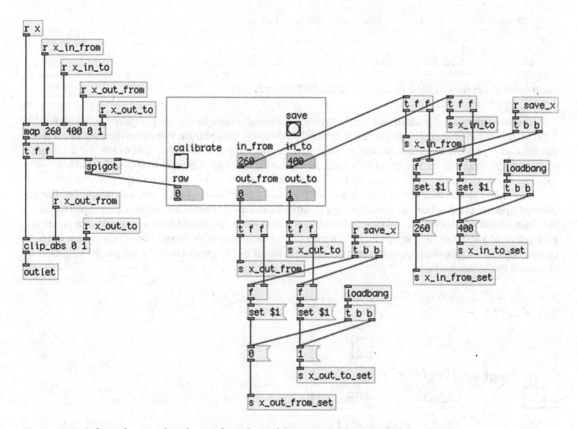

Figure 6-5. *Subpatch to read and map the values of the X axis of the accelerometer*

How to Use the Graph-On-Parent Feature

The middle part of the window in Figure 6-6 is labeled **Appearance of parent patch**. In that field, select the **Graph-On-Parent** tick box, but leave the **Hide object name and arguments** unselected. As soon as you select that tick box, the fields on the **Range and size** part of the window become highlighted (they are grayed out when **Graph-On-Parent** is not selected). By default, the value for X are 0, 1, 85, 100, and for Y –1, 1, 60, 100. What we care about are the last two numbers for each dimension.

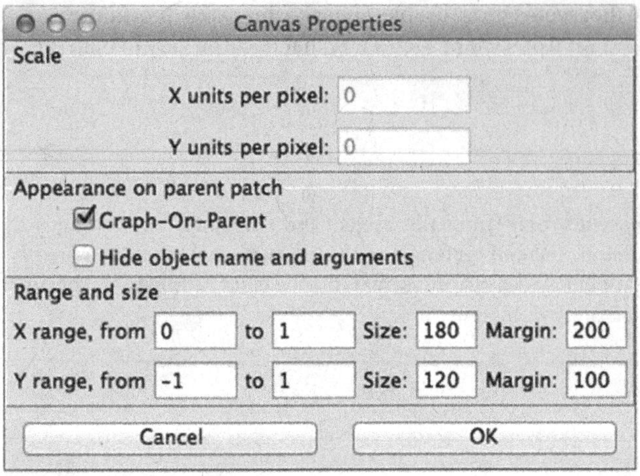

Canvas Properties

Scale

X units per pixel: 0

Y units per pixel: 0

Appearance on parent patch

☑ Graph–On–Parent

☐ Hide object name and arguments

Range and size

X range, from 0 to 1 Size: 180 Margin: 200

Y range, from -1 to 1 Size: 120 Margin: 100

Cancel | OK

Figure 6-6. *Properties window of a subpatch, called Canvas Properties*

The **Size** values are the width and height in pixels of the red rectangle you see in Figure 6-5 (width for the X/horizontal dimension and height for the Y/vertical dimension). The **Margin** values are the position of the top-left corner of the red rectangle inside the subpatch, in pixels. If you leave them both to 100, the rectangle will appear 100 pixels to the right, and 100 pixels down. I've changed the X margin to 200, so I could have some space on the left side of the rectangle to put some object that shouldn't appear in the parent patch.

If you take a look at your parent patch, you'll see that the subpatch you have opened, with **Graph-On-Parent** selected, appears as a gray box. Go ahead and put a number atom inside the red rectangle in the subpatch and any object outside the rectangle, and close it. Now in the parent patch you should have a rectangle with the name **pd x_axis** and a number atom inside it, but not the other object. Graph-On-Parent allows you to have only certain objects of a subpatch appear in the parent patch. Figure 6-7 illustrates how the subpatch in Figure 6-5 appears in the parent patch.

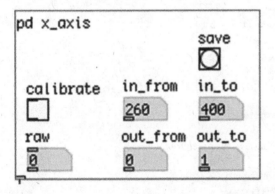

Figure 6-7. *Figure 6-5 subpatch appearance on parent patch*

If you compare the two figures you'll see that all objects inside the red rectangle in Figure 6-5 appear inside the rectangle in the parent patch shown in Figure 6-7, and all other objects are excluded. Notice also that the subpatch has an outlet, which you can see in Figure 6-7. Graph-On-Parent helps us create simple, yet functional interfaces in Pd, which make our patches more user-friendly.

244

Setting the Properties of the GOP GUIs

Go ahead and build the patch in Figure 6-5 inside your x_axis subpatch. Take care to distinguish the messages from the objects (everything that has a value, apart from the number atoms, is a message). Once you've done that, you'll need to change the properties of all the GUIs inside the rectangle (change the **Size** values in Figure 6-6 so that all the GUIs fit in the rectangle. If you've close the Properties window, right-click an empty part of the subpatch).

First, right-click the toggle and select **Properties**. In the **Label** field, type **calibrate**, and set the **X offset** to 0, and the **Y offset** to –7. Click **Apply** and **OK** (or simply hit **Return**). Then select the properties of the bang. Label it as **save** and set the same X and Y offsets for the label. Before you close it, in the **Send symbol:** field type **save_x**. This is like connecting the bang to [s save_x], so any [r save_x] will receive that bang. Click **Apply** and **OK** so that the new properties will take effect. Mind that the **Send symbol:** or **Receive symbol:** of bangs, number atoms, sliders, and so forth, are nice features, but also makes a patch not so obvious when trying to understand what's happening, since the symbols they use for sending and receiving data are hidden. I suggest using these features with care.

Now open the properties of the number atom below the toggle and in the **Label** field, type **raw**, and click **Top** to display the label on the top part of the number atom. As before, click **Apply** and **OK** (this goes for all Properties windows). Now open the Properties window of the top-left number atom, from the group of four number atoms, labeled **in_from** in Figure 6-5. In the Label field, type **in_from** and select **Top**. In the Receive symbol: field, type **x_in_from_set**. Open the Properties window of the number atom next to it, labeled **in_to** in Figure 6-7. Label it as **in_to**, set the label on **Top**, and set the **Receive symbol:** to **x_in_to_set**. The two number atoms below should take the same labels and receive symbols, only change the word "in" to "out", so the labels are **out_from** and **out_to**, and the receive symbols are **x_out_from_set** and **x_out_to_set**. These labels represent the range of a value going "from" some value "to" some other value.

What We Have Achieved So Far in this Project's Pd Patch

This procedure may seem a bit complicated for what this subpatch is supposed to do, but once you have the interface built, you'll see that it will be very helpful. The subpatch in Figure 6-5 takes the value of the X axis of the accelerometer and maps it from the range of 260-400 to the range of 0-1. The four number atoms we have just labeled are there to correct the incoming and outgoing range of the sensor values received. Since the values of the sensor are not exactly from 260 to exactly 400, we can use the top two number atoms to correct this range. The values of these number atoms are being stored in [f]s and when you click the bang, the values of [f]s will go to the messages [set $1(and from there they will be stored in the messages containing the values of [map]. Using "set" will store the value to the message but won't output it. If you save the patch, when you open it again, the values stored in the messages will be banged with [loadbang] and will go to the inlets of [map], and this way the calibrated mapping of the sensor range will be applied on load. Describing this procedure is not the same as seeing it in action. In the parent patch, duplicate the pd x_axis subpatch, rename it as **pd y_axis** (just click the subpatch once and its name will be editable), and change all *x*'s to *y*'s in all [send]s and [receive]s, plus in the bang's Properties window, and the four number atoms. Make sure that you don't miss any send/receive name because you'll get strange behavior that might be difficult to debug. Duplicate the subpatch once more and change its name to **pd z_axis** and also all the send/receive names accordingly. Connect the outlet of each subpatch to a number atom and put a [comport] connected to [serial_print_extended] in the parent patch. Figure 6-8 illustrates what your parent patch should look like.

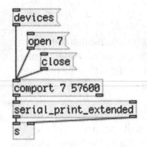

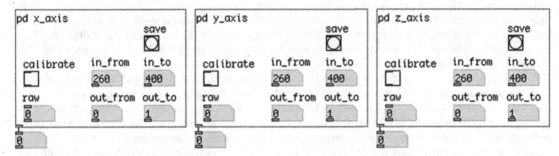

Figure 6-8. Parent patch of the x_axis, y_axis, and z_axis subpatches

If you've uploaded the Arduino sketch to your board, you should be able to use this patch immediately. Activating an axis of the accelerometer will display its values to the corresponding number atom connected to the subpatches. Click the **calibrate** toggle of the axis you've activated and the number atom labeled **raw** will display the same values. If these values go below 0, or above 1, the number atom connected to the subpatch will display either a 0 or a 1, whereas the "raw" number atom will display the values properly. This is because the subpatches use [clip_abs 0 1] (check Figure 6-5, which is an abstraction from the "miscellaneous_abstractions" GitHub page), right before [outlet]. The reason why [clip_abs] is preferred to [clip] is explained later.

The "raw" number atom helps us calibrate the mapping of these values. If the values of the sensor axis go below 0, raise the in_from value a bit, until it goes down to 0, and not lower (it might go down to –0.005, for example, which is good). If the value goes above 1, lower the in_to value until the sensor value goes up to 1, and not higher (again, something like 1.005 is good). This should make the sensor output much more stable. Do this for all three axes, until the sensor readings are stable within the 0-1 range. Then click all three **save** bangs and save your patch. Now if you close the patch and re-open it, the in_from and in_to number atoms should display the values you saved, and the sensor readings should be stable on load. For example, in my patch these values are 268, 403 for the X axis, 263, 398 for the Y axis, and 272, 403 for the Z axis, and these values have been saved with the patch, so every time I open the patch I get the same values. Using messages is a simple way to save the state of a patch, since number atoms and [f] will always be initialized to 0 when Pd launches.

Using the Canvas to Create Visual Feedback

There is one more piece of information the Arduino code is sending in the serial line, which we haven't used yet in our patch. This is the activity value of each axis, sent with the tags axis0, axis1, and axis2. It's very helpful to have some indication of the activity of each axis of the sensor when it comes to live performance. Since this project is not aimed at using a headless embedded computer, we can use the computer screen to provide the necessary visual feedback. The values sent along with the axis0 tags and so forth are either 0 or 1, depending on the activity of each axis. We'll use these values to create some sort of simple interface that will constantly inform us about the activity of each axis.

In the parent patch, put a *Canvas* (either from the **Put** menu, or with **Shift+Ctrl/Cmd+C**). A gray rectangle with a blue square on its top-left part will appear on the patch. This is a surface that we can edit in various ways to highlight certain parts of a patch, or to display a message that can change. Using a canvas is rather helpful because we can choose the font size (or even the font style), and we can even change the background or text color, so we can display messages in various ways. If you open the help patch of the canvas you'll see a [pd edit] subpatch on the top-right part. In there, you'll see various ways to edit a canvas, from which we're going to use a few. Right-click the blue square on the canvas and set its **Receive symbol:** to **x_canvas**. Place another two canvases and set their receive symbols to **y_canvas** and **z_canvas**. Create a new subpatch and call it **canvas_control** (or whatever else you like).

Figure 6-9 illustrates the contents in this subpatch. We're using four messages for each canvas, depending on the value received by [r axis0] and so forth. If the value is 0, we're setting the background color of the canvas to 10 (which is gray) and the font color to 12, which is black, using the "color" message. We're setting the size of the canvas to 130 and 30, for the X and Y dimensions, in pixels, using the "vis_size" message. Then we're setting the label position to 5 and 14 (this is pixels starting from the top-left corner) using the "label_pos" message, and finally we set the label to **X_axis_inactive**, using the "label" message. If the value received by [r axis0] is 1, we change the background color to red to indicate that the axis is active. We change the size to fit the label and we change the label to **X_axis_active**. The same applies to the other two canvases, only the [receive] name changes to [r axis1] and [r axis2], and the "label" messages change to Y_axis_inactive, Y_axis_active, Z_axis_inactive, and Z_axis_active. You can't include white spaces in the label. If you use white spaces instead of underscores in the properties of canvas, the message will be displayed with underscores anyway. If you use white spaces in messages, everything up to the white space will be executed and the rest will be discarded.

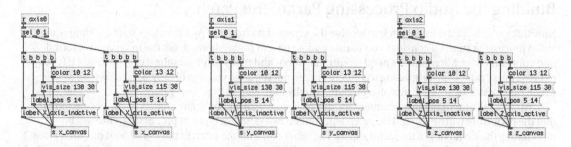

Figure 6-9. *Contents of the [pd canvas_control] subpatch*

Now whenever you activate an axis of the sensor, the corresponding canvas should change its color and message, making it clear to the performer that the axis is active. Figure 6-10 illustrates what you patch should look like with all of these additions. In this figure the Y axis of the sensor is active and you can see its output value displayed in the number atom connected to the pd y_axis subpatch. Notice the in_min and in_max values of all three subpatches in the figure, which are neither 260 nor 400. These are the calibrated values I used with my sensor, which are close to the initial values but not the same. After calibrating the sensor, I clicked all three save bangs and saved the patch. Now every time I open it, it launches with these values so I don't have to calibrate it again.

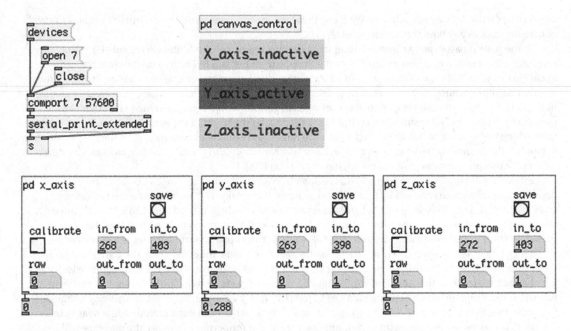

Figure 6-10. Parent patch with canvases indicating the activity of the sensor axes

Building the Audio Processing Part of the Patch

Now that we've built the basic patch to receive the sensor data from the Arduino, let's start building some audio processing stuff. I'll consider you're using some type of microphone to get the instrument's sound into your computer. Microphones need preamplification, and many external sound cards provide it (usually referred to as "gain"). If you're using a condenser type microphone you'll probably need *phantom power*, which many external sound cards also provide, usually indicated as "48V" or "phantom".

We'll use the input of the instrument for this project, which will be manipulated by the accelerometer that the performer will be using. The patch will contain two different types of processing. One will be modulating the frequency of the input signal. The other will shift the pitch of the signal, and output it along with the original signal, which creates a nice effect called beat frequency (when the frequency difference is rather small). We're going to use delay lines for both types of signal processing.

Pitch Shift Subpatch

Figure 6-11 illustrates the pitch shift part. Create a subpatch and name it **pitch_shift**, and put the objects in Figure 6-11 in it. What we're doing here is write the input signal to a short delay line (100 milliseconds length) and read it with an increasing or decreasing delay time, from 0 to 100 milliseconds. This is achieved by providing [phasor~] with either a positive or a negative frequency. If we provide a negative frequency to [phasor~], it will output its waveform inverted, so it will go from 1 to 0. Multiplying [phasor~]'s output by 100, will give a ramp either from 0 to 100 or from 100 to 0, depending on whether its frequency is positive or negative. On the right part of the patch in Figure 6-11 we're adding 0.5 to [phasor~]'s output and send that to [wrap~]. This gives an offset of half a period to [phasor~], since when it outputs 0, adding 0.5 will yield 0.5. When [phasor~] outputs 1, adding 0.5 will yield 1.5, but [wrap~] will wrap this value around 0, so it will again be 0.5, so the result of this is a [phasor~] with half a period offset. Below [wrap~] we have the same objects as the ones below [phasor~].

248

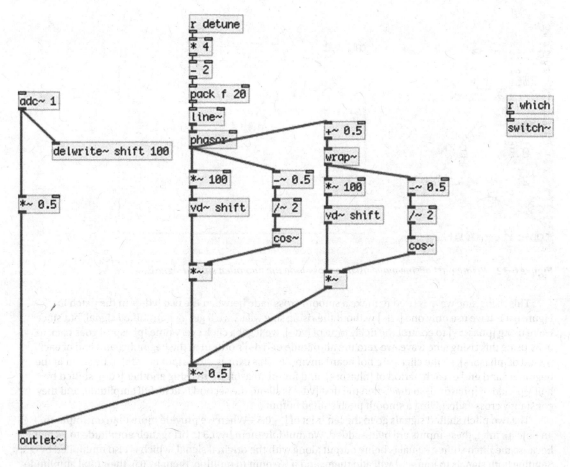

Figure 6-11. *Contents of the pitch_shift subpatch*

The part where we're subtracting 0.5 from [phasor~]'s output, then dividing by 2, and then feeding it to [cos~], produces half a sine wave starting from 0 and rising, which you can see in Figure 6-12. Try to understand how it works. You don't need to build the patch shown in Figure 6-12; it's there only to illustrate the specific part of the patch in Figure 6-11. The only thing you need to know about Figure 6-12 is that we're dividing the sampling rate by half the size of the array, so we can store two full periods of the waveform. We use [samplerate~], which gives the sampling rate we're currently using, when it's banged, and we're dividing it by 256, since the size of array1 is 512. When we bang [tabwrite~] we first send a 0 to the right inlet of [phasor~] to set its phase to the beginning, so the stored waveform will start at 0 (again, this is only for displaying the output of this part of the patch in Figure 6-11).

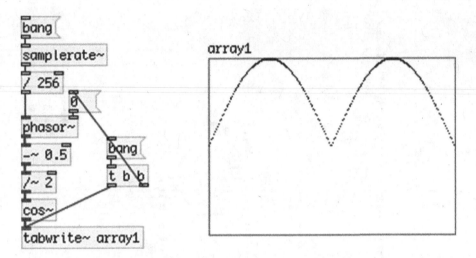

Figure 6-12. *Rising part of sine wave used to cross-fade the two pitch shifted signals*

This rising sine wave is used to make a smooth cross-fade between the two [vd~]s in the patch in Figure 6-11. If we use only one [vd~] without the rising sine wave, we'll get a pitch shifted signal, but since we're using [phasor~] to control the delay time of [vd~], we'll get a click every time [phasor~] goes back to 0. By using this rising sine wave, we zero the amplitude of [vd~]'s output at the beginning and end of each period of [phasor~], so the clicks are not heard anymore. This causes the output of [vd~] to be silent at the beginning and end of each period of [phasor~], and for this reason we're using another [vd~], shifted by half a period of [phasor~], so that when the first [vd~] is silent, the second is at its full amplitude, and they constantly cross-fade, giving a smooth pitch shifted output.

The two pitch shifted signals go to the left inlet of [*~ 0.5]. When we provide more than one input to an object's inlet, these inputs are being added. We multiply them by 0.5 to bring their amplitude to its half, because the pitch shifted signal is being output along with the original signal, which is also multiplied by 0.5. Sending both signals to [outlet~] will add them, and if we omit to multiply them by 0.5, their total amplitude will be 2, which is not good. By multiplying them by 0.5, we make their total amplitude go up to 1, which is what we want.

On the top part of the patch in Figure 6-11 we see a [r detune]. This object will receive values from a [s detune] in the parent patch, which will come out from one of the three axis Graph-On-Parent subpatches. This value will be used also in the frequency modulation subpatch, which we'll build afterward. For this reason, the value received from the Arduino is mapped to a generic range from 0 to 1 and then remapped in each subpatch separately. Here we want to have a range from –2 to 2. When we provide a 0 frequency to [phasor~], there will be no pitch shifting. When we provide a positive frequency, the signal's pitch will be shifted downward, and when we provide a negative frequency, it will be shifted upward. A range from –2 to 2 should be sufficient for this. You can fix this to a range that suits you most.

Finally, there's a [r which] going to [switch~] on the right side of the patch in Figure 6-11. [r which] will get a 1 or a 0 from the Arduino (we'll need to add one more push button for this), and we'll use it to control whether we want to use the pitch shift or the frequency modulation part of the patch. [switch~] turns the DSP of a subpatch (and all subpatches in this subpatch) on or off. It has some other features too, but we're using it here for this feature only.

Frequency Modulation Subpatch

Now let's build the frequency modulation subpatch, which is shown in Figure 6-13. There are a few new things in this subpatch. First of all, at the top we can see [sigmund~]. This object is used for sinusoidal analysis and pitch tracking. We're using it here for pitch tracking. With no arguments provided, we'll get the pitch of the input signal out the left outlet, and its amplitude out the right outlet (which we're not using in this project). We're connecting [sigmund~] to [moses 0], because it can output some negative values when you don't play. I'm not really sure why this is; probably the preamplification of the sound card is producing some artifacts. Using [moses 0], we get all values from 0 and above out of the right outlet, excluding all negative values. [sigmund~] outputs the pitch estimation of the input signal in MIDI note numbers, so we're using [mtof~] to convert them to frequency values. We don't need the [mtof_tune] abstraction here, because if your instrument is tuned at frequency other than 440 Hz, using [sigmund~] and [mtof] will yield the correct frequency. If your instrument is tuned at 442 Hz, and you use [mtof_tune], you'll get 444 Hz when you play the tuning A. The [r detune], [r index], and [r feedback] objects receive their values from the Graph-On-Parent subpatches. [r detune] will get values from 0 to 1, which are scaled to go from 0.5 to 2. Multiplying the output of [mtof] by 1, will yield the same frequency. Multiplying it by 0.5, yields half the frequency (one octave down), and by 2 yields double the frequency (one octave up). Multiplying it by values like 1.05, 0.55, or 1.95 will yield a slightly shifted frequency, which can give some nice results when this frequency is used to modulate the original signal. We're using [t b f] to get output from [*] whenever we provide a value with [r detune], even if the hot inlet of [*] doesn't receive input from [mtof].

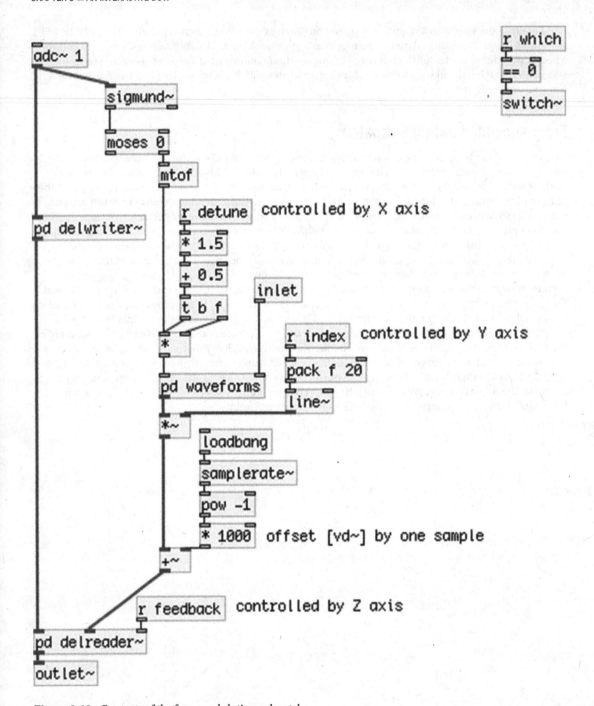

Figure 6-13. *Contents of the freq_modulation subpatch*

We're modulating the input signal with an oscillator, the waveform of which we can choose between two waveforms in the [pd waveforms] subpatch, which is shown in Figure 6-14. The two waveforms are a sine wave and a triangle, which we can choose by providing either a 0 or a 1 (which comes from [inlet], but is 0 by default, using [loadbang]), where with a 0 we'll choose the sine wave and with a 1 the triangle. [phasor~], which is common for both oscillators, takes in a signal to control its frequency, using [line~] to smooth out sudden changes that come both from the Arduino and the pitch estimation from [sigmund~]. Both waveforms go from 0 to 1, instead of –1 to 1. This is because we can't have a negative delay time with [vd~] (which is in the [pd delreader~] subpatch). [r index] controls the total amount of delay we'll use, by multiplying the output of the modulator oscillator with a value provided by one of the axes of the accelerometer.

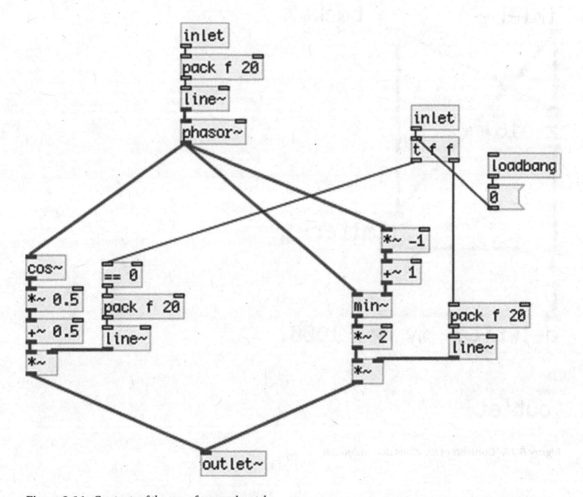

Figure 6-14. *Contents of the waveforms subpatch*

The part of the patch in Figure 6-13 with the comment "offset [vd~] by one sample" yields the time one sample takes, in milliseconds. Raising the sampling rate to the –1 power, will give the inverse of the sampling rate, since the result is like dividing 1 by the sampling rate. Multiplying this by 1000, gives the time in milliseconds between two consecutive samples. This is a tiny delay offset we give to [vd~], since the shortest delay it can provide is a one sample delay.

One thing that is different in the way we're using delay here is that we've placed both [delwrite~] and [vd~] into subpatches. Figures 6-15 and 6-16 illustrate these. If we don't put these two objects in subpatches, we won't be able to get delays shorter than one sample block. Pd runs with a 64 sample block by default, so if you don't place these objects in subpatches, the shortest delay that you'll be able to get is approximately 1.45 milliseconds (with a 44,100 sampling rate). To get shorter delays, writing to the delay line must be sorted before reading from it. Putting [delwrite~] and [vd~] in subpatches, we can force this order, by using a dummy [outlet~] in the subpatch of [delwrite~], and a dummy [inlet~] in the subpatch of [vd~], and by connecting the two subpatches.

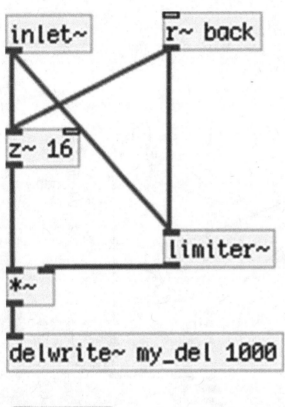

Figure 6-15. Contents of the delwriter~ subpatch

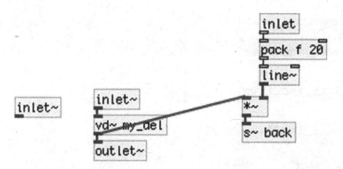

Figure 6-16. *Contents of the delreader~ subpatch*

DSP sorting in Pd follows the connection chain between objects, so connecting the two subpatches forces the order of the writing to and the reading from the delay line. This way we can get very short delays, down to the milliseconds between two consecutive samples, which with a sampling rate of 44,100, is approximately 0.0227 milliseconds. Since this is the shortest delay [vd~] can give, we provide this offset to the modulator oscillator, which is explained in the previous paragraph.

In the delreader~ subpatch we're sending the delayed signal to [s~ back], which goes back to the delwriter~ subpatch. We're controlling the amplitude of this delayed signal, so we can get a controlled feedback to the delay line. Since the original and the delayed signal will be added when they go into [delwrite~], we're using [limiter~] to make sure their total amplitude won't go above 1. When using [limiter~], we must delay the signal to be limited by a few samples, otherwise [limiter~] won't have enough time to apply the limiting envelope. [z~] is an object that delays a signal but takes its delay time in samples rather than milliseconds. Here we're delaying the signal by 16 samples, which should be enough for [limiter~] to apply the necessary amplitude changes. We've used this object the same way in Chapter 5.

Lastly, on the top-right part of Figure 6-13, there's a [r which] connected to [== 0] and then to [switch~]. Like the pitch_shift subpatch, [r which] receives either a 1 or a 0. Since this value controls the [switch~] of [pd pitch_shift] directly, we're inverting its value with [== 0] here, so that when [pd pitch_shift] is chosen, this subpatch will turn its DSP off, and when [pd pitch_shift] is deselected, this subpatch will turn its DSP on.

The Finished Patch

Figure 6-17 illustrates the finished parent patch. We've added a Vradio with two buttons to control the type of the modulator oscillator (which by default is a sine wave). We could have added another push button to our circuit for this, but when building such interfaces, you should hold back a bit, because if we start putting buttons for every single thing we might want to control, then our circuit would become a mess. Bearing in mind that the sensor circuit is supposed to be held in one hand, while holding a bow at the same time, we should be careful as to what decisions we take about the circuit. Controlling the waveform of the modulator doesn't seem to be of a very high priority to me, so I preferred to control it with a GUI instead.

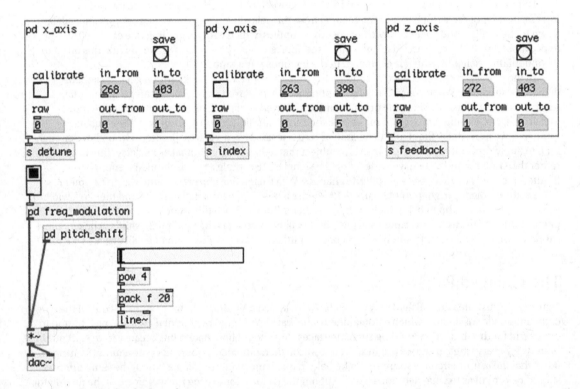

Figure 6-17. *The finished patch*

Another thing we've added is one more canvas to indicate what type of signal processing we're applying, whether it's frequency modulation of pitch shifting. Figure 6-18 illustrates the addition in [pd canvas_control] to control this canvas. As you can imagine, you should set the receive symbol of the canvas to **process_canvas**, in its properties. We've also added [pow 4] to the amplitude control to make it sound more natural by making the control range an exponential curve, like we did in Chapter 4.

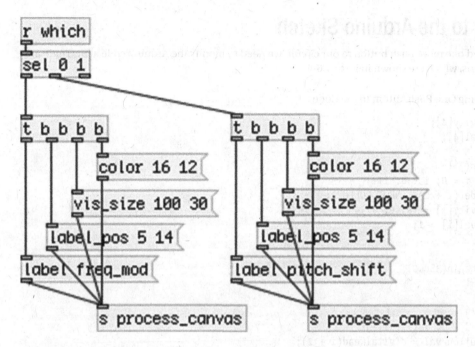

Figure 6-18. Addition to the canvas_control subpatch

In Figure 6-17, you can see the mapping values I've set for each axis of the accelerometer. The X axis is mapped to a range from 0 to 1, because it will control both the detune of the frequency modulation and the pitch shift. Each subpatch maps this range to its own one. The Y axis is mapped to a range from 0 to 5, and the Z axis to a range of 1 to 0. I've inverted the values of Z because I wanted to get 0 when the sensor is upright, and a 1 when it's flipped, and this is where the [clip_abs] abstraction used in Graph-On-Parent subpatches, came in handy. If we used [clip], in the case of the Z axis, we would have a problem, because with [clip] the first argument must always be smaller than the second one, and the same goes for the values overriding the arguments. In the Z axis, where the "from" value is 1 and the "to" is 0, we would have to set these values to [clip] manually to make sure we don't get unexpected behavior, and this would make the whole patch a bit not so functional. Using [clip_abs] helps make your interface more user-friendly and kind of plug-and-play. Check its help patch to see how this issue arises and how it is solved.

You'll have to experiment with which axis will control which aspect of the patch, and the value rages. Since the Y and Z axes are controlling one aspect each, I could map their range inside the [pd y_axis] and [pd z_axis] straight. Only the X axis is controlling two things, therefore I mapped it to the generic 0 to 1 range. Obviously, you're free to map them any way you like.

Additions to the Arduino Sketch

Since we've added one more push button to our circuit, we need to modify the Arduino code a bit. There are just a few additions, which are shown in Listing 6-4.

Listing 6-4. Adding One Push button to the Code

```
1.   bool activity[4];
2.   int old_val[4];
3.
4.   void setup() {
5.     for(int i = 0; i < 4; i++){
6.       pinMode((i + 2), INPUT_PULLUP);
7.         activity[i] = false;
8.         old_val[i] = 1;
9.     }
10.
11.    Serial.begin(38400);
12.  }
13.
14.  void loop() {
15.    for(int i = 0; i < 3; i++){
16.      int button_val = digitalRead(i + 2);
17.      if(button_val != old_val[i]){
18.        if(!button_val){ // if button is pressed
19.          activity[i] = !activity[i];
20.          // show on patch the on/off state of the axis
21.          Serial.print("axis");
22.          Serial.print(i);
23.          Serial.print("\t");
24.          Serial.println(activity[i]);
25.        }
26.        old_val[i] = button_val; // update old value
27.      }
28.    }
29.
30.    int button_val = digitalRead(5);
31.    if(button_val != old_val[3]){
32.      if(!button_val){
33.        activity[3] = !activity[3];
34.        Serial.print("which ");
35.        Serial.println(activity[3]);
36.      }
37.      old_val[3] = button_val;
38.    }
39.
40.    // call axes according to the button presses
41.    for(int i = 0; i < 3; i++) if(activity[i]) axes(i);
42.  }
43.
```

```
44.  void axes(int which){
45.    int value;
46.    switch(which){
47.      case 0:
48.        value = analogRead(which);
49.        Serial.print("x ");
50.        Serial.println(value);
51.        break;
52.
53.      case 1:
54.        value = analogRead(which);
55.        Serial.print("y ");
56.        Serial.println(value);
57.        break;
58.
59.      case 2:
60.        value = analogRead(which);
61.        Serial.print("z ");
62.        Serial.println(value);
63.        break;
64.    }
65.  }
```

In lines 1 and 2, we've increased the size of the arrays to 4. In line 5, we've increased the number of iterations of the for loop to 4 to include the new push button. In line 11, we've changed the baud rate of the serial communication to 38,400 because the 57,600 baud rate was very sluggish with the analog pins, while the 38,400 baud rate worked fine. Apparently this is because of the XBees, since the Arduino Pro Mini 5V/16 Hz works fine with the USB-to-serial board (which I'll explain later) with a 57,600 baud. If you choose the 38,400 baud rate, make sure that you set it in [comport] in the Pd patch as well; otherwise, you'll get weird behavior or the serial communication won't work at all.

In line 15, the for loop holds its 3 iterations, since it reads the push buttons that activate or deactivate the axes of the accelerometer, which are 3. Lines 30 to 37 read the new push button, which is on digital pin 5, hence the digitalRead(5) function call in line 30. The value of the push button is compared to its old value, the same way it's done with the other three push buttons. That's why the old_val array is increased by one element. Using the same technique as with the other push buttons, we detect if the push button is pressed, and if it is, we're swapping its activity value. That's why the activity array was increased by one element. When this is done, we print the new value of activity[3] with the "which" tag (again, the white space is used as a delimiter between the tag and the value). Notice that we've hard-coded the element number of both arrays old_val, and activity. Since this is a very specific project and there's no other process depending on the reading of the pin of the new push button, we can just as well use hard-coded numbers to indicate which element of the array we're accessing and which digital pin we're reading, instead of variables. The value we're using for the arrays is 3, as arrays start counting from 0, so the fourth element is 3. The rest of the code remains as is. The code in Listing 6-4 should be completely functional with the finished patch in this project.

Finalizing the Circuit

Now let's look at the new circuit, this time on a perforated board instead of a breadboard. It is illustrated in Figure 6-19. Since this is a rather simple circuit, luckily we won't need to cut the copper at any point of the board. In Figure 6-19 the board is shown with its solder lines on top. These are actually on the bottom side of the board, but Figure 6-19 illustrates them this way to make the circuit a bit easier to read. I've grouped the three push buttons that activate/deactivate the axes of the accelerometer together, and the push button that controls whether we're applying frequency modulation or pitch shift is a bit isolated, so it's easy for the performer to tell which push button is which. Take good care to place each component on a free line, so that you don't short any connections. If you take a close look, you'll see that the isolated push button is not aligned with the leftmost button of the group of three, so that they don't short, and we won't have to cut the copper of the board.

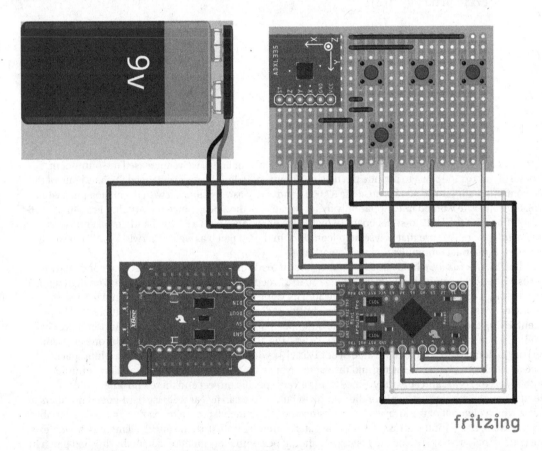

Figure 6-19. *Accelerometer circuit on a perforated board*

The Arduino (a Pro Mini in this case) gives ground to one pin of the perforated board, and all components get the ground from there, much like the way we do it with a breadboard, only now there's no clear indication that the specific strip is the ground. I have built this circuit on a perforated board with solder points instead of lines, because I had one lying around. Since the circuit is rather simple, it's not difficult to build it on a board with points as well. I already mentioned that both boards have their advantages and disadvantages, and it's up to you as to what kind of circuit board you choose. Figure 6-19 illustrates the circuit in this project on a perforated board with solder lines because it's easier to show the connections. Apart from that there's no real preference to one of the two kinds of boards.

The only component that needs voltage is the accelerometer (the push buttons use the internal pull-up resistor, so they're connected to 5V). The ADXL335 sensor is powered with 3.3V, so the Arduino Pro Mini can't provide the correct voltage for it. If used with an XBee, and an XBee Explorer, we can use the onboard 3.3V pin of the Explorer to power up the accelerometer (in Figure 6-19 the XBee is partly visible, in order for the board to be visible. On the actual circuit, an XBee is mounted on top of the XBee Explorer. The 3.3V pin is labeled, as all pins on the Explorer board). Mind that the Arduino connects to the XBee Explorer with angled headers and not jumper wires. It makes the connection more stable, Figure 6-20 shows it. The whole circuit is powered by a 9V battery, where the positive pole connects to the RAW pin of the Arduino, and the negative pole connects to ground. Then the Arduino provides power for the XBee, which provides power for the accelerometer. All components share the same ground (that's a rule in electronics, even if you have different voltages, the ground must be common for all components of the circuit).

Figure 6-20. The Arduino Pro Mini and the XBee mounted on the XBee Explorer on one side of the bow

The concept of this circuit is to be held with the same hand that holds the bow. The battery with the Arduino and the XBee can go on one side of the bow, and the perforated board can go on the other side. Make sure that the wires you use are long enough so that you're able to split the circuit in these two parts, but not very long, so that they hang from the performer's hand. Figures 6-20 and 6-21 illustrate the positioning of the Arduino and the sensor circuit. Figures 6-22 and 6-23 illustrate the bow held. I used some white glue and a rubber band to hold all the components of the circuit in place. Placing the circuit on the outer side of the bow, facilitates key presses. With a little bit of practicing, it should become rather easy to use the circuit.

To make this circuit work you need another XBee with a USB Explorer, connected to your computer. Make sure the XBees are properly configured, and they use the same baud rate with the Arduino (and [comport], of course). The two XBees must be able to talk to each other so that you can get the accelerometer data in Pd. Configuring the XBee was discussed in Chapter 3, so if you don't remember how to do it, go back to review that chapter.

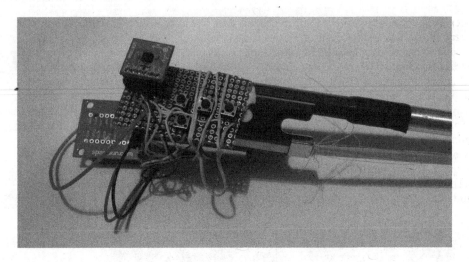

Figure 6-21. *The accelerometer circuit on the other side of the bow*

Figure 6-22. *The Arduino side of the bow held*

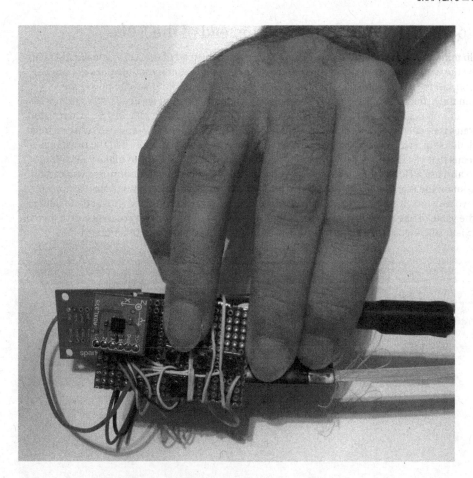

Figure 6-23. *The sensor side of the bow held*

The Arduino Pro Mini doesn't have a USB-to-serial converter, which is necessary to upload code to it from the Arduino IDE. The Arduino Uno has one on board, that's why we can interact with it from the computer right away. You need such a converter to upload code to the Pro Midi. There are a few different ones, so make sure the one you get has its pins broken out on its board in such a way that it will work with the Pro Mini. Consult your local electronics store as to which one to get.

The Arduino Pro Mini used in this project is 5V/16 MHz. There's also an Arduino Pro Mini 3.3V/8 MHz. I haven't used it so I don't really know if you can wire it with the XBee the way it is wired in the circuit in Figure 6-19, because the XBee Explorer pin connected to the VCC pin of the Arduino is a 5V pin, and 3.3V won't be sufficient. Also, I'm not sure if the 3.3V/8 MHz Arduino Pro Mini can handle the 38,400 baud rate with the analog pins, even with a USB-to-serial board. You'll have to test for yourself. An advantage of the 3.3V/8 MHz Arduino Pro Mini is that you can provide power from it for the accelerometer, since the accelerometer also works at 3.3V. Since the XBee Explorer provides this voltage from its breakout pins, this is not really a problem, when used with the 5V/16 MHz Arduino Pro Mini.

Using Bluetooth and a Voltage Divider Instead of the XBee

You should bear in mind that if you intend to use a Bluetooth module instead of the XBee, you should take care not to provide the wrong voltage to the accelerometer. The HC-06 Bluetooth module does not provide a voltage pin, so you can't use it to power up the accelerometer. In this case, you'll have to power the accelerometer from the Arduino. But I've already mentioned that the accelerometer needs 3.3V and not 5V. To solve this problem, you need to make a voltage divider. This is pretty simple—all it takes are two resistors. Figure 6-24 illustrates the schematic of a voltage divider that takes 5V and gives 3.3V. The two resistors must have the value relationship as in Figure 6-24. As shown in this figure, you should apply half the resistance of the other. This means that you don't necessarily need to use a 10K and a 20K resistor to make your voltage divider; you could also use a 1K and a 2K resistor, and you'd get the same results. I'm not sure if you can find a 2K or a 20K resistor on the market, but it's possible to build such a voltage divider with three resistors of the same value. Two resistors must be connected in series; this way, they will apply the sum of their values, which is double the value of one resistor. If you have three 10K resistors, then you can connect two of them in series and they'll yield 20K. Figure 6-25 illustrates how to build a voltage divider in a breadboard view.

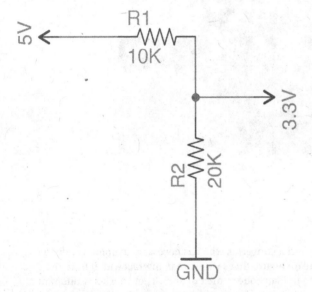

Figure 6-24. *Voltage divider schematic with two resistors*

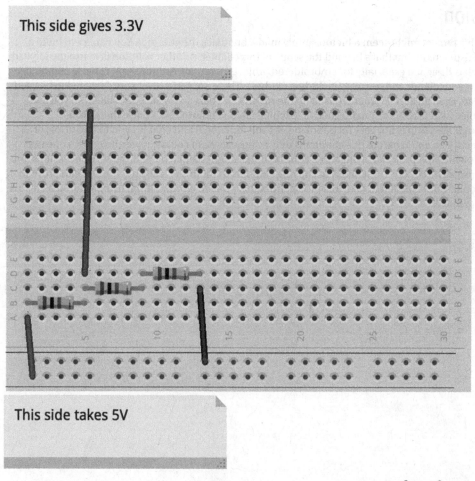

This side gives 3.3V

This side takes 5V

fritzing

Figure 6-25. *Voltage divider circuit with three resistors*

Another drawback of the HC-06 Bluetooth module is that its VCC and GND pins are inverted compared to those on the Arduino Pro Mini, so you can't really use headers like we did with the XBee; but you'll need to use wires, at least for these two pins. You could connect the RXD and TXD pins with headers on the Arduino since they are aligned (they are also inverted, but that's what we want—the RX1 pin of the Arduino to connect to the TXD of the HC-06, and the TX0 of the Arduino to connect to the RXD of the HC-06). But you'll need to use wires for power and ground; otherwise, it won't work and you might even damage your device.

Finally, once you've built the circuit, you might want to change what each axis of the accelerometer controls in the Pd patch. You may find it more appropriate that the X axis controls the feedback of the frequency modulation, and the Y axis controls the detune, for example. It's up to you (or the performer who you are building it for) how this circuit will behave; my input in this project are only suggestions.

Conclusion

The setup of this project might seem a bit too simply made, but that's the goal of it. You can try to build a case for the sensor circuit, but that's beyond the scope of this chapter. Another solution is to use the LilyPad Arduino, which is designed especially for embroidered projects. With this Arduino, you can use conductive thread and sew your circuit onto textiles (including clothes). There are various sensors designed to be used with the LilyPad, even an accelerometer based on the ADXL335. There's also an XBee shield, so you can make your project wireless. Since I'm no good at teaching anyone how to sew, I leave this to you.

Even though the setup is not very nice-looking, it's still functional and enables you to create lots of different projects. The audio part of this chapter is only a suggestion, of course; nevertheless, it's one that can be used as a finished project. Having three control inputs might seem a bit poor on some occasions, but being able to use all three controls without occupying your hands for this is really freeing and can prove to be very inspirational. You're invited to take this project from this point to another level, using your own creative thinking and practice.

■■■

## CHAPTER 7

An Interactive Drum Set

In this chapter, we're going to build an interactive drum set. We're not going to build the actual drum set, of course, but we're going to enhance a drum set by placing a few sensors on some of the drums, which will give us input on each drum hit. We'll use that input to trigger some samples in various ways. We're also going to use a few foot switches to change between various types of sample playback, but also to activate/deactivate each sensor separately.

The sensors we're going to use are simple and inexpensive piezo elements, the ones used for contact microphones. We're going to build four of these in such a way that it's easy to carry around and set up, so you can use this setup in gigs you're playing. This time, we'll use the Arduino Uno since this project is not going to be embedded anywhere, but we'll be using it with our personal computer. To make our lives easier, we'll use the Arduino Proto Shield, a board that mounts on top of the Arduino Uno with solder points that give easy access to all Arduino pins.

Parts List

Table 7-1 lists the components that we'll need to realize this project.

Table 7-1. Parts List

Arduino	Piezo elements	Foot switches	Connection Jacks	LEDs	Resistors	Proto Shield
1 (Uno or other type)	4 (with or without enclosure)	4	4 × 1/4-inch female (mono or stereo) 4 × 1/8-inch female (mono or stereo) 4 × 1/8-inch male	4	4 × 220Ω 4 × 1MΩ	1 (optional)

Other Things We'll Need

In this project, we'll use an abstraction I've made: [guard_points_extended]. You can find it on GitHub at https://github.com/alexdrymonitis/array_abstractions. All the rest will be explained in detail throughout the chapter.

First Approach to Detecting Drum Hits

Piezo elements can be used with the analog pins of the Arduino. They provide voltage that corresponds to the vibration it detects. Our sketch will be based on the Knock tutorial, found on the Arduino web site, but we're going to change it quite a lot to fit the needs of this project. What we essentially want is to detect drum hits. This can be easily achieved just by reading the analog pins of the Arduino and printing their values over to Pd. Since these sensors can be quite sensitive, we're going to apply a threshold value, below which nothing will be printed. Listing 7-1 shows a sketch very similar to the Knock tutorial sketch.

Listing 7-1. Basic Piezo Element Reading

```
1.   int thresh = 100;
2.   const int num_of_sensors = 4;
3.
4.   void setup() {
5.     Serial.begin(115200);
6.   }
7.
8.   void loop() {
9.     for(int i = 0; i < num_of_sensors; i++){
10.      int sensor_val = analogRead(i);
11.      if(sensor_val > thresh){
12.        Serial.print("drum"); Serial.print(i);
13.        Serial.print("\t");
14.        Serial.println(sensor_val);
15.      }
16.    }
17.  }
```

This code is rather simple and doesn't need a lot of explanation (or comments). All we do is set the number of sensors we're using (which must be wired from analog pin 0 and on, so that the for loop on line 9 will work as expected) and read through their analog pins. If the value of each pin is higher than the threshold value we've set on line 1, then we print that value with the tag "drum" followed by the number of the pin. Notice the high baud rate we're using in line 5. Since we want to detect every single hit on each drum, a high baud rate (the highest the Arduino Uno can handle) is desired here.

First Version of the Circuit

Figure 7-1 shows the circuit for the code in Listing 7-1. You don't really need to have four piezo elements; you can use as many as you have. Just make sure you set the correct number on line 1 of your code. Also, you can test the circuit in Figure 7-1 on some hard surfaces if you don't have a drum set available. Using drumsticks or something similar will work better than knocking on these surfaces with your hands.

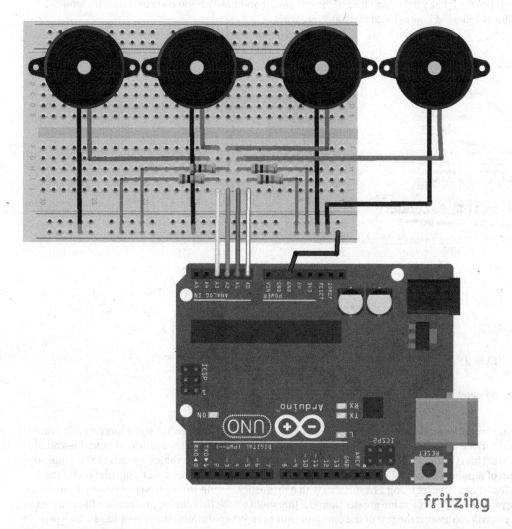

Figure 7-1. *Four piezo elements circuit*

The piezo elements shown in Figure 7-1 are enclosed in a hard case. Some electronics stores have these kinds of piezo elements, but you can also use the ones without an enclosure. Of course, the enclosure protects the sensor, which might be desired, especially during transportation (you don't want to find your sensor with its wires broken when you arrive at your gig venue for your setup and sound check). Both types of sensors—with and without the enclosure—come with wires soldered on them, so all you need to do is extend the wires to the length you want. Also, the enclosed sensors are easier to mount, since you can use duct tape straight on them without worrying that untaping them will rip off the wires. Even with the enclosure, these sensors are pretty cheap.

The resistors used in this circuit are 1MΩ resistors, which connect the positive wire of the sensor to ground. Other than that, the circuit is pretty simple and straightforward.

Read the Drum Hits in Pd

Figure 7-2 shows the Pd patch that you'll need for this circuit and the Arduino code to work with (you probably already guessed how to build this patch anyway).

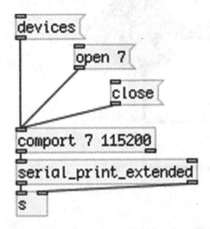

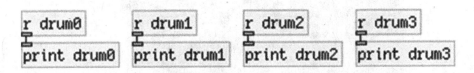

Figure 7-2. Pd patch that works with the Arduino sketch in Listing 7-1

Instead of number atoms, we're using [print] for each sensor for a reason. Try some hits on each drum (or any hard surface you're using) with a drumstick. What you'll probably get is a series of values instead of a single value. This is because the piezo element is an analog sensor, and the values it sends to the Arduino are a stream of numbers representing a continuously changing electrical current. Let's say that we hit the first drum (or whatever we're using as a surface) with a drumstick. As the for loop on line 9 goes through the analog pins, it will detect a value greater than the threshold on the first analog pin, and it will print that value. Then it will go through the rest of the pins and start over. When the loop starts over, the surface of the drum we've hit will still be vibrating, and pin 0 will still be giving values greater than the threshold, so the loop will again print these values to the serial line. Getting so many values per hit is rather messy, so it's better if you receive only one value, and preferably the highest value of the hit. We can do that in Pd or Arduino. I'm going to show this both ways. It will then be up to you what you choose.

Getting the Maximum Value in Arduino

To get the maximum value from all the values above the threshold is a rather easy task. All we need to do is check every value above the threshold if it's greater than the previous one. This is done by calling the max function and checking every new value against the previous one. Listing 7-2 shows the code.

Listing 7-2. Getting Maximum Sensor Value in Arduino

```
1.    int thresh = 100;
2.    const int num_of_sensors = 4;
3.    int sensor_max_val[num_of_sensors];
4.    int sensor_thresh[num_of_sensors];
5.    int old_sensor_thresh[num_of_sensors];
6.
7.
8.    void setup() {
9.      Serial.begin(115200);
10.   }
11.
12.   void loop() {
13.     for(int i = 0; i < num_of_sensors; i++){
14.       int sensor_val = analogRead(i);
15.       if(sensor_val > thresh){
16.         Serial.print("drum"); Serial.print(i);
17.         Serial.print("\t");
18.         Serial.println(sensor_val);
19.         sensor_thresh[i] = 1;
20.         sensor_max_val[i] = max(sensor_max_val[i], sensor_val);
21.         old_sensor_thresh[i] = sensor_thresh[i];
22.       }
23.       else{
24.         sensor_thresh[i] = 0;
25.         if(sensor_thresh[i] != old_sensor_thresh[i]){
26.           Serial.print("highest");
27.           Serial.print(i); Serial.print("\t");
28.           Serial.println(sensor_max_val[i]);
29.           sensor_max_val[i] = 0;
30.         }
31.         old_sensor_thresh[i] = sensor_thresh[i];
32.       }
33.     }
34.   }
```

Lines 3 to 5 define three new arrays. The first one, sensor_max_val, will be used to store the maximum value of each sensor, as its name suggests. The other two are used to see if we've just crossed the threshold from above to below. The loop function begins the same way it did in Listing 7-1. Again, we're printing all values above the threshold for comparison with the highest one to make sure that our code works as expected. In line 19, we set the current element of sensor_thresh to 1, denoting that we're above the threshold. In line 20, we test the value of the sensor to see if it's greater than the previous one.

We achieve this by comparing the current element of the sensor_max_val array against the value stored by the sensor, calling the max function. This function takes two values and returns the greater of the two. At the beginning, the sensor_max_val array has no values stored, so whatever value is read and stored in the sensor_val variable will be greater that the current element of the sensor_max_val array, so that value will be stored to the current element of the sensor_max_val array.

If the next time the for loop goes through the same sensor, its value is still above the threshold, if the new value is greater than the previous one, again line 20 will store the new value to the corresponding element of the sensor_max_val array, but if it's smaller, then sensor_max_val will retain its value. Even though we're assigning to the sensor_max_val array the value returned by the max function, during the test, the sensor_max_val array retains its value. Only if its value is smaller than the value of sensor_val will the sensor_max_val array change, otherwise max will assign to it its own value, since it was the greatest of the two. After we store the maximum value, we update the old_sensor_thresh array according to the corresponding value of the sensor_thresh array. We'll need these two values when we drop below the threshold.

Updating the sensor_thresh Array

If we drop below the threshold, we set the current element of the sensor_thresh array to 0, and then we test if its value is different from that of the old_sensor_thresh array, in line 25. If it's the first time we drop below the threshold after a drum hit, this test will be true, since the old_sensor_thresh array won't have been updated yet and its value will be 1. Then we print the maximum value stored and we set the current element of the sensor_max_val array to 0. If we don't set that to 0, the next time we hit the drum, the values of the sensor might not go over the last maximum value. And if sensor_max_val has retained its value, the max function in line 20 will fail to store the new maximum value to sensor_max_val, since sensor_max_val is part of the test.

After we've checked whether the current element of the sensor_thresh and the old_sensor_thresh arrays are different, we update the old_sensor_thresh array. This way, as long as we don't hit the drum, the test in line 25 will be false and we'll prevent the Arduino from printing the same maximum value in every iteration of the for loop.

Reading All Values Above the Threshold Along with the Maximum Value

Figure 7-3 illustrates the Pd patch for the code in Listing 7-3. Upload the code to your Arduino. (You'll need to send the message "close" to [comport] in Pd if you have the previous patch still open; otherwise, the Arduino's serial port will be busy and the IDE won't be able to upload code to the board.) Open its serial port in Pd. Now whenever you hit a drum, you'll get all the values that were above the threshold printed as drum0, drum1, and so forth, and the highest value of each sensor printed as highest0, highest1, and so on.

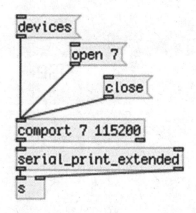

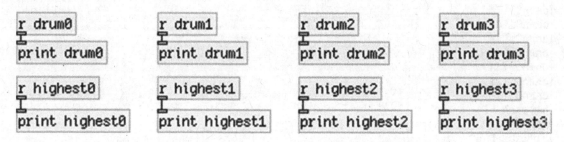

Figure 7-3. *Pd patch that works with the code in Listing 7-2, for testing the code and circuit*

What you'll probably notice is that you'll probably get more than one highest values with a single drum hit. Every time you hit a drum, the sensor values cross the threshold, and the code gives you the highest of the values, but the sensor values might drop below the threshold and then rise above it again, all within the same drum hit. Don't forget that this is an analog signal that fluctuates as long as the drum surface vibrates. Figure 7-4 illustrates this.

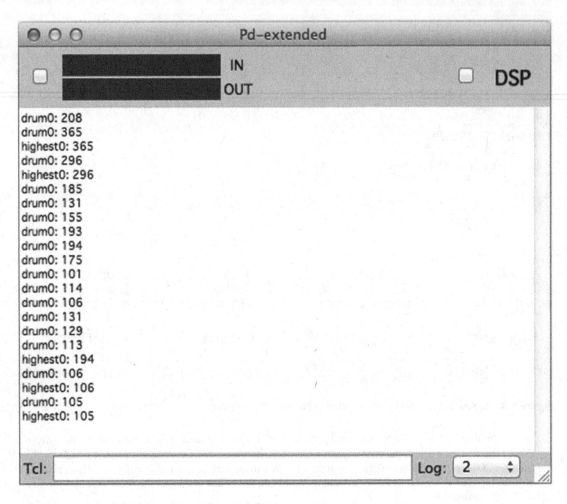

Figure 7-4. *Multiple highest values with a single hit*

Debouncing the Fluctuation Around the Threshold

This is a bit tricky to overcome. What we need to do is define a debounce time, before which the maximum value won't be printed to the serial line, but only saved. If the sensor values go above the threshold in a time shorter than the debounce time, then we'll keep on testing to get the maximum value. Only when the debounce time is over will we print that value to the serial line. Listing 7-3 shows the modified code that solves this problem.

Listing 7-3. Debouncing Sensor Readings

```
1.   // variables and constants that you might need to change
2.   int thresh = 100;
3.   const int num_of_sensors = 4;
4.   int debounce_time = 10;
5.
6.   // arrays for the sensors
7.   int sensor_thresh[num_of_sensors];
```

```
8.    int old_sensor_thresh[num_of_sensors];
9.    int sensor_below_new[num_of_sensors];
10.   int sensor_below_const[num_of_sensors];
11.   int sensor_max_val[num_of_sensors];
12.   int sensor_above_thresh[num_of_sensors];
13.
14.
15.   void setup() {
16.     Serial.begin(115200);
17.   }
18.
19.   void loop() {
20.     for(int i = 0; i < num_of_sensors; i++){
21.       int sensor_val = analogRead(i);
22.       if(sensor_val > thresh){
23.         Serial.print("drum"); Serial.print(i);
24.         Serial.print("\t");
25.         Serial.println(sensor_val);
26.         sensor_thresh[i] = 1;
27.         if(sensor_thresh[i] != old_sensor_thresh[i]){
28.           // count how many times the sensor rises above threshold
29.           sensor_above_thresh[i] += 1;
30.         }
31.         sensor_max_val[i] = max(sensor_max_val[i], sensor_val);
32.         old_sensor_thresh[i] = sensor_thresh[i];
33.       }
34.       else{
35.         sensor_thresh[i] = 0;
36.         // get a below threshold time stamp for the sensor
37.         sensor_below_const[i] = (int)millis();
38.         if(sensor_thresh[i] != old_sensor_thresh[i]){
39.           // get a time stamp every new time the sensor drops below threshold
40.           sensor_below_new[i] = (int)millis();
41.         }
42.         old_sensor_thresh[i] = sensor_thresh[i];
43.         if((sensor_below_const[i] - sensor_below_new[i]) > debounce_time){
44.           // make sure we print the values only once per hit
45.           if(sensor_above_thresh[i] > 0){
46.             Serial.print("highest");
47.             Serial.print(i); Serial.print("\t");
48.             Serial.println(sensor_max_val[i]);
49.             sensor_max_val[i] = 0;
50.             // zeroing sensor_above_thresh[i] will prevent this chunk of code
51.             // from be executed before we hit the drum again
52.             sensor_above_thresh[i] = 0;
53.           }
54.         }
55.       }
56.     }
57.   }
```

This code is a bit more complex than the code in Listing 7-2, as we need to overcome a pretty complex problem. In the beginning of the code, we define some variables and constants that we might need to change, but these are the only values; the rest of the code remains untouched. Lines 8 to 16 define some arrays for the sensors. All these arrays are necessary to detect whether the values rising above the threshold are within the same drum hit, or whether it's a new drum hit.

In the loop function, the lines 23 to 29 are identical to the beginning of the loop function in Listing 7-2. In line 27, we check whether we've risen above the threshold for the first time, by doing the reverse test we did in Listing 7-2, when we went below the threshold. If it's true, we increment the value of the corresponding element of the sensor_above_thresh array, which counts how many times we're going above the threshold when coming from below it. Then we go on and retrieve the maximum value of the sensor, like we did in Listing 7-2.

Getting Time Stamps

In line 34, the code of else starts. At the beginning, we set the current element of the sensor_thresh array to 0, like in Listing 7-2 and what we do then is take a time stamp of when the current sensor was measured to be below the threshold. We do that by calling the function millis(). This function returns the number of milliseconds since the program started (since we powered up the Arduino, or since we've uploaded the code to it). This function returns an unsigned long data type, but the sensor_below_const array is of the type int. To convert the data type that the millis function returns, we cast its returned value to an int, by calling it like this:

```
(int)millis();
```

This way the value millis returns will be converted to an int. It's probably OK not to do this casting at all, and call millis as if it returned an int, but casting its value is probably a better programming practice, and it's also safer.

Mind that the time stamp is not about when the sensor values drop from above the threshold to below it, but whenever the sensor is measured to be below the threshold, which will happen in every iteration of the for loop, if the sensor is below the threshold (meaning, for as long as the corresponding drum is not being hit).

In line 38, we're checking if the sensor values have dropped from above the threshold to below it. If it's true, we're taking a new time stamp, this time denoting the moment we dropped below the threshold. We're going to need these two time stamps for later. After taking the new time stamp, we update the old_sensor_thresh array.

In line 43, is where we're checking whether we have just dropped below the threshold (which means that we might go up again, and we don't want to print any values), or if we have passed the debounce time, so we need to print the maximum value to the serial line. We're checking if the difference between the two time stamps is greater than the debounce time we've set in line 5, which is 10 (milliseconds). Since this test will be made every time a sensor is measured below the threshold, which will happen for as long as a drum is not being hit, we need to take care that the code in this test is not being executed every single time, but only after a drum hit. This is where the sensor_above_thresh array is helpful. In line 45, we're checking whether the current sensor counter in the sensor_above_thresh array is above 0. If it is, it means that the sensor has gone above the threshold at least once, so there is some data we should print to the serial line. If the test of line 45 is true, we're printing the maximum value of all the values of the sensor, including those that went from below the threshold to above it, within the debounce time. After we print the value, we set to 0 the elements of the sensor_max_val and sensor_above_thresh arrays, so we can start from the beginning when a new drum hit occurs.

We have added only a few things to the code in Listing 7-2 and we have overcome the problem of multiple highest values with a single drum hit. All we needed was two time stamps and a debounce time. You can use the same Pd patch (the one in Figure 7-3) with this code. Compare the results with the code in

Listing 7-2 and the one in Listing 7-3. You'll see that with the latter you get only one "highest" value per hit, whereas with the former you might get more. If you still get more than one "highest" value with the code in Listing 7-3, try to raise the debounce time in line 5. Of course, we don't need the raw "drum" values; they're there only for testing. Experiment with the debounce time, try to set it to the lowest possible value, where you don't get multiple "highest" values with one hit, so that your system is as responsive as possible. Another thing to change can be the threshold value, which will determine how responsive to your playing the system will be. We'll leave that for later, as it's better if we control this from Pd instead of having to upload new code every time.

Getting the Maximum Value in Pd

Now let's see how we can achieve the same thing in Pd. We'll collect all the values printed by the Arduino, and when we're sure we're done with a drum hit, we'll get the maximum of these values. We'll take it straight from the second step, where we define a debounce time, so that we don't get multiple maximum values for one drum hit. Even though we're checking for the maximum value in Pd, we still have to start with the Arduino code. Listing 7-4 shows the code.

Listing 7-4. Getting Maximum Value in Pd

```
1.   // variables and constants that you might need to change
2.   int thresh = 100;
3.   const int num_of_sensors = 4;
4.   int debounce_time = 10;
5.
6.   // arrays for the sensors
7.   int sensor_thresh[num_of_sensors];
8.   int old_sensor_thresh[num_of_sensors];
9.   int sensor_below_new[num_of_sensors];
10.  int sensor_below_const[num_of_sensors];
11.  int sensor_above_thresh[num_of_sensors];
12.
13.  void setup() {
14.    Serial.begin(115200);
15.  }
16.
17.  void loop() {
18.    for(int i = 0; i < num_of_sensors; i++){
19.      int sensor_val = analogRead(i);
20.      if(sensor_val > thresh){
21.        Serial.print("drum"); Serial.print(i);
22.        Serial.print("\t");
23.        Serial.println(sensor_val);
24.        sensor_thresh[i] = 1;
25.        if(sensor_thresh[i] != old_sensor_thresh[i]){
26.          // count how many times the sensor rises above threshold
27.          sensor_above_thresh[i] += 1;
28.        }
29.        old_sensor_thresh[i] = sensor_thresh[i];
30.      }
```

```
31.      else{
32.        sensor_thresh[i] = 0;
33.        // get a below threshold time stamp for the sensor
34.        sensor_below_const[i] = (int)millis();
35.        if(sensor_thresh[i] != old_sensor_thresh[i]){
36.          // get a time stamp every new time the sensor drops below threshold
37.          sensor_below_new[i] = (int)millis();
38.        }
39.        old_sensor_thresh[i] = sensor_thresh[i];
40.        if((sensor_below_const[i] - sensor_below_new[i]) > debounce_time){
41.          // make sure we print the values only once per hit
42.          if(sensor_above_thresh[i] > 0){
43.            Serial.print("drum");
44.            Serial.print(i); Serial.print("\t");
45.            Serial.println("bang");
46.            // zeroing sensor_above_thresh[i] will prevent this chunk of code
47.            // from be executed before we hit the drum again
48.            sensor_above_thresh[i] = 0;
49.          }
50.        }
51.      }
52.    }
53.  }
```

This code is very similar to the code in Listing 7-3. What we don't do here is check the maximum of the values that are above the threshold since we'll do it in Pd, we only print them to the serial line. Again, we're taking the same time stamps and we're checking if we've crossed the debounce time so we can get the maximum value in Pd. If we've crossed the debounce time, we send a bang to Pd under the same tag, "drum" followed by the number of the sensor. This bang will force the maximum value to be output. Figure 7-5 shows the Pd patch for this code.

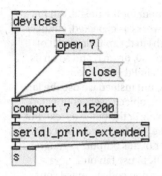

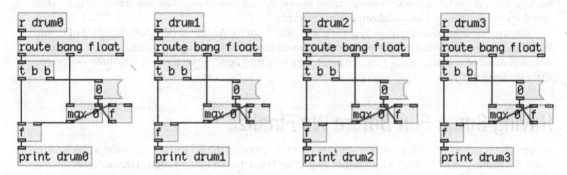

Figure 7-5. *Pd patch that calculates the maximum value of each sensor*

This patch is slightly more complex than the patch in Figure 7-3. Since both the values of the sensor, but also the bang are printed to the serial line under the same tag, we'll receive them with the same [receive]. We use [route bang float] to distinguish between floats (remember, in Pd all values are floats, so even though we're printing integers from the Arduino, Pd receives them as floats), and bangs. Each float goes to a [max 0], which outputs the maximum value between its arguments and the value coming in its left inlet to two [f]s. [max]'s output goes to the left (hot) inlet of the [f] on the right side, which will immediately output that same value and send it to the right (cold) inlet of [max]. This way the argument of [max] will be overridden by the output of the right [f], which is essentially the output of [max] itself. The same output goes to the right inlet of the [f] to the left side. This way, we store the maximum of all incoming values. It is the same as the following line of code in Arduino:

```
sensor_max_val[i] = max(sensor_max_val[i], sensor_val);
```

When we receive a bang, we first set 0 to [max], so we reset its initial argument for the next drum hit, and then we bang the left [f] to force it to output its value. To test whether this works as expected, connect a [print raw0] (type the number of each sensor, such as **raw1**, **raw2**, and so forth) to the middle outlet of [route], which outputs all the incoming floats, and see if the value printed as drum0, drum1, and so on, is equal to the maximum value of all the values printed as raw0, raw1, and so forth in Pd's console.

Mind that we don't use [trigger] to send the output of [max] to the two [f]s, but instead we use the fan out technique. The order these values are being output depends on the order the objects were connected, and there's no way to really tell how these objects were connected (only by looking at the patch as text, but this is not something very helpful anyway, plus its rather confusing). In this case, the fan out doesn't affect the patch at all, since there's no other calculation made depending on the order of this output. In general, it's better to use [trigger], but in such cases (like this or the case of a counter), it's OK to use fan out.

Again, you might want to experiment with the debounce time set in the Arduino code to make your system as responsive as possible. Make sure that you get only one value printed as drum0, drum1, and so forth for every drum hit. If you start having more than one, then you should raise the debounce time a bit. Test this until you get the lowest debounce time possible.

What we've done here is take away some complexity from the Arduino code and add it to the Pd patch. This shows us that the same thing is possible in different programming languages (well, they're not so different, Arduino is written in C++ and Pd in C). Now let's start building on top of this to make our project more interesting.

Having Some Fun Before We Finalize

This project is aimed at playing various samples stored in arrays in the Pd patch, every time we hit a drum. There are various ways to play back a sample. Before we build the circuit and code of the Arduino, let's first have some samples play back in reverse every time we hit a drum.

I have downloaded some drum samples from freesound.org. Since it's a drum set, it could be nice to have a drum sound corresponding to each kind of drum (kick drum, snare drum, and so forth), playing backward every time we hit it. Figure 7-6 illustrates the Pd patch.

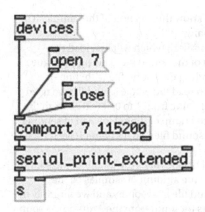

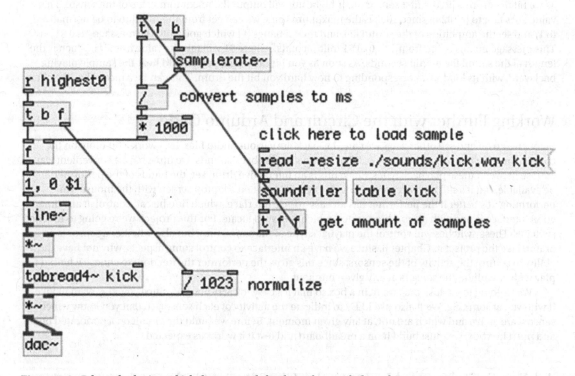

Figure 7-6. Pd patch playing a kick drum sample backward on each drum hit

I've used the code in Listing 7-3 instead in Listing7-4, where the maximum value is being calculated in the Arduino code and not Pd, just to make the patch a bit simpler. If you want to calculate the maximum value in Pd, replace [r highest0] in Figure 7-6 with the part of the patch in Figure 7-5 that calculates the maximum value ([r highest] should be replaced by [r drum0], and [print drum0] should not be included). Of course, now we don't need to print the raw values that are printed with the "drum" tag, and we're going to change that as we develop the Arduino code even more. For now, we use the code in Listing 7-3 as is to make some sounds with Pd.

The first thing you need to do is click the "read -resize ./sounds/kick.wav kick" message, which will write the sample called kick.wav to the kick table. This sample is in a directory (folder) inside the directory of the patch, hence the path ./sounds/kick.wav. We could also use [openpanel] instead which would open a dialog window with which we could navigate to the sample, but if we know the location of the sample, it's easier if we load it this way, since we can also automate it with a [loadbang].

Once we click the message, the kick.wav file will go through [soundfiler], which will first load the sound file to the specified table and will output the number of samples of the sound file. This number goes to two locations: [t f b] on the right side of the patch and [*~]. We're reading the sound file with [line~], which makes a ramp from 1 to 0 (we'll read the sound file backward). To read the whole sound file we need to multiply this ramp by the number of samples of the file, so instead of going from 1 to 0, it will go from the total length of the sound file to 0. We could have used this value to set the ramp of [line~] instead of sending it the message "1, 0 $1"; but further on, we'll use different ways to read sound files, so it's better if we have a generic ramp and multiply accordingly.

The part where the number of samples of the sound file goes to [t f b] calculates the length of the sound file in milliseconds. This is a pretty straightforward equation. We divide the amount of samples by the sampling rate (using [samplerate~]), which yields the length of the sound file in seconds, and we multiply by 1000 to convert it to milliseconds. We store that value in [f], which waits for input from the Arduino. As soon as we hit the drum with the first sensor on, [r highest0] will output the maximum value of the sensor. This value goes first to [/ 1023], since 1023 is the maximum value we can get from the Arduino to be normalized to 1, and set the amplitude of the sound file, and then it bangs [f] which goes into the message "1, 0 $1". This message means to "go from 1 to 0 in $1 milliseconds," where $1 will take its value from [f], which is the length of the sound file in milliseconds. As soon as you hit the drum, you should hear the sample playing backward, with its loudness corresponding to how hard you hit the drum. This can be a nice echo-like effect.

Working Further with the Circuit and Arduino Code

Before we go on and see other ways we can play back the various audio files, let's work a bit more on the circuit. Since we're going to have a few different ways to play back sounds, we must have a convenient way to choose between these playback types. Since this is an interactive drum set, the hands of the performer might be available, but it's still not easy to navigate to a certain point on a laptop screen with the mouse while performing. It's better if the performer has an easy-to-use interface, which he/she can control at any time. A good interface is one with quite a big surface, which is easy to locate. For this project, we're going to use foot switches. These switches will control the playback type of the sound files, but also whether a sensor is active or not. Like the project in Chapter 6, since we have an interface to control some aspects, why not have the ability to control the activity of the sensors, since this gives the performer the freedom to choose whether to play with or without the sensors at any given moment.

We're going to enclose the circuit in a box so that it's easy to carry around, since you may want to take it with you at some gig. We'll also use LEDs to indicate the activity of each sensor, so that you know which sensors are active and which are not, at any given moment. Before we build the circuit on a perforated board and put it in a box, we must build it on a breadboard and test if it works as expected.

Adding Switches and LEDs to the Circuit and Code

Since we have four sensors in our circuit, we're going to use four foot switches to control them. This applies some limitation to how many different kinds of playback we'll have, but when building interfaces we must set strict priorities, which concern functionality, flexibility, ease of use and transport, among others. The more control we want over a certain interface, the more the circuit will grow, and this can lead to rather complex circuits, which may not be desirable. If you want to add more foot switches or piezo elements to your project, you're free to do it. This chapter focuses on four of each.

To test the circuit and code we'll first use tactile switches (breadboard-friendly push buttons). Once it's all functioning properly, we'll go on and enclose the circuit in a box, using foot switches. Figure 7-7 shows the circuit with four push buttons and four LEDs.

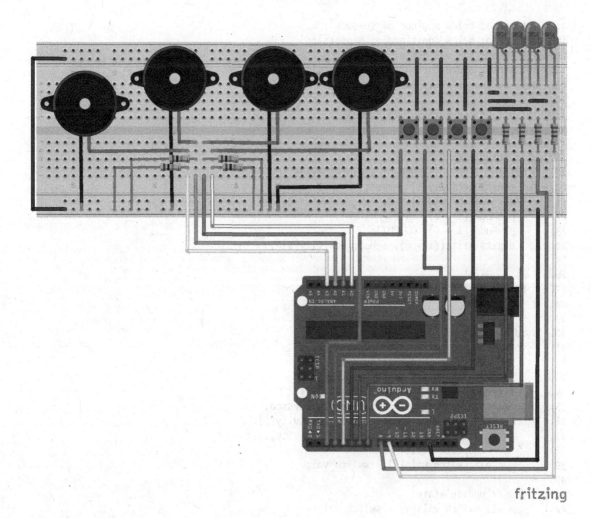

Figure 7-7. *Full test circuit*

There's nothing really to explain about the circuit. We'll use the Arduino's integrated pull-up resistors for the switches, and we're using 220Ω resistors for the LEDs, and 1MΩ resistors for the piezo elements. What I need to explain is the code that will work with this circuit. First, let's try to make the switches control the activity of the piezo elements. The code is shown in Listing 7-5.

Listing 7-5. Controlling Piezo Activity with Switches

```
1.  // variables and constants that you might need to change
2.  int thresh = 100;
3.  const int num_of_sensors = 4;
4.  int debounce_time = 10;
5.
6.  // arrays for the sensors
7.  int sensor_thresh[num_of_sensors];
8.  int old_sensor_thresh[num_of_sensors];
```

```
9.    int sensor_below_new[num_of_sensors];
10.   int sensor_below_const[num_of_sensors];
11.   int sensor_max_val[num_of_sensors];
12.   int sensor_above_thresh[num_of_sensors];
13.   bool sensor_activity[num_of_sensors];
14.
15.   int all_switch_vals[num_of_sensors];
16.
17.
18.   void setup() {
19.     for(int i = 0; i < num_of_sensors; i++){
20.       sensor_activity[i] = true;
21.       all_switch_vals[i] = 1;
22.       pinMode((i + 2), INPUT_PULLUP);
23.       pinMode((i + 6), OUTPUT);
24.       digitalWrite((i + 6), sensor_activity[i]);
25.     }
26.
27.     Serial.begin(115200);
28.   }
29.
30.   void loop() {
31.     for(int i = 0; i < num_of_sensors; i++){
32.       int switch_val = digitalRead(i + 2);
33.       // check if switch state has changed
34.       if(switch_val != all_switch_vals[i]){
35.         if(!switch_val){ // if switch is pressed
36.           sensor_activity[i] = !sensor_activity[i];
37.           digitalWrite((i + 6), sensor_activity[i]);
38.           // update state
39.           all_switch_vals[i] = switch_val;
40.         }
41.         // update state
42.         all_switch_vals[i] = switch_val;
43.       }
44.     }
45.
46.     for(int i = 0; i < num_of_sensors; i++){
47.       if(sensor_activity[i]){ // check sensor activity
48.         int sensor_val = analogRead(i);
49.         if(sensor_val > thresh){
50.           sensor_thresh[i] = 1;
51.           if(sensor_thresh[i] != old_sensor_thresh[i]){
52.             // count how many times the sensor rises above threshold
53.             sensor_above_thresh[i] += 1;
54.           }
55.           sensor_max_val[i] = max(sensor_max_val[i], sensor_val);
56.           old_sensor_thresh[i] = sensor_thresh[i];
57.         }
```

```
58.        else{
59.          sensor_thresh[i] = 0;
60.          // get a below threshold time stamp for the sensor
61.          sensor_below_const[i] = (int)millis();
62.          if(sensor_thresh[i] != old_sensor_thresh[i]){
63.            // get a time stamp every new time the sensor drops below threshold
64.            sensor_below_new[i] = (int)millis();
65.          }
66.          old_sensor_thresh[i] = sensor_thresh[i];
67.          if((sensor_below_const[i] - sensor_below_new[i]) > debounce_time){
68.            // make sure we print the values only once per hit
69.            if(sensor_above_thresh[i] > 0){
70.              Serial.print("drum");
71.              Serial.print(i); Serial.print("\t");
72.              Serial.println(sensor_max_val[i]);
73.              sensor_max_val[i] = 0;
74.              // zeroing sensor_above_thresh[i] will prevent this chunk of code
75.              // from be executed before we hit the drum again
76.              sensor_above_thresh[i] = 0;
77.            }
78.          }
79.        }
80.      }
81.    }
82.  }
```

The additions are very few, we've only added two arrays sensor_activity, and all_switch_vals. The first determines whether each sensor is active or not, as its name suggests. Since we only need this array to determine whether a state is true or false, we define it as a bool. The second array is used to test whether a switch has changed its state, otherwise the code of the for loop of line 31 would be executed constantly.

In the setup function, we set values for the two arrays and the mode for the digital pins. We also set the state of the LEDs according to the values of the sensor_activity array. By default, all sensors are active, so all the LEDs will turn on as soon as the Arduino is powered.

In the loop function, we run through all the switches and test if their state has changed (whether they are being pressed or released). If it has, we'll check if the switch is pressed in line 35, and if it is, we'll switch the corresponding value of the sensor_activity array and update the state of the corresponding LED. Then in line 39, we update the corresponding value of the all_switch_vals array so that the test in line 34 will work properly when the for loop goes through the same switch pin again. This update happens only if a switch is pressed, so in line 42, we update the all_switch_vals array, in case the state of a switch has changed, but we're releasing it.

Then in line 46, we go through the sensor pins, and in line 47, we check the activity of each sensor. If the sensor we're at is active, then the code that reads its values, gets the maximum, and prints it is executed. If the sensor is not active, then we go on to the next sensor. The code inside the if of line 47 is the same as with Listing 7-3, only this time we're not printing every value that's above the threshold, but only the maximum of all the values with the "drum" tag instead of "highest". Figure 7-8 illustrates the Pd patch for this code. It's essentially the same as with Figure 7-3, only there are not [r highest] objects, since we're not printing any values with this tag.

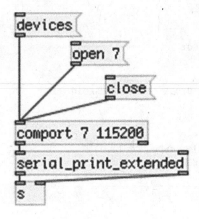

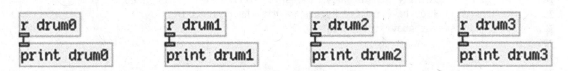

Figure 7-8. Pd patch for the Arduino code in Listing 7-5

With this code and patch, you have control over to which sensor is active and which is not, having direct visual feedback from the corresponding LEDs. Try to deactivate a sensor and hit its drum. Check the Pd console to see if you get any values printed, you should not. This gives us great flexibility, as we might not want to trigger samples all the time, but at certain moments during our playing.

Using the Switches to Control Either Playback Type or Sensor Activity

Now to the next step, where we'll use the switches to control two things, playback type or sensor activity. To do this we'll use the first switch to determine which of the two aspects we'll control. We'll do this by checking for how long we keep the switch pressed. If we keep it pressed for more than one second, we'll switch the type of control. If we keep it pressed for less than one second, the switch will control whatever the other switches control. (If the control type is set to sensor activity, it will control the activity of the first sensor; otherwise, it will control the type of playback.)

To do this, we again need to take time stamps of when we press and when we release the first foot switch. We'll then compare the difference between the two time stamps; if it's more than 1000 (milliseconds), we'll switch the type of control the switches will have. I'm not going to show all the code because the greatest part of it is the same as before. I'll only show the additions we need to make. Listing 7-6 shows some new variables and arrays we need to create. Of course, all these strings should be written in one line in the Arduino IDE, they just don't fit in one line in a book.

Listing 7-6. Additional Arrays for Switch Control Types

```
1.   // variables and arrays for the switches
2.   int all_switch_vals[num_of_sensors];
3.   int control = 1;
4.   String control_type[2] = { "playback", "activity" };
5.   String playback_type[num_of_sensors] = { "ascending", "descending", "backwards",
     "repeatedly" };
```

The first array all_switch_vals, was there in Listing 7-5 as well, but we're including it here because these variables and arrays all concern the switches. The second variable is an int that sets the type of control. You'll see how this is done in Listing 7-7. Lines 4 and 5 define a new type that we haven't seen before: String. String is a built-in class in the Arduino language, which manipulates strings, therefore its name. This class enables us to easily create string objects, even arrays of strings. Line 4 defines the strings "playback" and "activity", which we'll use in Pd to label the type of control the foot switch currently has. Line 5 defines four strings that have to do with the type of playback we'll use. "ascending" means that we'll play a sound file a few times repeatedly, where each time it will be played faster and at a higher pitch. "descending" means the opposite. "backwards" means just that, and "repeatedly" means that we'll play a short fragment of the sound file many times repeatedly, creating a sort of tone. We'll see all these types of playback in practice when we build the Pd patch. Place this chunk of code at line 15 in Listing 7-5, where we set the all_switch_vals array.

Listing 7-7 shows the for loop for the switches. This must replace the code in Listing 7-5 from line 31 to line 44. Mind that if you have added the code in Listing 7-6, these line numbers will have changed.

Listing 7-7. Modified for Loop to Select Control Type for the Switches

```
1.   for(int i = 0; i < num_of_sensors; i++){
2.       int switch_val = digitalRead(i + 2);
3.       // check if switch state has changed
4.       if(switch_val != all_switch_vals[i]){
5.         if(!i){ // check if it's the first switch
6.           static int press_time, release_time;
7.           if(!switch_val){
8.             press_time = (int)millis();
9.             all_switch_vals[i] = switch_val;
10.          }
11.          else{
12.            release_time = (int)millis();
13.            all_switch_vals[i] = switch_val;
14.            if((release_time - press_time) > 1000){
15.              control = !control;
16.              Serial.print("control ");
17.              Serial.println(control_type[control]);
18.            }
19.            else{
20.              if(control){
21.                sensor_activity[i] = !sensor_activity[i];
22.                digitalWrite((i + 6), sensor_activity[i]);
23.              }
```

```
24.              else{
25.                Serial.print("playback ");
26.                Serial.println(playback_type[i]);
27.              }
28.            }
29.          all_switch_vals[i] = switch_val;
30.        }
31.      }
32.      else{
33.        if(switch_val){
34.          if(control){
35.            sensor_activity[i] = !sensor_activity[i];
36.            digitalWrite((i + 6), sensor_activity[i]);
37.          }
38.          else{
39.            Serial.print("playback ");
40.            Serial.println(playback_type[i]);
41.          }
42.        }
43.      }
44.      // update state
45.      all_switch_vals[i] = switch_val;
46.    }
47.  }
```

This is quite more complex than the for loop for the switches in Listing 7-5. The first thing we do after we read each digital pin is check whether its state has changed, which is the same as in Listing 7-5. Then in line 5, we're checking if the switch that is being changed is the first one, by using the variable i with the exclamation mark. If it's the first one, i will hold 0, and the exclamation mark will make it 1, which is the same as true. If i holds any other value (1, 2, or 3), the exclamation mark will make it 0, which is the same as false. In line 6, we define two local variables for the two time stamps we need to take, which are both static, so even though they're local, even when the function in which they're defined is over, they will retain their values.

In line 7, we check if the first foot switch is being pressed, and if it is, we take a time stamp of when we pressed it, and then update all_switch_vals so testing if the switch's state has changed will work properly. else in line 11 will be activated if the first switch is being released, where we're taking another time stamp of when that happened, and we again update all_switch_vals. Then in line 14, we're testing the time difference between the two time stamps. If it's greater than 1000 we switch the value of the control variable, and we print the corresponding string from the control_type string array, where if control is 0, we'll print "playback", and if it's 1, we'll print "activity". By default, control is 1, so the first time we'll hold the first switch pressed for more than a second, the control type will change to "playback".

If the difference between the two time stamps is less than 1000, we first check the type of control the switches have. If it's 1, releasing the foot switch will switch the state of the first element of the sensor_activity array, and this value will be written to the pin of the first LED. If control is 0, we'll print the first element of the playback_type string array, which is "ascending".

Afterward, we're updating the all_switch_vals array right before else of line 11 has ended, which is called if the first switch is being released. It's very important to place the updates of the all_switch_vals array in the right places, otherwise the code won't work properly. In this code, we're separating the first foot switch from the rest, plus we're separating what happens when we press the switch from what happens when we release it. For the rest of the switches we only care about releasing the switch. So we need to update the all_switch_vals array every time we press or release the first switch and every time the state of any of the other switches changes. That's why this array is being updated in one more place in line 45.

In line 32, we have an else that will be executed if a state change has been detected, but it was not the first switch. In line 33, we're checking if the switch is being released. In Listing 7-5, we were checking when a switch was being pressed. Since for the first switch we can't determine what we want to do when we press it, but only when we release it, because we're counting the amount of time we held the switch pressed, it's better to use the release for the rest of the switches, for the compatibility of the interface. Depending on the type of control the switches have, we'll switch the current element of the sensor_activity array and write that value to the corresponding LED, or we'll print the corresponding element of the playback_type string array, which will be "descending", "backwards", or "repeatedly". After updating the all_switch_vals array, the loop ends and we go on to read the pins of the sensor. This is not included in Listing 7-7 because, as I already mentioned, it is exactly the same in Listing 7-5.

Building the Final Pd Patch

We're now at a point where we can start building the actual patch for this project. This patch will consist of three different levels, the audio file playback, the Arduino input, and the main patch. We'll first build the patch that reads the audio files, and then the patch that receives data from the Arduino and triggers the playback. Then we'll go on and put everything together in the main patch for this project.

Building the Audio File Abstraction

Since the only difference between each audio file playback for each sensor will be the table where we'll load the audio file to, it's better if we first make an abstraction that we can duplicate for each sensor. We can set the name of the table via an argument, along with some other data that we'd like to pass, but different for each abstraction. This abstraction should take a bang and play back the audio file loaded on the table specified via an argument, in a way set by the switches. Figure 7-9 illustrates this abstraction.

The heart of this abstraction is [line~], which is the index for the table of the sound file, but also for the window that smooths out the beginning and the end of the file. [loadbang] goes first to [del 100], because in the parent patch we first want to load the audio files to the tables and then do any calculations depending on those files, so we give some short time to Pd when the patch loads to first load these files to the tables. After this 100 milliseconds delay, we get the size of the table each abstraction reads with [arraysize $1]. $1 will be replaced by the first argument of the abstraction, which will be the name of the table with the audio file we want to read. The value output by [arraysize] is the number of samples of the audio file, since the table where it will be loaded will be resized to a size that will fit all the samples of the file. This value goes to two locations, first it goes to [*~] next to which there is the comment "this is the total length of the table". We'll set [line~] to go from 0 to 1, so we'll need to multiply it by the number of samples of the audio file so that [tabread4~] can output the whole file.

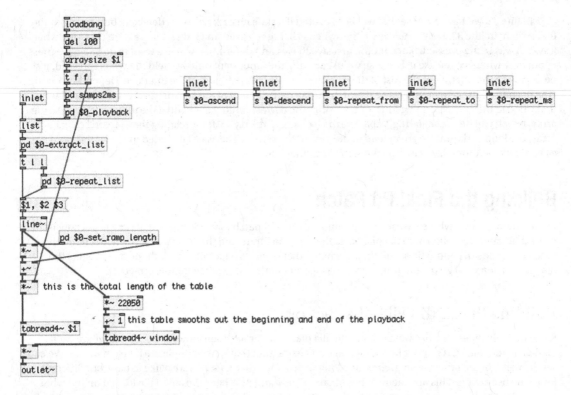

Figure 7-9. The read_sample abstraction

Converting Samples to Milliseconds and Dealing with the Sampling Rate

The second location the size of the table goes is the [pd samps2ms] subpatch, which calculates the amount of milliseconds it will take to read the whole audio file. This is very often useful; you might want to consider making an abstraction so that you can use it frequently. The contents of the subpatch are shown in Figure 7-10. The calculations in this subpatch are pretty simple, so I won't explain it. One thing that I need to mention at this point is that you should take care of the sampling rate your audio files have been recorded in. If you download sounds from freesound.org, the sampling rate of each sound is mentioned on the sound's web page. Otherwise, you can get this information from your computer, if you have stored sound files in your hard drive. All the audio files that you load in your patch should have the same sampling rate. Pd must run at the same sampling rate too for the audio files to be played back properly. To change Pd's sampling rate, go to **Media ➤ Audio Settings…**.

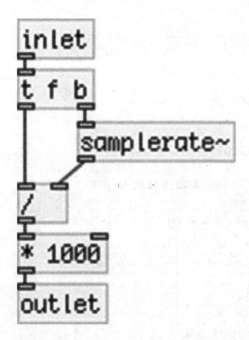

Figure 7-10. *Contents of the samps2ms subpatch*

Creating Different Types of Playback

After [pd samps2ms] has calculated the amount of milliseconds an audio file takes to be read, it sends its value to [pd $0-playback], the contents of which are shown in Figure 7-11. In this subpatch, we're creating lists according to the string that sets the type of playback, sent by the Arduino. The only straightforward playback type is "backward" which outputs the list "1 0 $1", where $1 is replaced by the milliseconds it takes to read the audio file. This list will be sent to [line~] which will make a ramp from 1 to 0, lasting as long as it needs to read the audio file at normal speed, which will cause [tabread4~] to read the audio file backward.

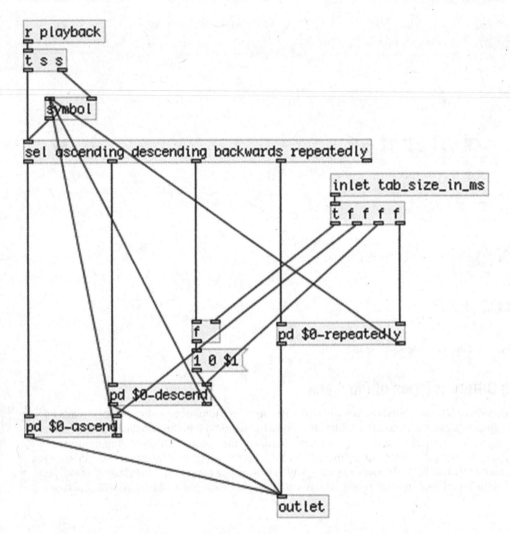

Figure 7-11. *Contents of the $0-playback subpatch*

The rest of the playback types are inside subpatches because their calculations are not so simple, and putting them in subpatches makes the patch a bit tidier. They are illustrated in Figures 7-12 through 7-15.

The "$0-ascend" subpatch

In the "ascend" subpatch we receive, store, and output the audio file length in milliseconds in [f], and we run a loop for as many times set by the second argument to the abstraction, which is stored in [$2]. We need to store the value in case we want to set a different number of repetitions for this playback type with a value entered in the second [inlet] of the abstraction. By sending the milliseconds length in the left inlet of [f], we both output it and save it at the same time, so we can use it again even if [inlet tab_size_in_ms] doesn't output a new value.

[until] will bang [f] below it as many times the loop runs. At the beginning, [f] holds the length of the audio file in milliseconds, and every time it is banged, it outputs it and it also gets divided by 2, so it stores half that value. This values goes to the message "0 1 $1", which is the list that will be sent to [line~] to make a ramp from 0 to 1 in the time specified by $1. (Remember that $1 in a message means the first value of a list—in this case, it's just one value list—that arrives in the inlet of the message. $1 in an object contained in an abstraction means the first argument given to the abstraction. The interpretation of $1—or $2, $3, and so forth—is totally different between messages and objects). The list of this message goes to [list prepend] which outputs the lists it gets in its left inlet to [t l l] ([trigger list list]). [t l l] will first send the list to the right inlet of [list prepend], and the latter will store that list, and then it will send it to the right inlet of [list] at the bottom, which will also store it.

This way we can create a growing list, since [list prepend] will prepend anything stored in its right inlet to anything that arrives in its left inlet. So, if the length of the audio file is 1000 milliseconds, the message that goes to [list prepend] will first output "0 1 1000". [list prepend] will output this list and it will also store it in its right inlet. The second time, the message will send "0 1 500", which will go to the left inlet of [list prepend] and the previous list will be prepended to it, resulting in [list prepend] storing the list "0 1 1000 0 1 500". The third time the message will output "0 1 250", so the list stored in [list prepend] will be "0 1 1000 0 1 500 0 1 250", and so on. This will go on for as many times as we'll set via the second argument to the abstraction. All these lists will go to the right inlet of [list] at the bottom of the subpatch, and every time [list] will replace whatever it had previously stored with the new list that arrives in its right inlet. When [r playback] in the subpatch in Figure 7-11 receives the "ascending" symbol, it will bang [pd ascend] which will output the final list stored in [list], which will be a set of lists of three elements, 0, 1, and the decreasing time in milliseconds. This full list will go to the right inlet of [list] in the top-left part of the parent patch of the abstraction, shown in Figure 7-9, which will be banged whenever the corresponding sensor will output a value. We'll see later on how this works.

Notice that in Figure 7-12 there's a [r $0-ascend] which receives values from [s $0-ascend] from the parent patch of the abstraction (Figure 7-9). If we want to change the number of repetitions of the ascending playback on the fly, we can send a value to the second inlet of the abstraction and it will arrive in [pd $0-ascend]. When we receive this value we'll first bang the right inlet of [list prepend]. This will cause the object to clear any stored list. Afterward, we'll store the value in [$2], so its value will be replaced. Then we'll bang [f], which holds the duration of the audio file in milliseconds. And finally we'll send a bang to the outlet of the subpatch, which will go to [pd $0-playback] (Figure 7-11). This bang is sent to the left inlet of [symbol], which has stored the symbol arriving from the Arduino ("ascending", "descending", "backwards", or "repeatedly"). We do that so we can output the list generated by the subpatch, which is stored in [list] at the bottom of [pd $0-ascend]. If we don't bang it, [list] won't output its list and we won't be able to hear the new setting we made by sending a value in the inlet of the abstraction.

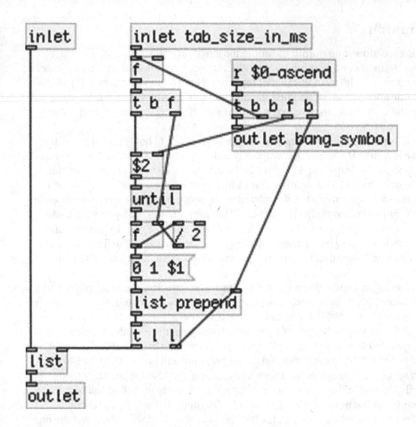

Figure 7-12. *Contents of the $0-ascend subpatch*

The $0-descend Subpatch

The next one is [pd $0-descend], shown in Figure 7-13. This is very similar to [pd $0-ascend]. Instead of playing the audio file twice as fast (and at a higher pitch) every time, we start by playing it fast, and every time we play it twice as slow, end it at the normal speed (starting at a high pitch and ending at the normal pitch). In this subpatch, we calculate the shortest duration according to the length of the audio file and the number of repetitions, which is set via the third argument, and then we run a loop starting with this shortest duration and every time we double it. Apart from that the rest of the patch works the same way as [pd $0-ascend].

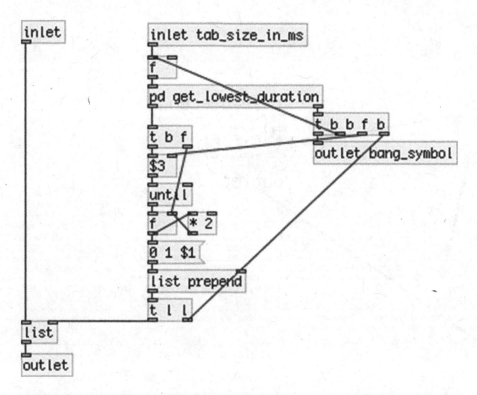

Figure 7-13. *Contents of the $0-descend subpatch*

I need to explain how the get_lowest_duration subpatch works, which is shown in Figure 7-14. This subpatch runs a loop similar to that of [pd $0-ascend]. We start with the normal duration of the audio file and every time we divide it by two. When we've gone through all the iterations of the loop, [sel $3] will bang the final value, which is stored in [f] at the bottom of the subpatch. This value will then go in [pd $0-descend] and another loop will run, now starting with the shortest duration, every time multiplying by two, ending at the normal duration. Again, we have a long list stored in [list], which we bang whenever we send the symbol "descending" from the Arduino. Notice that [r $0-descend] is in [pd get_lowest_duration], as if we changed the number of repetitions. We'll need that number in there too to calculate the shortest duration for the audio file. Mind the order of sending that value or a bang and try to understand why we need things to go in this order, so that everything works as expected.

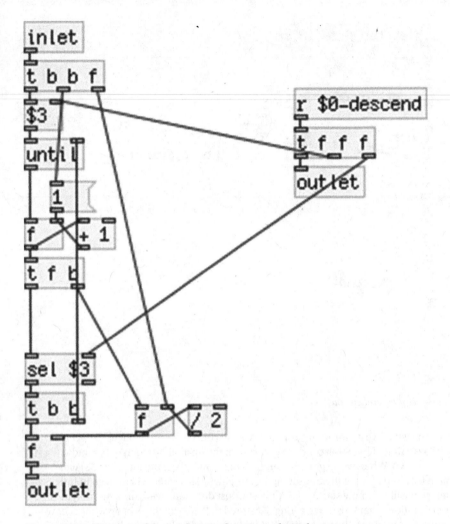

Figure 7-14. Contents of the get_lowest_duration subpatch

The $0-repeatedly Subpatch

Then we go to [pd $0-repeatedly], which is shown in Figure 7-15. This is quite different than the other
two subpatches that I've explained. In this subpatch, we're using the fourth and fifth arguments of the
abstraction, which are the points of the audio file where we want our repeated playback to start and end. We
don't need to know the length of the audio file, either in milliseconds or samples, to set these two arguments.
They are set in some kind of percentage, from 0 to 1, where 0 will be the beginning of the file and 1 its end.
What we do in [pd $0-repeatedly] is subtract $4 from $5 to get the duration of the file we want to play back,
in a 0 to 1 scale, and then we multiply the result by the length of the audio file in milliseconds. This will tell
[line~] to go from 0 to 1 in the time it takes to play the portion of the file we want. In [pd $0-repeatedly] we
can see a [s $0-repeat_length] with a comment stating, "goes into [pd set_ramp_length]". This value (the
difference between $4 and $5) will set the amount of scaling and an offset to the ramp of [line~]. If [line~]
went from 0 to 1 in the time it takes to read only a portion of the file, which is set via $4 and $5, the result
would be that [tabread4~] would read the whole file much faster that its normal speed. Since we're using the

ramp of [line~] to control a smoothing window, in the parent patch of the abstraction, we're telling [line~] to go from 0 to 1, so that the smoothing window will be read properly, and then we scale its ramp and we give it an offset before if goes to [tabread4~]. All this might start to get a bit complex, but if you go back and forth from subpatch to parent patch, and try to keep track of what's going on, you'll understand how things work in this patch. In [pd $0-repeatedly] we have two [receive]s, a [r $0-repeat_from] and a [r $0-repeat_to], which take values from the fourth and fifth inlet of the abstraction. If we want a different portion of the audio file, we can change the values with these two inlets, and the necessary calculations will be done. Again, we store the resulting list to [list], which we'll bang if the Arduino sends the symbol "repeatedly". And that's all about [pd $0-playback], now back to the parent patch.

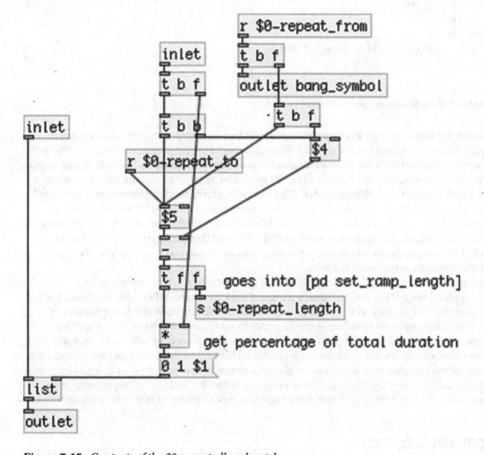

Figure 7-15. *Contents of the $0-repeatedly subpatch*

The $0-extract_list Subpatch

In the parent patch we see another subpatch, [pd $0-extract_list], which is shown in Figure 7-16. This subpatch takes input from [list] at the top-left part of the parent patch, which holds the list of the type of playback we want. If we've set the playback type to "ascending" or "descending", then the corresponding subpatches will output a long list, which should be split to groups of three, the beginning of the ramp of [line~], the end of the ramp, and the duration of the ramp. In [pd $0-extract_list] we do exactly that. [list split 3] splits the incoming list at the point specified by its argument. So this object will output the first three elements of its list out its left outlet, and the rest will come out the middle outlet.

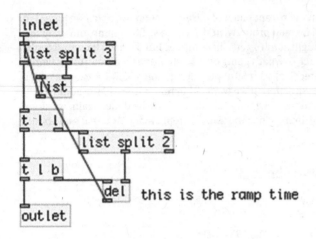

Figure 7-16. *Contents of the $0-extract_list subpatch*

Since Pd has the right-to-left execution order, [list split] will first give output from the middle outlet and then from the left one. The middle outlet receives the remaining of the list and stores it in [list]. The left outlet outputs the three element list to [t l l], which first outputs this list to [list split 2]. [list split 2] will output a list with the first two elements of the list it received out its left outlet, and the remaining third element out its middle outlet, which goes to the right inlet of [del]. This last value stored in [del] is the ramp time of the first three elements, in milliseconds. After the ramp time has been stored in [del], [t l l] will output the three element list out its left outlet, which goes to [outlet]. From there, it will go to the message "$1, $2 $3" in the parent patch, which tells [line~]: "Jump to the first value of the list, and then make a ramp to the second value which will last as many milliseconds as the third value." So [line~] will make a ramp from 0 to 1 that will last as many milliseconds as the audio file lasts.

At the same time [t l l] outputs the list through [outlet] in [pd $0-extract_list], it also bangs [del] (actually it first bangs [del] since [t l b] will first send a bang out its right outlet). When [del] receives a bang, it will delay it for as many milliseconds as the value it has stored in its right inlet, which is the amount of milliseconds the ramp takes. So as soon as [line~] finishes with its ramp, [del] will bang [list], which has stored the remaining list, and it will send it to [list split 3], and the whole procedure will start over, every time taking the first three elements of the list, until the list is finished. This way we can have repeated ramps, each lasting half the time of the previous one, in case we're playback in the "ascending" type, or twice as long, in case we're playing back in the "descending" type. If we playback in the "backwards" or "repeatedly" type, [pd $0-extract_list] will output one list only, since these two types produce a list of three elements only.

The $0-repeat_list Subpatch

The output of [pd $0-extract_list] goes to [t l l] in the parent patch, which outputs the incoming lists first to [pd $0-repeat_list] and then to the message sent to [line~]. [pd $0-repeat_list] is activated only if we receive the symbol "repeatedly" from the Arduino. What it does is take the list generated by [pd $0-repeatedly] and repeat it for an amount of time set in milliseconds by the sixth argument to the abstraction. It is shown in Figure 7-17. If the Arduino sends the symbol "repeatedly", we'll receive it in Pd via [r playback] and [sel repeatedly] will send a bang out its left outlet, which will send 1 to the two [spigot]s. As soon as this subpatch receives a list, it first bangs [del $6], which will have its delay time set via the sixth argument. Then, it will bang [f] containing 1. Then, it will send the list to [list split 2], which will output the ramp time of the list out its middle outlet into [pipe 0 0 0 0]'s rightmost inlet, and then the same list to the leftmost inlet of [pipe].

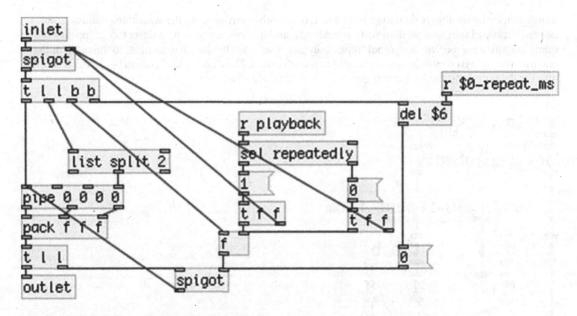

Figure 7-17. *Contents of the $0-repeat_list subpatch*

[pipe] is like [del], only instead of outputting bangs delayed by a specified time, it outputs numeric values entered in its inlets. It can delay whole lists, where the arguments initialize the lists' values, and the very last argument is the delay time. Its inlets correspond to its arguments, so sending a value to the rightmost inlet will set a new delay time; sending a list in the leftmost inlet will replace as many arguments as the elements of the list with these elements. If [pd $0-repeatedly] generates the list "0 1 50", [list spit 2] will send "50" to the rightmost inlet of [pipe], which will be the delay time, and in the leftmost inlet, [pipe] will receive the whole list, which it will delay for 50 milliseconds.

This list goes out three different outlets ([pipe] creates one outlet for each element of the list, the delay time is not included), which are [pack]ed to be sent as a list to [line~]. Since the [spigot]s are open, [t l l] will first send the list back to [pipe], which will again be delayed for 50 milliseconds, and this will go on until [del $6] bangs the message "0", which will close the lower [spigot]. This way we can send the same list over and over for an amount of time that we specify either via an argument or through the sixth inlet of the abstraction (which goes to [s $0-repeat_ms], received by [r $0-repeat_ms] in [pd $0-repeat_list]). As soon as we send a symbol other than "repeatedly" from the Arduino, the top [spigot] in [pd $0-repeat_list] will close, and incoming lists won't go through anymore. And this covers how the four different ways of playback are achieved.

The $0-set_ramp_length Subpatch

There's one more subpatch that I need to explain and we're done with the [read_samp] abstraction, which is [pd $0-set_ramp_length], shown in Figure 7-18. If you see Figure 7-9 you'll see that the output of [line~] goes to two destinations, [*~] and [*~ 22050]. The first multiplication object ends up in [tabread4~ $1], which reads the audio file stored in the table the name of which is set via the first argument to the abstraction. [*~ 22050] ends up in [tabread4~ window] which reads a smoothing window function which we'll place in the parent patch of our project. This window will smooth out the beginning and end of each audio file since it may not begin or end smoothly and cause clicks, or in the case of playing back the sound repeatedly, since we won't be reading the file from beginning to end, even if it's smooth at its edges, we'll get clicks for sure. You'll see this window further on. The point here is that we need the same ramp to control two different

299

things, that's why we always set it to go from 0 to 1, or the other way around (the smoothing window can be read backward too), and we then scale accordingly, and if necessary, give it an offset too. In [pd $0-set_ ramp_length], if we receive the symbol "repeatedly", we'll send $4 (the fourth argument) to the right [outlet] and the value received by [r $0-repeat_length] to the left outlet. The right [outlet] goes to [+~], which sets the offset of the ramp, and the left [outlet] goes to [*~] which scales the ramp.

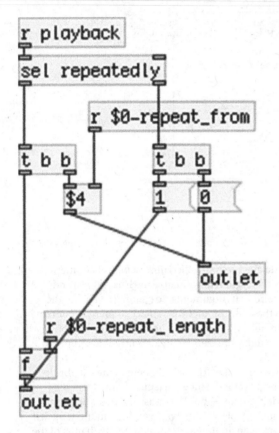

Figure 7-18. *Contents of the $0-set_ramp_length subpatch*

If we want to play a portion of the audio file, which we set with values from 0 to 1, we need to offset the ramp with the lowest value, and scale it with the difference between the two values, which is the value received by [r $0-repeat_length]. If, for example, we set to read the file from 0.25 to 0.60, we need our ramp to start at 0.25 and go up to 0.60, giving the ramp a total length of 0.35, which is the difference between the two values. This way we set the correct ramp for our sample without affecting the ramp that reads the smoothing window function. Of course, if we change the start and end point of the portion of the audio sample we want to play, [r $0-repeat_from] will receive the beginning point and store it in [$4]. If the Arduino sends any symbol other than "repeatedly", we'll send a 1 to [* ~], and a 0 to [+~], so we'll multiply the ramp by 1, which gives the ramp intact, and we'll add 0, which gives no offset.

This concludes the [read_samp] abstraction. It is a bit complex, but we'll use it for all four sensors as is, the only thing we'll change to each instance of the abstraction is its arguments. The next thing we need to do is build an abstraction to receive the input from the sensors, and that will be enough for our patch to work with the Arduino code. There will be only a minor addition to the Arduino code, which will enable us to control the threshold and debounce time from Pd, so we don't need to upload the code every time we want to change one of the two values.

In the directory where you'll save your main patch, make a directory called abstractions, and save [read_samp] in there. This abstraction is project-specific, so it's better if you save it in a directory that's not in Pd's search path. (Don't confuse the directory you'll make and call abstractions with the abstractions directory that you might have already made and stored generic abstractions, like the ones from my "miscellaneous_abstractions" GitHub page. They should be two different directories in different places, even though they share the same name). In the main patch, we'll use [declare] to help Pd find the abstraction.

Building the Abstraction to Receive Input from the Arduino

This abstraction is much simpler than the [read_samp] abstraction. It is shown in Figure 7-19. All we do here is receive input from a sensor and normalize its value to a range from 0 to 1, using the [map] abstraction. The values from the Arduino are printed with the "drum" tag along with the number of the pin of each sensor. In [drum] we have a [r drum$1]. This means that for the first sensor, we must provide the value 0 as the first argument to the abstraction, so that [r drum$1] will receive the value of that sensor. This abstraction has three inlets, one to set a new threshold, one to set a new debounce time, and one for debugging. The first two will take number atoms, while the third will take a toggle. When the toggle outputs 1, the value received by the sensor won't go out the abstraction's outlet, but will be printed as drum0, drum1, and so forth. This is very helpful as we can calibrate our patch on the fly. We'll cover that a bit further on. When the abstraction outputs its value, it will send the normalized value out its right outlet and a bang out its left outlet, so it can bang [read_samp] and the patch will play the audio file.

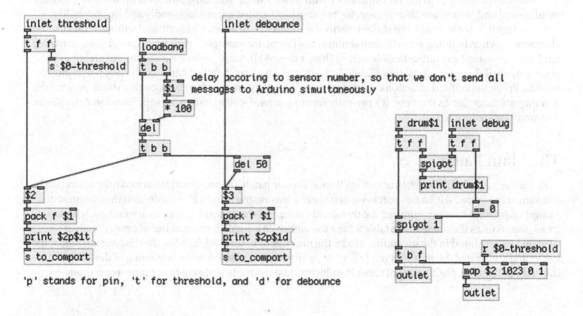

Figure 7-19. The "drum" abstraction

301

Sending the Threshold and Debounce Values to the Arduino

One thing we need to explain is the [loadbang] of the abstraction. Since we might want to have different thresholds and different debounce times for each drum, we need to set these as arguments and send them to the Arduino on load. To avoid sending all of these messages at the same time, we set a delay for each abstraction. This delay depends on the first argument of the abstraction, which is the number of the pin of the sensor each abstraction is listening to. So for the first abstraction, we'll bang 0 × 100 = 0 milliseconds, so there's no delay, and we'll bang the threshold argument first, and the debounce argument delayed by 50 milliseconds. The second abstraction will have its arguments delayed by 1 × 100 = 100 milliseconds for the threshold argument, and another 50 milliseconds for the debounce. The third will be delayed 2 × 100 = 200 milliseconds, and so on. This way we avoid sending a bunch of messages all at the same time.

Lastly, I must explain the syntax of the messages sent to the Arduino. You've seen this in Chapter 2, but I'll explain it shortly here as well. The message "print" sent to [comport] will convert all characters of the message to their ASCII values, which makes it easy to diffuse values in the Arduino code. To set the threshold for a specific sensor, we must define both the sensor pin and the threshold. The same applies to the debounce time. The message "print $2p$1t" will print the second value stored in [pack f $1], which is the number of the sensor pin, then "p", then the value of the second argument, or the value sent in the leftmost inlet, and lastly, "t". All numeric values will be assembled from their ASCII values in the Arduino code. When the Arduino receives "p", it will set the value it assembled as the number of the pin. When it receives "t", it will set the value it assembled as the value for the threshold for that pin. You'll see that in detail when we go through the additions to the Arduino code.

Save this abstraction to the same directory with [read_samp]. You must have noticed that even though [read_samp] and [drum] are abstractions, we're using hard-coded names for [send]s and [receive]s, like [s to_comport] or [r playback]. We should clarify the difference between a generic and a project-specific abstraction. When building generic abstractions (like [loop], for example), you should avoid using anything hard-coded; instead use rather flexible names (like [r drum$1]). Since these abstractions are meant to be used in many different projects, they should be built in such a way that they can adapt to any patch they're used in. Project-specific abstractions are different because they are meant to be used in a single project, the one they are made for. In this case, it's perfectly fine to use hard-coded names in [send]s and [receive]s, since that won't create problems.

The Main Patch

Now that we've built the two abstraction we'll need for our patch, it's pretty simple to make the main patch. It is shown in Figure 7-20. In this patch, we can see the two abstractions we've made with their arguments. [drum] takes the sensor pin number, the threshold value, and the debounce time as arguments; whereas [read_samp] takes the name of the table where the audio file is loaded, the number of times the file is repeated when played in the ascending mode, the number of times it is played in the descending mode, the beginning and end of the repeated mode (in a scale from 0 to 1, where 0 is the beginning of the file and 1 is the end), and finally, the amount of time (in milliseconds) the file is repeated in the repeatedly mode.

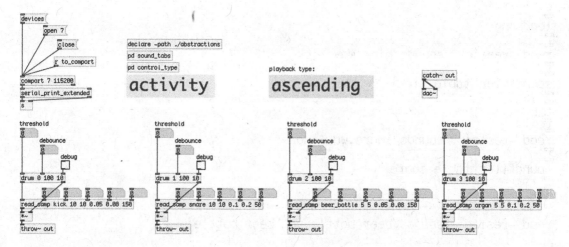

Figure 7-20. *The main patch*

Each abstraction takes a number atom for each argument to experiment with the playback settings and to calibrate each sensor. Notice that there's a [r to_comport] which takes input from [s to_comport] which is in [drum], so we can send messages to the Arduino. Also notice that [drum] outputs the sensor value from its right outlet, which goes to [*~] to control the amplitude of the audio file playback, and it then bangs [read_samp] to read the file. We don't need a [line~] to smooth the amplitude changes, because they occur before the file is triggered.

The sound_tabs Subpatch

Apart from that we have a [declare -path ./abstractions], which tells Pd to look into the abstractions/ directory for the two abstractions. There's also a [pd sound_tabs] subpatch. This is similar to the subpatch with the same name in Chapter 4, and it is shown in Figure 7-21. In this subpatch, we're loading our audio files to the tables on load of the patch. The message to load an audio file is the same as already used in this book a few times. When [soundfiler] loads the file to the specified table, we convert its output (which is the length of the file in samples) to a bang so we load the next audio file. As you can see, I've made a directory called sounds in the directory of the main patch, and I've stored the audio files in there. Making separate directories for the audio files and the abstractions—instead of having everything together (along with the main patch) in the same directory—keeps things tidy and easy to understand.

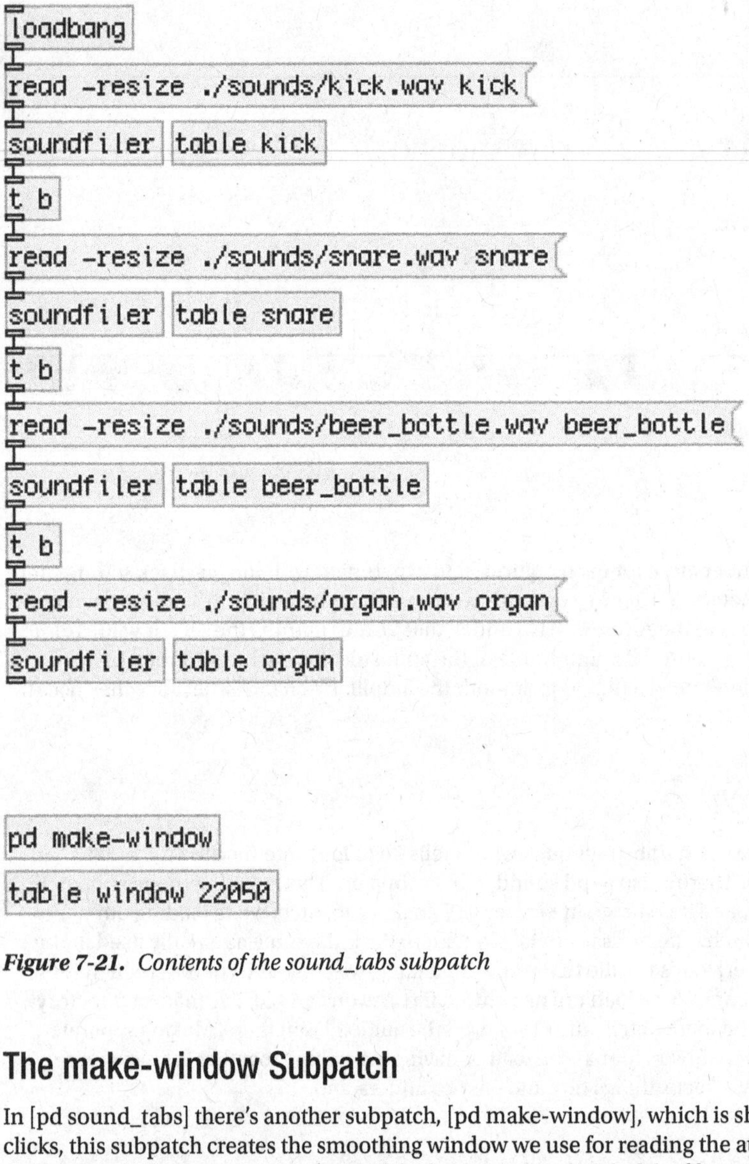

```
loadbang

read -resize ./sounds/kick.wav kick

soundfiler   table kick

t b

read -resize ./sounds/snare.wav snare

soundfiler   table snare

t b

read -resize ./sounds/beer_bottle.wav beer_bottle

soundfiler   table beer_bottle

t b

read -resize ./sounds/organ.wav organ

soundfiler   table organ
```

```
pd make-window

table window 22050
```

Figure 7-21. *Contents of the sound_tabs subpatch*

The make-window Subpatch

In [pd sound_tabs] there's another subpatch, [pd make-window], which is shown in Figure 7-22. To avoid clicks, this subpatch creates the smoothing window we use for reading the audio files when the beginning and ending points are not zero. (Even if the beginning and end of the file are zero, when reading the file in the repeatedly mode, the two points will most likely not be zero, since we'll be reading from some point other than the beginning to some point other than the end.) What this subpatch does is create a rapidly rising ramp from 0 to 1, taking 500 samples, then it stays at 1 for 21050 samples, and finally it makes a rapidly falling ramp from 1 to 0, which takes another 500 samples. You can see its output in Figure 7-23. We're using [loop] to iterate through all the points of the table, and [moses] to separate the two ramps from the full amplitude value, 1. [moses] separates two streams of values at the value of its argument, or a value in its right inlet. Take a minute to try to understand how it works. The only thing that might be a bit confusing is [+ 499]. Since we start counting from 0, after 22050 iterations we'll reach 22049.

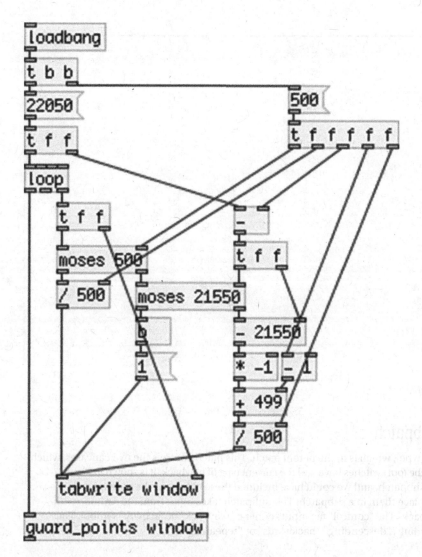

Figure 7-22. *Contents of the make-window subpatch*

Figure 7-23 illustrates what is stored in the window table. Its frame shows values between –1 and 1, in the Y axis, that's why the window starts from the middle of the graph. You can see the rapidly rising ramp, which hits the ceiling of the graph when it reaches 1. It stays there for the greatest part of the table and it then goes back down to 0, making a sharp ramp. When we read each audio file, we're reading this table at the same time and we're multiplying the two tables. This results in beginning with silence, then rapidly rising to full amplitude, staying there for almost the whole table, and finally going back to silence. This prevents any clicks at the beginning and end points of the table, since they will always be silent.

Figure 7-23. *The window table*

The control_type Subpatch

Now I'll explain [pd control_type], which is in the parent patch. This patch controls the two canvases, which label the kind of control that the foot switches have and the current type of playback. It is shown in Figure 7-24. This is an extremely simple subpatch, and we could have included these objects in the parent patch. It's tidier and looks better if we place them in a subpatch. This subpatch takes input from the Arduino with the tags "control" and "playback". The "control" tag inputs either a "playback" or "activity" symbol. The "playback" tag inputs "ascending", "descending", "backwards" or "repeatedly" .

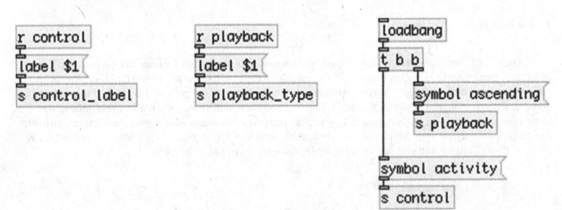

Figure 7-24. *Contents of the control_type subpatch*

The subpatch controls the label of the canvases in the parent patch, which on load are labeled as "activity" and "ascending". "activity" is the default control of the switches when the Arduino starts. We explicitly set "ascending" as the default playback type. You need to set some of their properties to make it look like this. The Properties window of the "control" (the yellow) canvas is shown in Figure 7-25.

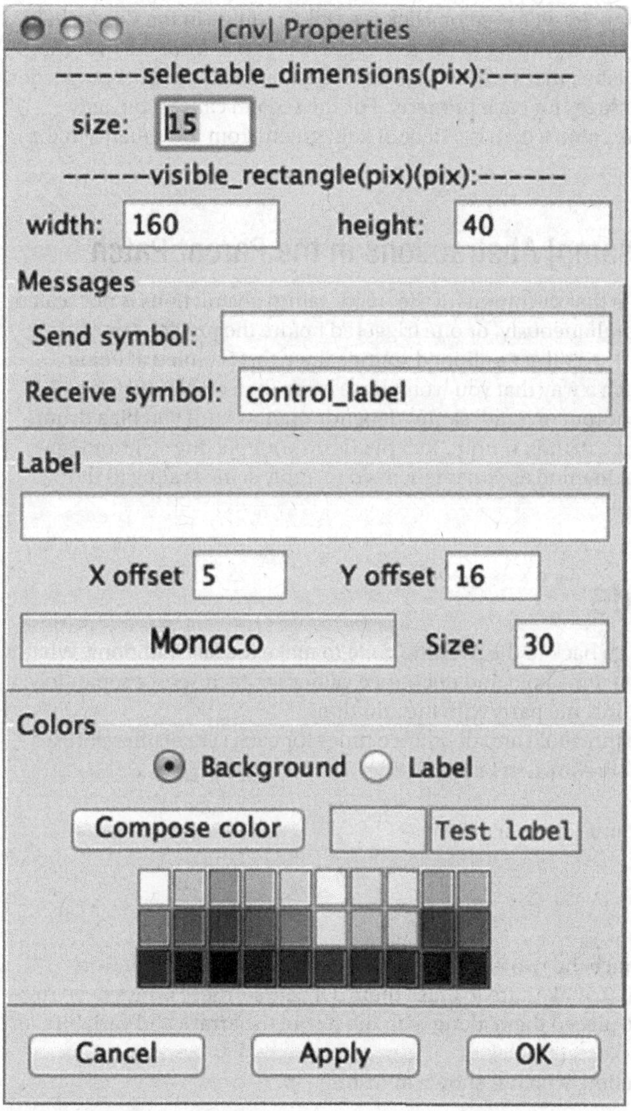

Figure 7-25. *Canvas Properties*

Change the Properties of the Playback Type Canvases

To open its properties right-click the top-left corner of the canvas, or left-click that point, and you'll see a blue rectangle. That's the control point of the canvas, which we can use to move the canvas around, or to open its properties. For this canvas, I've set its width to 160, its height to 40, its **Receive symbol:** field to control_label, which is the same name [send] has in Figure 7-24, its **X offset** to 5, its **Y offset** to 16, its **Size:** (this if the font size) to 30, and its color to yellow. To set the color, click the yellow square in the **Colors** field. This time we're not controlling its color, or its size according to the messages we receive from the Arduino, like we did in Chapter 6. We're only setting its label; that's why we can set everything else using its properties, which is more user-friendly, than sending messages for each property. For the second canvas, the only differences are the width, which is 190, and the color, which is a default light green, from the squares in the Properties window.

Scaling the Output of the [read_samp] Abstractions in the Parent Patch

One thing that you might need to take care of is that the output of the [read_samp] abstractions is not scaled, so if you get two audio files being triggered simultaneously, or one triggered before the previous one has ended, and their amplitude sum is more than 1, you'll get a clipped sound. Since this is aimed at being used with a drum set, you might set it up in such a way that you won't have more than one audio file being triggered at a time. Also, the amplitude of the output of [read_samp] depends on how hard you hit a drum. For these reasons, I haven't included any scaling to their output. It depends on your playing style and how you're going to utilize this project, so have that in mind as you might need to apply some scaling to the output audio.

Finalizing the Arduino Code

Now that the Pd patch has been covered, let's go back to the Arduino code to make the last additions. What we haven't done yet is to allow Pd to control the threshold and debounce values for each sensor separately. Again, I'm not going to show the whole code, only the parts with the additions.

First of all, we now want to have separate threshold and debounce times for each sensor, therefore we need to create two new arrays for that, which are shown in Listing 7-8.

Listing 7-8. Adding Arrays for the Threshold and Debounce Values

```
1.   int threshold_vals[num_of_sensors];
2.   int debounce[num_of_sensors];
```

As you can imagine, these arrays will replace the two variables thresh, and debounce, which were defined in the beginning of the previous code, so make sure to erase them. Of course, these arrays need to be global, so place them before any function. I've placed them along with the rest of the arrays and variables for the sensors.

The next change occurs in the setup function, which is shown in Listing 7-9.

Listing 7-9. Initializing the New Arrays

```
1.   void setup() {
2.     for(int i = 0; i < num_of_sensors; i++){
3.       sensor_activity[i] = true;
4.       all_switch_vals[i] = 1;
5.       threshold_vals[i] = 100;
```

```
6.        debounce[i] = 10;
7.        pinMode((i + 2), INPUT_PULLUP);
8.        pinMode((i + 6), OUTPUT);
9.        digitalWrite((i + 6), sensor_activity[i]);
10.   }
11.
12.   Serial.begin(115200);
13. }
```

Since it's not big, I've included the whole function to avoid any errors that could occur while adding things to the existing code. Here we're just initializing the two new arrays with default values; all the values of the threshold_vals array are set to 100 and the debounce are set to 10.

The next addition is in the loop function, where we're checking for input in the serial line and diffusing the values to their destinations. This is shown in Listing 7-10.

Listing 7-10. Receiving Data from Pd Through the Serial Line

```
1.  if(Serial.available()){
2.    static int temp_val, pin;
3.    byte in_byte = Serial.read();
4.    if((in_byte >= '0') && (in_byte <= '9'))
5.      temp_val = temp_val * 10 + in_byte - '0';
6.    else if(in_byte == 'p'){
7.      pin = temp_val;
8.      temp_val = 0;
9.    }
10.   else if(in_byte = 't'){
11.     threshold_vals[pin] = temp_val;
12.     temp_val = 0;
13.   }
14.   else if(in_byte = 'd'){
15.     debounce[pin] = temp_val;
16.     temp_val = 0;
17.   }
18. }
```

If there is data in the serial line, we're creating two static ints, temp_val and pin, and then we're checking what kind of input we got. If it's numeric values we're assembling them and store them to temp_val. If it's 'p', we're copying the content of temp_val to pin and resetting temp_val to 0. If it's 't', we're assigning the value of temp_val to threshold_vals, to the element set by pin. Finally, if it's 'd', we're assigning the value of temp_val to debounce, to the element set by pin. So if we send the message "0p150t" from Pd, we'll store the value 150 to threshold_vals[0]. If we send the message "3p5d", we store the value 5 to debounce[3]. This way we can dynamically set a new threshold and debounce time for every sensor individually. This means that whenever you use this patch and code, you should upload the Arduino code only once to your board, and then enable "debug" with the toggle in the [drum] abstraction. Then you should check each sensor by hitting the corresponding drum and make sure that you don't get more than one value for each hit. If you get more than one, raise the debounce time. If you get only one, bring the debounce time even lower, until the lowest value that prints only one value per hit. The threshold will set how sensitive the sensor will be. This will also be helpful to avoid getting input from a sensor when you hit another drum, as your whole drum set may be vibrating and all the sensors might be giving output. You should bring the threshold to the lowest value where you get input from the sensor only when you hit the drum of that sensor. If you want your sensor to be less responsive, you should bring the threshold even higher. This type of control is very helpful because it enables the system to adapt to many different situations (different drum sets, different drum sticks, and so forth), without needing to tweak the code at all. If you want to save your new settings, all you need to do is change the arguments to the Pd abstractions.

The last addition to the code is in the for loop that goes through the sensors. The whole loop is shown in Listing 7-11 to avoid potential errors.

Listing 7-11. Final Version of the for Loop That Reads the Piezo Elements

```
1.   for(int i = 0; i < num_of_sensors; i++){
2.     if(sensor_activity[i]){ // check sensor activity
3.       int sensor_val = analogRead(i);
4.       if(sensor_val > threshold_vals[i]){
5.         sensor_thresh[i] = 1;
6.         if(sensor_thresh[i] != old_sensor_thresh[i]){
7.           // count how many times the sensor rises above threshold
8.           sensor_above_thresh[i] += 1;
9.         }
10.        sensor_max_val[i] = max(sensor_max_val[i], sensor_val);
11.        old_sensor_thresh[i] = sensor_thresh[i];
12.      }
13.      else{
14.        sensor_thresh[i] = 0;
15.        // get a below threshold time stamp for the sensor
16.        sensor_below_const[i] = (int)millis();
17.        if(sensor_thresh[i] != old_sensor_thresh[i]){
18.          // get a time stamp every new time the sensor drops below threshold
19.          sensor_below_new[i] = (int)millis();
20.        }
21.        old_sensor_thresh[i] = sensor_thresh[i];
22.        if((sensor_below_const[i] - sensor_below_new[i]) > debounce[i]){
23.          // make sure we print the values only once per hit
24.          if(sensor_above_thresh[i] > 0){
25.            Serial.print("drum");
26.            Serial.print(i); Serial.print("\t");
27.            Serial.println(sensor_max_val[i]);
28.            sensor_max_val[i] = 0;
29.            // zeroing sensor_above_thresh[i] will prevent this chunk of code
30.            // from be executed before we hit the drum again
31.            sensor_above_thresh[i] = 0;
32.          }
33.        }
34.      }
35.    }
36.  }
```

The only change is in line 4, where we now use threshold_vals[i] instead of thresh, and in line 22, where we use debounce[i] instead of debounce. Using arrays enables us to have a different value for each sensor, which gives us great flexibility. And this concludes the code in this project. Make sure that you've written everything correctly, and that your code and patch work properly. Once you have everything working as expected you can start building your circuit on a perforated board and enclose it in a box. Next, I'll provide some suggestions on how to make an enclosure to keep your project safe for transporting, and easy to carry and install anywhere.

Making the Circuit Enclosure

Bringing around a circuit on a breadboard is not really a good idea, because every time you'll probably need to build the circuit from scratch, as the breadboard is not stable and the circuit can get destroyed during transfer. The breadboards are designed only for testing circuits. What's best is to build the circuit on a perforated board and close it in a box so you can easily carry it around. Another good idea is to build your sensors in such a way that it's easy and safe to carry them around. Figure 7-26 shows a piezo element connected to a mono 1/8-inch jack. This way it's very easy to transport them and use them.

Figure 7-26. *Piezo element wired to a 1/8-inch jack*

Figure 7-27 illustrates a box that I have made for this project. On the front side, there are four quarter-inch female jacks for plugging in the foot switches, which usually come with a 1/4-inch male jack extension. Since the switches have two pins, these jacks are mono. On the top side of the box, there are the four LEDs that indicate which sensor is active and which is not, and the four female 1/8-inch jacks for the sensors. This setup makes it extremely easy to plug in your switches and sensors, and play.

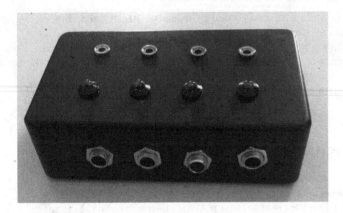

Figure 7-27. *Box containing the circuit of the project*

This box doesn't contain an Arduino, but it has a IDC male angled connector on its back side, which you can see in Figure 7-28. I've used a Proto Shield along with the Arduino, which makes prototyping very easy, giving easy access to all the pins of the Arduino.

Figure 7-28. *Back side of the box*

To connect the box to the Arduino I just used a ribbon cable (the kind of cable used with the IDC connectors) between it and the Proto Shield. You can see that connection in Figure 7-29.

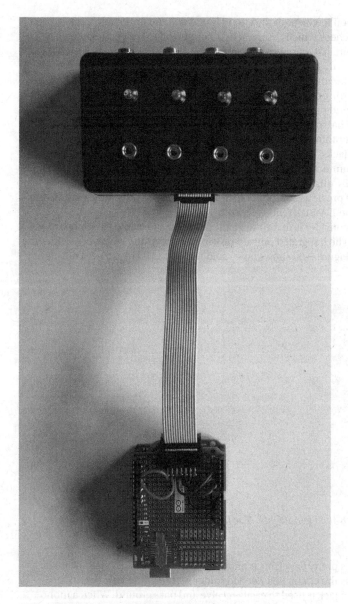

Figure 7-29. Connecting the circuit box to the Arduino

What you need to take care of is how you'll wire both the circuit board and the Proto Shield. First, connect the two IDC connectors and then check which component is connected to which pin of the IDC, so that every component goes to the correct pin. IDC connectors may be a bit confusing, and you'll probably need a multimeter to use its continuity to verify your connections. You'll also probably need to solder some wires on the back side of the circuit board and the Proto Shield.

I've already mentioned that when building circuits, it may be preferable to have only one component connected to ground on the circuit board, and pass it to the other components (the same goes for voltage, but we're not passing voltage here). Here, I have connected the first 1/4-inch jack to ground and I have then daisy chained all components. Another thing I have used is stereo female 1/8-inch jacks for the piezo elements. Even though they require mono jacks, using stereo in the box makes it easier to pass the ground to the other components. As soon as you connect the male mono jack, the first two pins of the stereo female jack will be connected (it's pretty simple why this happens, test for yourself with a multimeter). This way you can use one pin to receive ground from the previous component, and another pin to pass the ground to the next component. You could apply this to the 1/4-inch female jacks too.

Another thing that you might want to consider is to minimize the components on the circuit board as much as you can. In the same line of using the integrated pull-up resistors of the Arduino, you can solder the other resistors right on their components. Figure 7-30 shows how I soldered the 1MΩ resistors of the piezo elements on the female 1/8-inch jacks.

Figure 7-30. *1MΩ resistor soldered on a 1/8-inch female jack. The first two pins are used to receive and pass ground*

Soldering these resistors on the jacks is pretty easy (this is another advantage of using stereo female jacks with mono male jacks) and minimizes the circuit you need to solder on the perforated board. In Figure 7-30, you can also see how the stereo jack is used to easily receive and pass ground. When a mono male jack is inserted, the two wires on top are both grounded, since they both touch the same area of the male jack. The left one receives ground from the previous component, and the right one passes it to the next component. Mind, though, that this is true only when the male mono jack is plugged in. If you don't plug in a mono male jack, the two pins won't be connected. If you plug in a stereo male jack, again the two pins won't be connected, because they will be touching different parts of the male jack. If you omit connecting a piezo element, make sure you don't have any other components in the chain waiting to receive ground; otherwise, any components after the unconnected female jack won't be grounded and won't work. You can also wire the 220Ω resistors of the LEDs on their long legs, but I didn't do that. I prefer to solder these resistors on the perforated board instead.

On the one hand, this kind of enclosure makes the circuit very stable, easy to use, and safe to transfer; on the other hand, you don't need to "sacrifice" your Arduino since you're not enclosing it along with the circuit, enabling you to use it for other projects any time.

Conclusion

We have created a robust and flexible combination of hardware and software that enables us to extend the use of the drums. As with all projects, the Pd patch is only a suggestion as to what you can do with these sensors. You can try out different things with this patch and build something that suits your needs. The Arduino code is not so much of a suggestion, as it provides a stable way for these sensors to function as expected. This also applies to the part of the Pd patch that reads the sensor data (the [drum] abstraction).

The threshold values used here are most likely very low, and you'll probably find that you need much higher values to avoid getting input from one sensor when you hit the drum of another sensor. A very low threshold might also result in getting more than one value per hit, even if you use a rather high debounce time value. Setting these two values is a matter of personal playing style, what you actually want to achieve, and the drum set you use.

Another thing to bear in mind is that you might not want to use sensors on drums that share the same base (tom drums, for example, which are based on the kick drum). Also, depending on the tuning of the drums, hitting a drum might cause another drum to resonate, which can also be a factor you'd like to experiment with. Regardless of these aspects, your sensor setup should bring in a lot of inspiration and fun.

CHAPTER 8

■ ■ ■

A DIY Theremin

In this chapter, we're going to build an interface that's very similar to the very popular *theremin*, an electronic musical instrument originally built by Léon Theremin in the 1920s. The theremin allows the performer to have no physical contact with the instrument; he or she can control both the pitch and amplitude of the instrument with two antennae that sense the distance of the performer's hands from it. The instrument uses oscillators to produce sound, making it a purely electronic instrument. In this chapter, we'll replicate the theremin by using two proximity sensors for the controls, and a few potentiometers for some other controls that are available in today's version of the instrument.

Parts List

To build this project we'll need all the components listed in Table 8-1.

Table 8-1. Theremin Parts List

Part	Quantity
Arduino (Uno or Nano will do)	1
Proximity infrared sensors (I've used the Sharp 2Y0A21)	2
10KΩ potentiometers	4
Push buttons to mount on a panel	4
Switches to mount on a panel	2
LEDs	4
LED holders	4
Raspberry Pi (or other embedded computer, if enclosing the computer)	1
Proto Shield (if you're using an Arduino Uno)	1
Perforated board (if you're using an Arduino Nano, or other microcontroller)	1
1/4-inch female jack (if the computer will be embedded into the enclosure)	1
Wire to make all the circuit connections	Length of your choice

Using a Proximity Sensor with the Arduino

Before we start building this project let's take a first look at the proximity sensors we'll be using. There are various types of such sensors, including infrared and ultrasound, but we'll use infrared because they are wired to the analog pins of the Arduino and are more straightforward in their use. There are some libraries that can be used to convert the output of the sensor to a scale that is easier to read, like centimeters, but in our case, it's not really necessary, so we won't bother to use any of them.

The infrared proximity sensors have an LED that emits infrared light that we humans can't see, and a receiver that detects the reflection of this light. According to the time spent for the light to be reflected back to the device, the sensor outputs values to express this amount of time in distance. The sensors that I'm using for this project are rather inexpensive, but you might find even cheaper ones. Let's take a look at the code that reads one proximity sensor, which is shown in Listing 8-1.

Listing 8-1. Reading One Proximity Sensor

```
1.    // we'll use Serial.write so we need to split analog values to two
2.    // we also need a unique byte at the beginning of the data stream
3.    const int num_of_data = 3;
4.    byte transfer_data[num_of_data] = { 192 };
5.
6.    void setup() {
7.      Serial.begin(57600);
8.    }
9.
10.   void loop() {
11.     int index = 1;
12.
13.     int sens_val = analogRead(0);
14.     transfer_data[index++] = sens_val & 0x007f;
15.     transfer_data[index++] = sens_val >> 7;
16.
17.     Serial.write(transfer_data, num_of_data);
18.   }
```

Here we're using the write function of the Serial class, instead of print, which writes raw bytes to the serial line. This means that any value above 255 (which is the highest value of a byte) will be wrapped back to 0. In order to obtain the 10-bit resolution of the analog pins of the Arduino we must split any analog value to two, and reassemble it back in Pd. This is done in lines 15 and 16; it was explained in Chapter 2.

With the write function, it's complicated to send values to Pd tagged with a string, because the string will arrive in ASCII in the serial line, but the numeric value will arrive as a raw byte. [serial_print] takes only ASCII values, so it won't work with write. What's easy to do with this function, and quite effective, is to send data in one piece, meaning an array. In this case, write can prove to be simpler than print, because we just call it once and send a whole array to the serial line. In the code in Listing 8-1 we're sending one value only, but split in two bytes, plus one byte to denote the beginning of the data stream (192 which is initialized in line 4). For this piece of code we could have used print instead, but later on we'll send more values, all as one piece because we won't really have to distinguish between these values, so we'll see the advantage of write in such cases.

Refreshing Our Memory on Serial.write

We'll go through the code in Listing 8-1 to refresh our memory a bit. The first two lines contain a comment saying how many bytes we need to send to the serial line. In line 4, we define the array that will hold these values, initializing its first element to 192, which will denote the beginning of the data stream. Then in line 11, we define a local variable called index, initializing it to 1, which will set the index to write values to the transfer_data array. In line 13, we're reading the first analog pin. In lines 14 and 15, we're disassembling this value to two bytes, stored in two sequential locations in the array, using the post-increment technique. This means that in line 14, index will hold 1 (the value that has been assigned to it in line 11) until this line of code has been executed. Then it will be incremented by one, so in line 15, it will hold 2, after which it will again be incremented by one, but we don't care anymore because we stop writing values to the array. In line 15, we could have omitted the two plus signs, but when we'll be reading more analog pins, we'll use this technique inside a loop, so it will be necessary that both of these lines use the post-increment technique. In line 17, we're writing the whole array to the serial line, which we'll receive in Pd and assemble using the [serial_write] abstraction.

Proximity Sensor Circuit

Figure 8-1 shows the circuit with the proximity sensor. It is actually an extremely simple circuit. It's just the proximity sensor with its red wire connecting to Arduino's 5V, the black wire to GND, and the yellow wire (it might be white on some) to analog pin 0. Make these connections and upload the code in Listing 8-1 to your Arduino board.

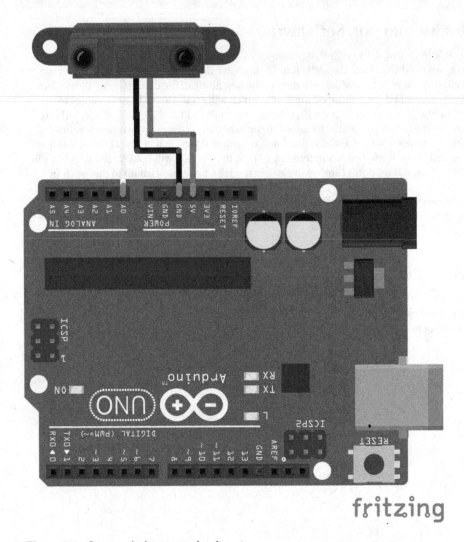

Figure 8-1. *One proximity sensor circuit*

Pd Patch That Reads the Proximity Sensor Data

Figure 8-2 illustrates the Pd patch that receives data from the proximity sensor. As with the code and circuit, it's a very simple patch. We're using the [serial_write] abstraction with the arguments "analog 1", which means that we're expecting one analog value. Then [serial_write] outputs this value assembled back to its 10-bit resolution out its left outlet. As you move your hand closer to the sensor, you should see the values in the number atom rising, and as you move it further away, these values should be falling.

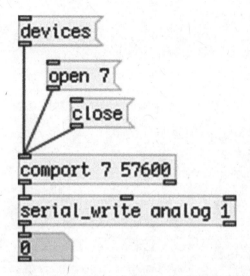

Figure 8-2. *Pd patch that receives data from the proximity sensor*

One thing you might have noticed is that the sensor is quite noisy, meaning that its values are bouncing up and down a bit. There are a couple of things we can do to remedy this. First of all, we can set a low and high threshold, so we can have a steady floor and ceiling. We do this with the object [clip]. In the case of the infrared sensor, a good low threshold could be 40, and a good high threshold could be 600. So connect [serial_write] to [clip 40 600] and then to the number atom. Now when your hand is far from the sensor, the value in the number atom should more or less stay at 40, and when you get close to the sensor it should stay at 600. If you get very close to it, almost touching it, the values start falling again. This is probably due to light reflections, and the sensor is not receiving correct data. Mind the distance where you get the highest value and use that as the minimum distance you can have from the sensor.

Smoothing out the Sensor Values

The first thing we can do to reduce the noise of the sensor is to use a technique provided from the Arduino web site, which smooths out noisy analog input. Listing 8-2 shows this code modified for this project.

Listing 8-2. Smoothing out the Values of the Sensor

```
1.   byte transfer_data[3] = { 192 };
2.
3.   const int num_of_readings = 20;
4.   int readings[num_of_readings];
5.   int read_index = 0;
6.   int total = 0;
7.   int average = 0;
8.
```

```
9.  void setup() {
10.    for(int i = 0; i < num_of_readings; i++)
11.      readings[i] = 0;
12.    // put your setup code here, to run once:
13.    Serial.begin(57600);
14.
15.  }
16.
17.  void loop() {
18.    // put your main code here, to run repeatedly:
19.    int index = 1;
20.
21.    // subtract the last reading:
22.    total = total - readings[read_index];
23.    // read from the sensor:
24.    readings[read_index] = analogRead(0);
25.    // add the reading to the total:
26.    total = total + readings[read_index];
27.    // advance to the next position in the array:
28.    read_index++;
29.
30.    // if we're at the end of the array…
31.    if (read_index >= num_of_readings) {
32.      // …wrap around to the beginning:
33.      read_index = 0;
34.    }
35.
36.    // calculate the average:
37.    average = total / num_of_readings;
38.
39.    transfer_data[index++] = average & 0x007f;
40.    transfer_data[index++] = average >> 7;
41.
42.    Serial.write(transfer_data, 3);
43.  }
```

The comments in the lines where the smoothing occurs are taken from the Arduino web site, where this technique is introduced. Each reading of the sensor is added to a table. Before getting a new reading, we subtract the old reading at the current position of the table from the total, and then store the current reading in that position of the table. We add it to the total, and then divide the total by the number of readings set in line 3.

If you upload this code to your board and open the serial port in the patch in Figure 8-2 (which should now include a [clip 40 600] between [serial_write] and the number atom), you should see that the floor and ceiling values are more stable than before. What we want to do with these values is to control the frequency of an oscillator. Figure 8-3 illustrates a basic test, where the incoming values are being mapped to a range from 45 to 69 (MIDI notes from A to A, three octaves in total ending in the tuning A). We take the integral part of these values with [i], and then we're converting them to the respective frequencies of the MIDI notes using the [mtof_tune] abstraction. We're using [line~] to smooth the changes between successive notes.

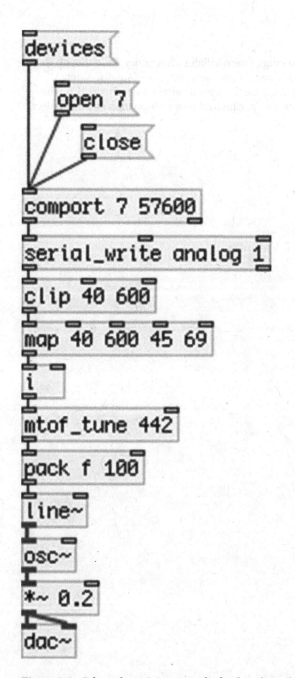

Figure 8-3. Pd patch receiving smoothed values from the infrared sensor

Smoothing Out Even More

Still with this setup you'll probably get quite some bouncing of the oscillator's frequency. Even though we're restricting the value stream to integers of the MIDI scale, the oscillator still gets a lot of small frequency changes that are quite audible and probably annoying. To make this sound better we need to work on the value smoothing even more. Figure 8-4 illustrates the next step, which adds one abstraction and one object.

Figure 8-4. Smoothing out the infrared sensor data even further

The abstraction [smooth_out] is an abstraction I built especially for this project, hence I've not included it in my "miscellaneous_abstractions" repository in GitHub. Figure 8-5 illustrates it. What it does is store the incoming values to a table, the size of which is determined by the first argument. The second argument sets the initial value, which is the lowest value [map] will output, and we're using it here to initialize its process. When a new value comes in, [smooth_out] will iterate through all the values stored in its table and will test for equality with the initial value. If a value is equal to the initial value, it will bang the counter on the bottom-left part of the abstraction. This counter needs to be banged as many times as the third argument sets (here it's 3). If it's banged three times, the initial value, which has been stored in [f] on the bottom side of the abstraction, will pass.

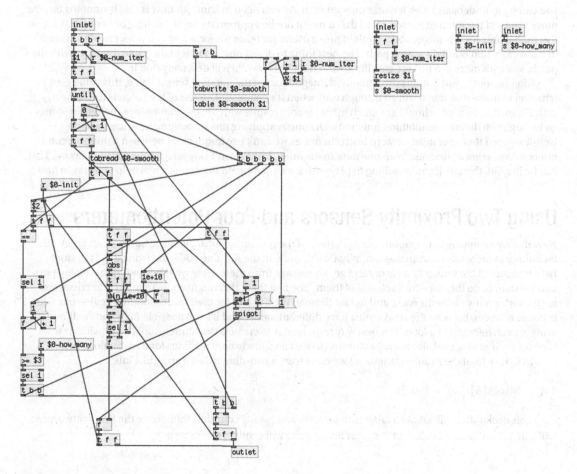

Figure 8-5. *The [smooth_out] abstraction*

At the same time it will subtract its initial value from each value of its table, taking the absolute of this difference and it will store the minimum of these values. This is achieved by initializing [min] with a very large value, 1e+10, which means 1×10^{10} (one times ten to the tenth power). Any value that will come in [min] will definitely be smaller than this value. If the abstraction fails to find the initial value at least three times, the minimum difference between the initial value and all the values of its table will pass, and will be saved as the value to compare the next incoming value against. On the top side of the abstraction we're using [% $1], which makes sure that we won't go over the bounds of the table as we store each incoming value to it. There are also inlets for each argument, so you can experiment with these three values to see what works best for you.

This abstraction is somewhat complex, but it's not really necessary to explain it thoroughly. If you want, you can try to understand how it works on your own. All you need to know for now is that it smooths out the noisy values of the infrared sensor. Mind that it might not be appropriate for smoothing out any noisy input, I've built it only for this project. Still it might prove useful for other projects that you work on. Make sure it's in a directory that is in Pd's search path. The best thing to do for now is put it in the same directory with the patch, since it's more of a project-specific abstraction. This way Pd will definitely find it.

After [smooth_out] outputs the smoothed integral part of the mapped sensor value, this value goes to [change] to make sure that it will go through only when it's changed. The rest of the patch is the same as with Figure 8-3. Now you should get much more smooth results with the oscillator, since the sensor values go through two different smoothing functions. Of course, applying this smoothing, restricts us to the well-tempered tuning of most western instruments, as we can't use frequencies between a minor second interval. We do get a glissando from one note to the other, but we can't stay on a frequency in between. This can be helpful, though, if you're willing to play with a well-tempered tuning, as it will help you stay in tune.

Using Two Proximity Sensors and Four Potentiometers

Now that we've managed to smooth out the values of the proximity sensor, let's see how we can apply this technique to more than one analog pin. What we've done in the Arduino code, in Listing 8-2, is to store twenty values of the sensor to an array and get an average from there. Now we'll be reading six analog pins and we want to do the same for each one of them. The first thing that might come to mind is to create six arrays with twenty elements each, and iterate through their values for their corresponding analog pin. This is not so efficient, because the arrays must have different names and it's not possible to include a different array at each iteration of a loop. The best way to go here is to create a two-dimensional array with six rows (one for each analog pin) and twenty columns (so we can store twenty readings for each analog pin).

In C/C++ (and the Arduino language) we can create a two-dimensional array like this:

```
int readings[6][20] = { 0 };
```

This declaration will create a table with six rows and twenty columns. Initializing the first value to zero will actually initialize all values to 0, so this line of code will result in the Table 8-2.

Table 8-2. *A Two-Dimensional Array with Six Rows and Twenty Columns with All Values Initialized to 0*

0	0	0	0	0	0	0	0	0	0	0	0	0	0	0	0	0	0	0	0
0	0	0	0	0	0	0	0	0	0	0	0	0	0	0	0	0	0	0	0
0	0	0	0	0	0	0	0	0	0	0	0	0	0	0	0	0	0	0	0
0	0	0	0	0	0	0	0	0	0	0	0	0	0	0	0	0	0	0	0
0	0	0	0	0	0	0	0	0	0	0	0	0	0	0	0	0	0	0	0
0	0	0	0	0	0	0	0	0	0	0	0	0	0	0	0	0	0	0	0

To access an element of a two-dimensional array we need to first provide the row of the array and then the column. For example, the following line of code would result in Table 8-3.

```
readings[1][5] = 20;
```

Table 8-3. *Assigning a Value to the Second Row and Sixth Column of Table 8-2*

0	0	0	0	0	0	0	0	0	0	0	0	0	0	0	0	0	0	0	0
0	0	0	0	0	20	0	0	0	0	0	0	0	0	0	0	0	0	0	0
0	0	0	0	0	0	0	0	0	0	0	0	0	0	0	0	0	0	0	0
0	0	0	0	0	0	0	0	0	0	0	0	0	0	0	0	0	0	0	0
0	0	0	0	0	0	0	0	0	0	0	0	0	0	0	0	0	0	0	0
0	0	0	0	0	0	0	0	0	0	0	0	0	0	0	0	0	0	0	0

With this line of code we have assigned the value 20 to the element of the second row and sixth column of the two-dimensional array. We use the same approach to read from a two-dimensional array, where the following line of code would assign the value 20 to the variable var (if we have already assigned 20 to that position of the two-dimensional array).

```
int var = readings[1][5];
```

The Arduino Code

Now that I've explained the two-dimensional array we can have a look at the code that reads six analog pins and performs smoothing to these readings. It is shown in Listing 8-3.

Listing 8-3. Smoothing the Values of Six Analog Pins

```
1.   // analog values array size, must be constant
2.   const int num_of_analog_pins = 6;
3.
4.   // assemble number of bytes we need
5.   // analog values are being split in two, so their number times 2
6.   // plus a unique byte to denote the beginning of the data stream
7.   const int num_of_data = (num_of_analog_pins * 2) + 1;
8.
```

```
9.    // array to store all bytes
10.   byte transfer_array[num_of_data] = { 192 };
11.
12.   // set the number of readings to smooth out the noise
13.   const int num_of_readings = 10;
14.   // set a two dimensional array to store 10 readings for each analog pin
15.   int readings[num_of_analog_pins][num_of_readings] = { 0 };
16.   // set variables to accumulate and average readings
17.   int total[num_of_analog_pins] = { 0 };
18.   int read_index = 0;
19.   int average = 0;
20.
21.
22.   void setup() {
23.     Serial.begin(57600);
24.   }
25.
26.   void loop() {
27.     int index = 1; // index offset
28.
29.     // store the analog values to the array
30.     for(int i = 0; i < num_of_analog_pins; i++){
31.       total[i] -= readings[i][read_index];
32.       readings[i][read_index] = analogRead(i);
33.       total[i] += readings[i][read_index];
34.       average = total[i] / num_of_readings;
35.       // split analog values so they can retain their 10-bit resolution
36.       transfer_array[index++] = average & 0x007f;
37.       transfer_array[index++] = average >> 7;
38.     }
39.
40.     read_index++;
41.     if(read_index >= num_of_readings) read_index = 0;
42.
43.     // transfer bytes over serial
44.     Serial.write(transfer_array, num_of_data);
45.   }
```

In line 2, we define the number of analog pins we'll use. In line 7, we define the number of bytes we'll be transferring to Pd over serial. The comments above this line explain why we need this number of bytes. In line 15 we define the two-dimensional array that will store the readings for each analog pin. We use the two constants num_of_analog_pins, and num_of_readings to define the number of rows and columns for the array. Lines 17, 18, and 19 define the three variables we used in Listing 8-2 for the smoothing process, only this time the variable that will accumulate these readings is defined as an array total[num_of_analog_pins], since we need one accumulated value for each analog pin.

The next new thing in this code is the for loop in line 30. This loop creates a variable called i, which is used to iterate through the six analog pins. At the first iteration of the loop in line 31, we subtract the value stored at the first row and first column of the two-dimensional array readings (since both i and read_index are 0) from the first element of the total array, which stores the accumulated values. In line 32, we assign to that position of the two-dimensional array the value read by the first analog pin (pin 0). In line 33, we add the value we just read to the first element of the total array. In line 34, we divide the first element of the total array by the number of readings we've set and assign the result to the average variable.

Both total and readings are arrays since they need a separate value or set of values for each analog pin. average is a single variable since at every iteration of the loop it will be assigned the average value of the readings of each pin. We don't need that average for anything else other than writing it to the serial line, whereas each element of the total array must be maintained throughout the program to perform the smoothing we want. The same applies to the readings array, which needs to maintain all its elements so that they can be accumulated in the total array.

The rest of the code is the same with Listing 8-2, so I'm not going to explain it here. This time, of course, we're transferring more bytes than what we were transferring in Listing 8-2, since we're reading more pins. Still the code is the same.

The Circuit

Now let's take a look at the circuit for this code (see Figure 8-6). This circuit is pretty straightforward. The only thing that might look a bit strange is that the first infrared sensor is wired to analog pin 0, whereas the second is wired to analog pin 2. This is just to make the Pd patch a bit nicer as we'll be using the first potentiometer to control the sensitivity of the sensor, and this wiring will make the patching a bit easier (it's actually a minor detail, I just like to have it like that, you'll see it in the Pd patch).

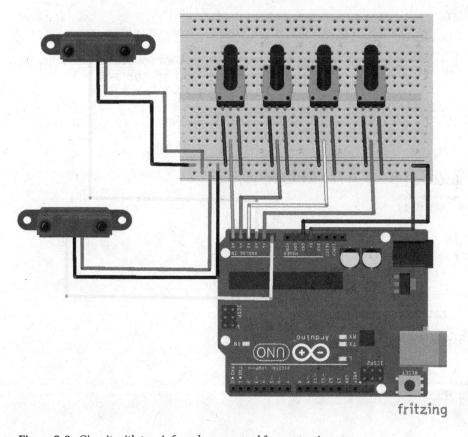

Figure 8-6. *Circuit with two infrared sensors and four potentiometers*

The Pd Patch

For now let's take a quick look at the Pd patch that reads these six values. Figure 8-7 illustrates it. Again, we're using the [serial_write] abstraction but this time with the arguments "analog 6", since we'll be reading six analog values. I've only added the mapping and smoothing of the values just to show how we're going to use our sensors at the final stage of the project. The first potentiometer, which is the second analog value of the Arduino, controls the clipping and mapping of the first infrared sensor. At its lowest position, the potentiometer will set the sensor to its full sensitivity, like we've already used it. At its highest position, it will constrain this sensitivity to a smaller range, as the sensor values above 500 will be clipped, resulting in a smaller range (clipping will occur at a position further away from the sensor, resulting in less responsiveness in short ranges).

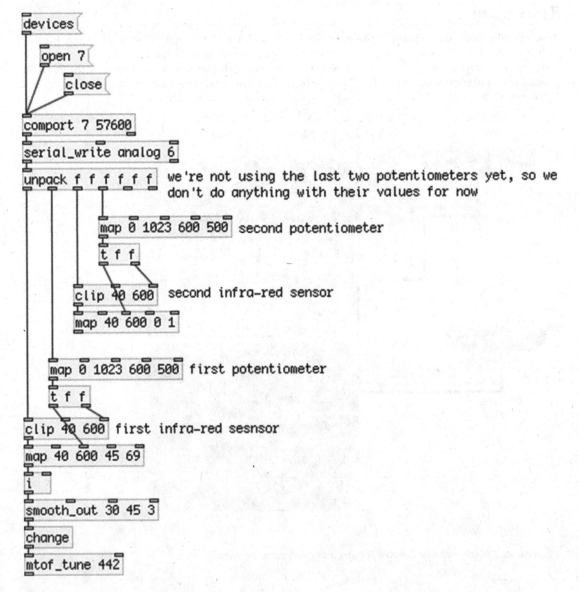

Figure 8-7. *Reading six analog pins of the Arduino in Pd*

The same technique is applied to the second infrared sensor. This patch should make clear why we've wired the second sensor to pin 2. If we had wired it to pin 1, the connection of the first potentiometer would have to go over the connection of the second infrared sensor, making the patch not very "clean."

Notice that we're not using [smooth_out] with the second sensor, as mapping its values to a range from 0 to 1 makes the sensor's noise less audible.

Building the Oscillators for the Theremin

Let's start building the audio part of the patch. We'll use two oscillators where one will be modulating the phase of the other. This way, we'll be able to control the brightness of the sound, which is one aspect the theremin has. For this project we'll use band-limited waveforms. To do this that we'll use a native Pd feature that calculates various sine waves as partials and sums them to create a more complex waveform. Let's take a look at this feature first.

Creating Band-Limited Waveforms

To create a sine wave using this feature, we have to create an array where the waveform will be stored, and then we must send the following message to the array: "sinesum 2048 1".

"sinesum" is the feature that sums sine waves to create waveforms. The first number in this message is the size of the array, which must be a power of 2. If it's not, Pd will round it to the closest power of 2 that's smaller than this value (for example, if this number was 2000, Pd would round it to 1024). The second value is the amplitude of the first harmonic of the waveform. In this case, we create one sine wave only, with full amplitude. Figure 8-8 illustrates how this works. This message will automatically resize the array to the size we've specified, it will then create the waveform resulting from the sum of the partial amplitudes (where in this case, it's only one partial) and then it will add the *guard points*, which are necessary for applying cubic interpolation when reading this table. This last bit is not necessary to grasp; it's only mentioned because if you click the array's properties, you'll see that its size is 2051 instead of 2048 (due to the guard points, which are three). Explaining how the guard points are being used is beyond the scope of this book. If you want to learn more about it, look for information on cubic interpolation.

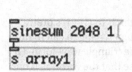

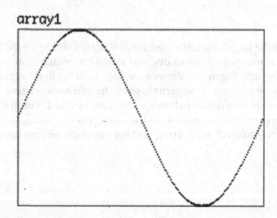

Figure 8-8. Using sinesum to create a sine wave

Of course, making a sine wave is not very helpful, since Pd already has another two ways to create sine waves ([osc~] and [phasor~] connected to [cos~]). What this feature is really helpful for, is creating other types of waveforms.

Creating a Triangle Waveform with sinesum

Let's first build the other three standard waveforms we've already been using, the triangle, the sawtooth, and the square wave.

What we need to do is provide the partial strengths (the amplitude of each harmonic) for each waveform. Let's start with the triangle. The strength of each partial is defined by the following equation:

```
partials[i] = 1 / i^2
```

This means that we divide 1 by the number of the partial raised to the second power. That's not all for this waveform. We also need to use only odd harmonics, and all even harmonics must have 0 amplitude. Lastly, every odd harmonic we must change the sign, so the first is positive, the second is negative, and they keep on alternating. If we were to write code in C++ to obtain these harmonics, we should write the code in Listing 8-4.

Listing 8-4. C++ Code to Obtain the Partial Strengths of a Triangle Waveform

```
const int num_of_partials = 40;
float partials[num_of_partials];
int partial_sign = 1;

for(int i = 1; i <= num_of_partials; i++){
        if((i % 2) == 0) partials[i - 1] = 0;
        else{

                partials[i - 1] = partial_sign * (1 / (pow(i, 2)));
                partial_sign *= -1;
        }
}
```

We're starting the loop from 1, because the first partial is considered to have index 1. In C++ it's pretty simple to create the array with these values, and for sure it's much more preferable than having to calculate them by hand every time. Figure 8-9 shows a message with the first ten partials for the triangle waveform. Of course, we could have just used nine partials, since the odd ones are zero. You can see that the waveform has gone over the boundaries of the digital audio, which are –1 and 1. This happens with both the triangle and the sawtooth, whereas with the square wave the result is below the audio boundaries. To fix this we need to send the message "normalize" to the array, and the waveform will get back to the proper range.

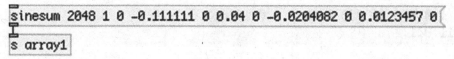

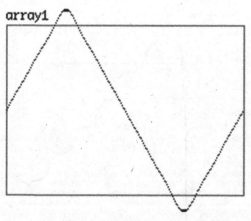

Figure 8-9. *A non-normalized band-limited triangle waveform with ten partials*

Creating a Sawtooth Waveform with sinesum

To create a sawtooth waveform is much simpler. The equation to obtain the partial strengths is this:

partials[i] = 1/i

This means that all we need to do is divide one by the number of the partial. Figure 8-10 illustrates a sawtooth waveform with ten partials that hasn't been normalized.

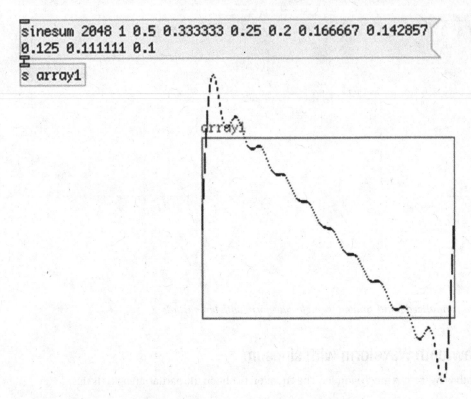

```
sinesum 2048 1 0.5 0.333333 0.25 0.2 0.166667 0.142857
0.125 0.111111 0.1
```

```
s array1
```

Figure 8-10. *A non-normalized band-limited sawtooth waveform with ten partials*

Creating a Square Waveform with sinesum

Lastly, the equation to obtain the partial strengths of a square wave is the same as with the sawtooth, only this time we're using only the odd partials. Figure 8-11 illustrates a square waveform with ten partials that hasn't been normalized. You can see that its amplitude is below the digital audio range.

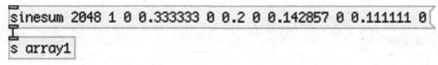

```
sinesum 2048 1 0 0.333333 0 0.2 0 0.142857 0 0.111111 0
s array1
```

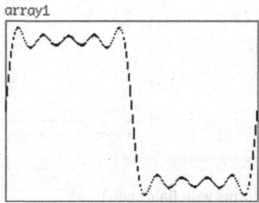

array1

Figure 8-11. *A non-normalized band-limited square waveform with ten partials*

It can be a bit cumbersome to calculate these partial strengths, but I've made some abstractions that do this automatically. You can find them on GitHub at https://github.com/alexdrymonitis/band_limited_ waveforms. These are [bl_trianble], [bl_sawtooth], [bl_squarewave], [bl_impulse], and [bl_fibonacci]. We haven't talked about the last two: impulse and fibonacci. An *impulse waveform* has all its partial strengths set to 1, and [bl_fibonacci] uses the Fibonacci numbers to set the partial strengths. We're going to use these abstractions (except from [bl_fibonacci]) to easily create four different waveforms for out theremin. Check their help patches for more information.

These waveforms might not look exactly like the shape their names imply, but that's only because we've been using very few partials. In our project we'll use more partials, 50 or more, and the waveforms will look much more like the shape they're supposed to have.

Reading the Stored Band-Limited Waveforms

There are two ways to read these waveforms, once we have stored them in an array. The easiest way is to use [tabosc4~], which is built to read such waveforms. Figure 8-12 illustrates a part of the help patch of [bl_ triangle], where this object is used.

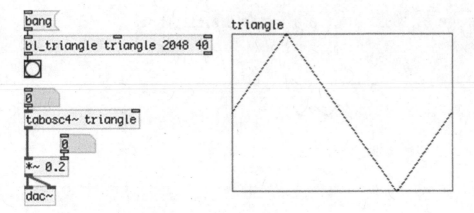

Figure 8-12. Part of the help patch of [bl_triangle]

Reading Waveforms with [tabosc4~]

As you can see, all we need to do to use [tabosc4~] is to provide a frequency to its left inlet and the table to read the waveform from as an argument. Also the waveform looks more like a triangle, since we're using 40 partials (the third argument of [bl_triangle]) and it's also normalized, so it stays within the digital audio range. The number 4 in the object's name stands for the cubic interpolation, also called *four-point interpolation.*

There are two disadvantages with [tabosc4~]. First, the table it will read from must have a size of a power of two plus three (for the three guard points). If the table doesn't have such a size, Pd will throw an error to its console and [tabosc4~] won't produce any output. The second disadvantage is that we don't have access to its phase, so if we want to do phase modulation, we can't use this object.

Reading Waveforms with [tabread4~]

The other way to read stored waveforms is to use [tabread4~] (again, 4 stands for *four-point interpolation*). Figure 8-13 illustrates how to read a waveform with [tabread4~], in the same way as with Figure 8-12. Not much complexity has been added. To read the waveform correctly, we need to use a [phasor~], which will be the driving force for our oscillator. We multiply its output by the size of the table before the guard points were added (so 2048 and not 2051), add 1, so we start reading from the second index (the first index is now part of the guard points and we shouldn't read it with [phasor~]) and send the result to [tabread4~]. Since [tabread4~] applies cubic interpolation, it must read from index 1 to last index, minus 2. The very first and the last two indexes of the table are the guard points.

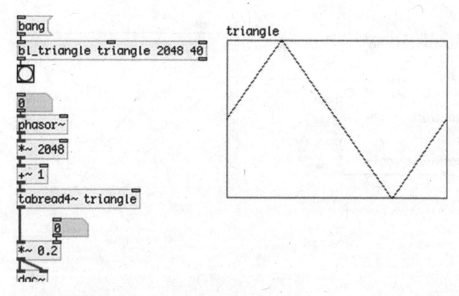

Figure 8-13. *Reading a stored waveform with [tabread4~]*

This way of reading waveforms is a bit more complex, but now we have access to the phase of the band-limited oscillator (which is actually the [phasor~]). Another advantage of [tabread4~] compared to [tabosc4~] is that the former can read tables that don't have a size of a power of two plus three. It is independent of specific table sizes. This doesn't really matter in the case of waveforms built with the sinesum feature of Pd, since this feature needs to use sizes that are a power of 2 and adds the three guard points.

Applying Phase Modulation to a Band-Limited Oscillator

Before we move on to build the rest of the patch, let's look at how we can apply phase modulation to a band-limited oscillator. We'll use [tabread4~] for the carrier oscillator, and [tabosc4~] for the modulator, since we don't need to have access to its phase. To refresh our memory, modulating the phase of an oscillator means to interfere between its phase and the waveform table (the kind of oscillator that reads the waveform out of a table is called a *table lookup oscillator*). If we use the patch in Figure 8-13, what we need to do is to place a [+~] between [phasor~] and [*~ 2048], where the right inlet of [+~] will get input from [tabosc4~] (after it's been scaled). Figure 8-14 illustrates this technique.

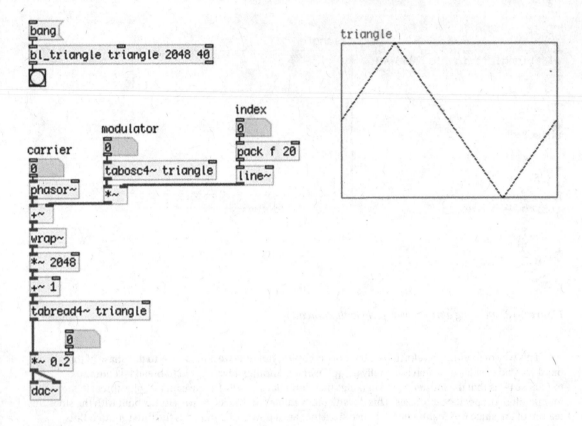

Figure 8-14. Applying phase modulation to a band-limited oscillator

Notice that after [+~] we've placed [wrap~]. Since the signal of an oscillator is bipolar, adding it to a [phasor~] might result in negative values (if [phasor~] outputs 0.1 and [tabosc4~] outputs –0.5, the result will be –0.4). These negative values will be fed to [tabread4~], which will look for negative indexes in the waveform table, which don't exist. In general, in programming, when trying to read a value from an array beyond its bounds (by providing an index that's either greater than its maximum index, or smaller than its minimum, which is 0) will yield unexpected results or will even lead to a crash of the program. Many programming environments provide safety checks for such occasions. Pd clips incoming index values to the maximum index of the table. So if we send the value 3000 to [tabosc4~], it will output the value at index 2050 (this is the last index after the guard points have been added, which resize the array to 2051). If we send the value –2, [tabread4~] will output the value at index 0. Still, even though there are safety checks so that we don't get unexpected results or crashes, the right thing to do is wrap the addition of [phasor~] and [tabosc4~] around 0 and 1 to make sure that the indexing of the table, after being multiplied by the size of the waveform, will always be between the waveform bounds.

Finalizing the Interface

We've now built our two oscillators and combined them. What's left to do is build all five waveforms so that we can choose between them. Before we move on to that part, we must first finalize our interface. Since this project is aimed at using either the Arduino Uno or the Nano, we must restrict ourselves to the number of analog pins of the Uno (the Nano has eight analog pins, whereas the Uno has only six). We must think of what we want our instrument to be capable of, in combination with the capabilities of the Arduino.

Adding Push Buttons to the Arduino Code to Control the Waveforms of the Oscillators

Since we'll be using two infrared sensors, we're left with four analog pins and all the digital ones. We've already written code that uses the rest of the analog pins with potentiometers, so that's what we're going to use. What we haven't really decided yet is what the fourth potentiometer will control. The first one controls the sensitivity of the pitch sensor, the second controls the sensitivity of the amplitude sensor, and the third controls the overall brightness, which will actually be the index of the modulation. What about the last potentiometer? Since this is an instrument aimed at being used in various situations, and since we needed to constrain it to the MIDI notes to eliminate the noise of the sensor, it is a good idea to use the last potentiometer for tuning the instrument. We can use it to send the frequency of the tuning A to the [mtof_tune] abstraction. But if we do that, how can we control the waveform of the carrier oscillator and the modulator independently?

We're left only with digital pins, so this is what we must use. There are 12 digital pins. The first thing that might come to mind is to use four push buttons for the carrier and another four for the modulator. If we do that, we won't have any visual feedback as to what waveform each oscillator has, and we're left with only four digital pins, which are not enough for us to use LEDs for this task. One solution is to use four push buttons, one for each waveform, for both oscillators, four LEDs that will indicate which waveform has been chosen, and a switch to determine whether we'll be choosing a waveform for the carrier or for the modulator. This switch also enables you to visualize which waveform is chosen for which oscillator at any time, since we can use it only to display the two choices without using the push buttons. With this setup, we're using 9 digital pins, so we're good to go.

The code for this task might be a little bit tricky, but if explained well, it should be easy to understand. It is shown in Listing 8-5.

Listing 8-5. Using Switches and LEDs to Control the Waveform of the Two Oscillators

```
1.   const int num_of_switches = 4;
2.   const int num_of_leds = 4;
3.   const int num_of_data = 3;
4.
5.   byte transfer_data[num_of_data] = { 192 };
6.
7.   int old_switches[num_of_switches];
8.   int car_or_mod[2] = { 0, 0 };
9.   int old_car_or_mod[2] = { 0, 0 };
10.  int old_ctl_switch = 1;
11.  bool button_pressed = false;
12.  bool ctl_switch_changed = false;
13.
```

```
14.  void setup() {
15.    for(int i = 0; i <= num_of_switches; i++){
16.      pinMode((i+2), INPUT_PULLUP);
17.      // the old_switches array has four elements, not five
18.      // so make sure you don't exceed its bounds
19.      if(i < num_of_switches) old_switches[i] = 1;
20.    }
21.
22.    for(int i = 0; i < num_of_leds; i++) pinMode((i+7), OUTPUT);
23.    // triangle oscillator is the default one, so turn the first LED on
24.    digitalWrite(7, HIGH);
25.
26.    Serial.begin(57600);
27.  }
28.
29.  void loop() {
30.    int index = 1;
31.
32.    // use the switch to control what the push buttons will set
33.    int ctl_switch = digitalRead(2);
34.    if(ctl_switch != old_ctl_switch){
35.      ctl_switch_changed = true;
36.      old_ctl_switch = ctl_switch;
37.    }
38.
39.    for(int i = 0; i < num_of_switches; i++){
40.      int switch_val = digitalRead(i + 3);
41.      if(switch_val != old_switches[i]){
42.        if(!switch_val){ // if button is pressed
43.          // save its number to either the carrier or modulator element
44.          car_or_mod[!ctl_switch] = i;
45.          button_pressed = true;
46.          old_switches[i] = switch_val;
47.        }
48.        else old_switches[i] = switch_val;
49.      }
50.    }
51.
52.    if(ctl_switch_changed){
53.      // if we change the switch, show which LED of the current position is on
54.      digitalWrite((car_or_mod[ctl_switch] + 7), LOW);
55.      digitalWrite((car_or_mod[!ctl_switch] + 7), HIGH);
56.      ctl_switch_changed = false;
57.    }
58.    if(button_pressed){
59.      digitalWrite((old_car_or_mod[!ctl_switch] + 7), LOW);
60.      digitalWrite((car_or_mod[!ctl_switch] + 7), HIGH);
61.      old_car_or_mod[!ctl_switch] = car_or_mod[!ctl_switch];
```

```
62.     button_pressed = false;
63.     index += !ctl_switch;
64.     transfer_data[index] = car_or_mod[!ctl_switch];
65.   }
66.
67.   Serial.write(transfer_data, num_of_data);
68. }
```

The Global Variables

The first lines don't really need explanation; only line 3, which sets the number of bytes that we'll be transferring to Pd. There are three because we want to transfer the waveform value for each oscillator, plus the unique byte at the beginning of the data stream to let Pd know that we're starting to send bytes.

In line 7, we define an array to hold the old values of the push buttons so that we can compare them to the value we'll be getting in every iteration of the loop function for each push button to execute code only when we press one. Line 8 defines an array that will hold the waveform values for each oscillator. For example, if we want the carrier oscillator to have a square waveform and the modulator to have an impulse waveform, then this array will hold these values:

```
car_or_mod[2] = { 2, 3 };
```

Where 2 is the square waveform value and 3 the impulse. Line 9 defines an array to hold the old waveform values. This time we don't need this array for comparison, but to know which LED to turn off when we change a waveform. You'll see how to do this a bit later. Both arrays of lines 8 and 9 are initialized with all their values to 0.

Line 10 defines a variable to hold the old value of the switch that will control which oscillator's waveform the push buttons will set. This one is needed for comparison to the current value of the switch; it is initialized with the value 1 because we're using the internal pull-up resistors for all digital input pins. Finally, lines 11 and 12 define two variables of the bool type (which is the same as boolean) to state whether we have pressed a push button and whether the switch has changed position. Both these variables are initialized as false.

The setup Function

In the setup function we set the mode of the pins we use (in this sketch we're using only digital pins). Notice that in line 1 we have defined that we'll be using four switches (actually push buttons, but they are also considered to be switches, called *momentary switches*). That number corresponds to the number of push buttons only, and does not include the switch we're using. For this reason, in line 15, the test condition of the for loop checks whether the incrementing variable i is less than or equal to the num_of_switches. This way the loop will iterate five times instead of four, and we'll be able to set the mode of all five input pins. If we do that, we must make sure we won't try to assign a value to the old_switches array in an index beyond its bounds. That's why in line 19 we test if i is less than the num_of_switches, which is 4, and only as long as it's less, will a value be assigned to the array. If we omit this test, the code will go on and try to assign a value to the index 4 of the array, which doesn't exists (it has four elements, counting from 0, so the last element has the index 3). This will probably not throw an error, but the code won't work. You should always be careful with array indexing, as it's a very common mistake programmers do.

In line 22, we set the mode of the LED pins, and in line 24, we turn the first LED on to indicate that the triangle waveform has been chosen. This is done because this waveform will be used as the default one, so we don't need to choose a waveform as soon as the program boots.

Notice the addition to the i variable in all pin mode settings. In the case of the switches, we add 2, because we start using digital pins from 2 onward, and in the case of the LEDs we add 7, because we have 5 switches, plus 2 for the first two digital pins that we don't use.

Using the Push Buttons to Select a Waveform

In line 29, we start the loop function. The first thing to do is to assign the value 1 to the index offset variable index, which we're using when we store data to an array combined with the Serial.write function. In line 33, we read the pin of the control switch and check if its state has changed, by comparing the value we just read to the value stored in the old_ctl_switch variable. If I has changed, we set the ctl_switch_changed variable to true, and we update the old_ctl_switch variable. Notice that the names of the variables used here are as self-explanatory as possible.

In line 39, we read through the pins of the push buttons. This time we add 3 to the i variable, because the switch is wired to pin 2, so the push buttons start from pin 3. Again, we check if the state of a push button has changed by comparing the current value to the one stored in the corresponding element of the old_switches array. If the state of a push button has changed, in line 42, we check whether the push button is being pressed instead of being released.

If the test of line 42 passes, we store the value of i to the car_or_mod array, using the inverse value of the ctl_switch variable as the index. ctl_switch has stored the value read from the pin where the switch is attached to. Since we're using pull-up resistors, when the switch is in the off position, the variable will hold 1, and when the switch is in the on position, it will hold 0. By inverting this value we get the first index of the car_or_mod array, when the switch is in the off position, and the second index when the switch is in the on position. This means that when the switch is in the off position, we can use the push buttons to set the waveform for the carrier oscillator (it makes more sense to place the waveform value of the carrier to the first element of the car_or_mod array). When the switch is in the off position, we can use the push buttons to set the waveform for the modulator oscillator. Depending on which push button we have pressed, i will hold the corresponding value. So if we press the first push button, i will hold 0, if we press the second, i will hold 1, and so forth. This way, depending on the position of the switch, we store the number of the push button to either the first or second element of the car_or_mod array, enabling the control of the waveforms of both oscillators, with the same four push buttons.

In line 45, we set the button_pressed variable to true, and then we update the corresponding element of the old_switches array. In line 48, we update the corresponding element of the old_switches array when a push button is being released (since the test of line 42 will be false, and line 46 won't be executed).

Controlling the LEDs According to the Switch and the Push Buttons

Lines 52 and 58 are where the two booleans come to play. In line 52, we check if ctl_switch_changed is true, and if it is, we switch LEDs using the car_or_mod array. If the switch goes to the on position, the ctl_switch variable will hold 0, which is the index of the waveform of the carrier oscillator in the car_or_mod array. Since we've put the switch to its on position, we want the LEDs to display which waveform is chosen for the modulator oscillator instead. For this reason, we turn off the LED in line 54, which indicates the waveform of the carrier by using the first element of the car_or_mod array (remember, ctl_switch now holds 0). We turn on the LED that indicates the waveform of the modulator by using the inverse of ctl_switch to read the second element of the car_or_mod array. This is done so that we can check which waveform both oscillators have, without needing to select a new one (without needing to press any push button). After this is done we set ctl_switch_changed to false, so that this chunk of code doesn't get executed again, before we change the position of the switch again.

In line 58, we check whether a push button has been pressed. If it has, we'll turn off the LED that indicates the previous waveform of the oscillator selected (depending of the position of the switch) and we'll turn on the LED of the push button that has just been pressed. Here we need the old_car_or_mod array, because it has stored the previous waveform value. In both lines 59 and 60 we use the inverse of ctl_switch, because now we're controlling the LEDs of one oscillator only. If the switch is in the off position, ctl_switch will hold 1, but we want to use the first element of the two arrays, so we invert the value. After we set the LEDs, we update the element of the old_car_or_mod array and we set button_pressed to false.

Writing the Data to the Serial Line

In line 63, we add the inverse of the ctl_switch value to index, so that we can set the correct index of the transfer_data array. In line 30, we initialize index to 1. If we set a waveform for the carrier oscillator, it means that the control switch is at the off position, and ctl_switch holds 1. We still want the carrier waveform value to come before the modulator waveform value. Inverting ctl_switch will yield 0, and adding this to index (which holds 1) will yield 1. This is the index of the transfer_data array we want to write to, because the first index (index 0) holds the unique byte of the data stream (192), so the next available index is index 1. If we set a waveform for the modulator, then the switch will be at the on position, and ctl_switch will hold 0. Inverting it will yield 1, and adding it to index will yield 2, which is the index of the transfer_data array we want to write to. In line 64, we write to the correct index of the transfer_data array the value stored at the correct index of the car_or_mod array. Finally, in line 67, we transfer the stored data to the serial line with the Serial.write function. As already mentioned, this function can't really tag data with strings, so we're sending all data, even if something hasn't changed. We'll deal with the unchanged data in the Pd patch that we're building for this project.

Making the Circuit

The best way to understand the Arduino code is always when tested with its circuit (and when other software is involved, combined with it). This is shown in Figure 8-15. It's pretty straightforward. I'm using the components mentioned when explaining the code, which are one switch, four push buttons, and four LEDs.

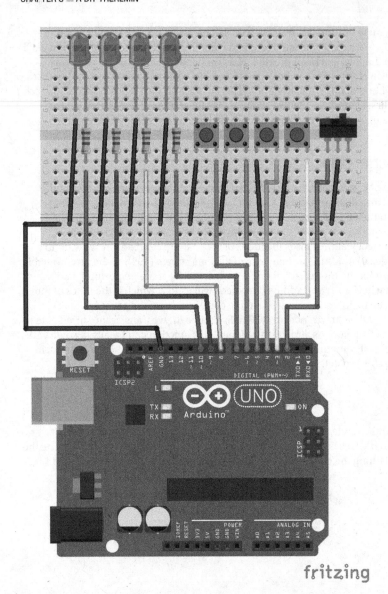

Figure 8-15. *Circuit with the push buttons controlling the waveform of the two oscillators*

Figure 8-16 shows the Pd patch that receives data from this Arduino code. It is only for data visualization, so it's a very simple, stripped to the only necessary patch. Once you build the circuit and upload the code to your Arduino board, open the patch and test it. When the switch is in the off position, you'll be able to set the waveform of the carrier. If you press a push button, you see the corresponding value being printed to the Pd console with the name *car* (for carrier). The corresponding LED lights up, whereas the previously lit LED goes off. Since we're using [change] in the Pd patch, only when a value is changed will it go through and it will be printed to the console. The two [print] objects have arguments to set the name of the printing values, so we can distinguish one from the other. If you put the switch to the on position, the LED of the waveform of the modulator will go on, and the previously lit LED (which was the LED of the selected carrier waveform) will go off. Now pressing a push button will print its corresponding value to the Pd console under "mod," and the corresponding LED will light up. Switching between the two positions of the switch will control the LED waveform display of the two oscillators, without needing to press any push button.

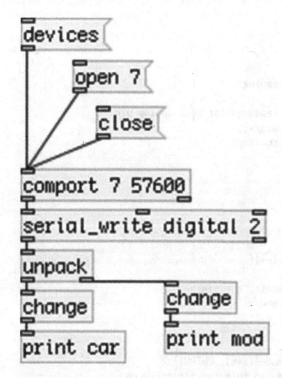

Figure 8-16. *Pd patch to receive the waveform data from the Arduino*

Putting It All Together

We can finally put all the circuit components together and test the whole circuit along with the finished patch. We'll first finalize the code and then the circuit on a breadboard, as we need to test it to make sure everything works as expected. Afterward, we're going to build the final Pd patch. Lastly we're going to put everything in an enclosure along with a Raspberry Pi and finalize our theremin. Let's first look at the code, which is shown in Listing 8-6.

Listing 8-6. The Final Version of the Code

```
1.   // this constant should exclude the DSP and control switch
2.   const int num_of_switches = 4;
3.   const int num_of_leds = 5;
4.   const int num_of_analog_pins = 6;
5.   // number of bytes to send are the number of analog pins times 2
6.   // one byte for the waveform of the carrier, one for the waveform of the modulator
7.   // one byte for the DSP, and one for the beginning of the data stream
8.   const int num_of_data = (num_of_analog_pins * 2) + 4;
9.
10.  byte transfer_data[num_of_data] = { 192 };
```

```
11.
12.  int old_switches[num_of_switches];
13.  int car_or_mod[2] = { 0, 0 };
14.  int old_car_or_mod[2] = { 0, 0 };
15.  int old_ctl_switch = 1;
16.  bool button_pressed = false;
17.  bool ctl_switch_changed = false;
18.
19.  // variables for smoothing the analog pin readings
20.  const int num_of_readings = 10;
21.  // set a two dimensional array to store 10 readings for each analog pin
22.  int readings[num_of_analog_pins][num_of_readings];
23.  // set variables to accumulate and average readings
24.  int total[num_of_analog_pins] = { 0 };
25.  int read_index = 0;
26.  int average = 0;
27.
28.  int dsp_led = 12;
29.
30.  void setup() {
31.    for(int i = 0; i <= num_of_switches + 1; i++){
32.      pinMode((i+2), INPUT_PULLUP);
33.      // the old_switches array has five elements, not six
34.      // so make sure you don't exceed its bounds
35.      if(i < num_of_switches) old_switches[i] = 1;
36.    }
37.
38.    for(int i = 0; i < num_of_leds; i++) pinMode((i+8), OUTPUT);
39.    // triangle oscillator is the default one, so turn the first LED on
40.    digitalWrite(8, HIGH);
41.
42.    Serial.begin(57600);
43.  }
44.
45.  void loop() {
46.    int index = 1;
47.
48.    // store the analog values to the array
49.    for(int i = 0; i < num_of_analog_pins; i++){
50.      total[i] -= readings[i][read_index];
51.      readings[i][read_index] = analogRead(i);
52.      total[i] += readings[i][read_index];
53.      average = total[i] / num_of_readings;
54.      // split analog values so they can retain their 10-bit resolution
55.      transfer_data[index++] = average & 0x007f;
56.      transfer_data[index++] = average >> 7;
57.    }
58.
59.    read_index++;
60.    if(read_index >= num_of_readings) read_index = 0;
61.
```

```
62.    // store the value of the DSP switch
63.    transfer_data[index++] = !digitalRead(2);
64.    // turn the DSP LED on or off, according to the DSP state in Pd
65.    if(Serial.available()) digitalWrite(dsp_led, Serial.read());
66.
67.    // use the switch to control what the push buttons will set
68.    int ctl_switch = digitalRead(3);
69.    if(ctl_switch != old_ctl_switch){
70.      ctl_switch_changed = true;
71.      old_ctl_switch = ctl_switch;
72.    }
73.
74.    for(int i = 0; i < num_of_switches; i++){
75.      int switch_val = digitalRead(i + 4);
76.      if(switch_val != old_switches[i]){
77.        if(!switch_val){ // if button is pressed
78.          // save its number to either the carrier or modulator element
79.          car_or_mod[!ctl_switch] = i;
80.          button_pressed = true;
81.          old_switches[i] = switch_val;
82.        }
83.        else old_switches[i] = switch_val;
84.      }
85.    }
86.
87.    if(ctl_switch_changed){
88.      // if we change the switch, show which LED of the current position is on
89.      digitalWrite((car_or_mod[ctl_switch] + 8), LOW);
90.      digitalWrite((car_or_mod[!ctl_switch] + 8), HIGH);
91.      ctl_switch_changed = false;
92.    }
93.  if(button_pressed){
94.    digitalWrite((old_car_or_mod[!ctl_switch] + 8), LOW);
95.    digitalWrite((car_or_mod[!ctl_switch] + 8), HIGH);
96.    old_car_or_mod[!ctl_switch] = car_or_mod[!ctl_switch];
97.    button_pressed = false;
98.    index += !ctl_switch;
99.    transfer_data[index] = car_or_mod[!ctl_switch];
100.  }
101.
102.  Serial.write(transfer_data, num_of_data);
103. }
```

I've already explained most of the code, all we've done is assembled the different bits we've been through in this chapter. One thing that has been added is the number of LEDs in line 3, which is now 5, because we want to have an LED indicating the DSP state. Lines 5 to 7 explain how many bytes we need to transfer over serial. We have added one byte for the DSP state, which is going to be controlled by a switch we'll put in our circuit. Lines 12 to 17 are the same with Listing 8-5, which are the arrays and variables for reading the digital pins and comparing these readings to their previous state. Lines 20 to 26 are the same with Listing 8-3, where we define arrays and variables for smoothing the analog pins readings. In line 28, we define the pin number where we'll wire the LED that will indicate the DSP state. That will be the last LED in our circuit.

The setup function is the same with Listing 8-5, only this time the offset given to the output digital pins is 8 instead of 7, because we've added one switch to control the DSP in Pd. The beginning of the loop function is the same with Listing 8-3, where we read and smooth out the readings of the analog pins. After we've done that, in line 63, we read the switch that controls the DSP state of Pd, which is wired to digital pin 2 of the Arduino, and we store its value to the next position in the transfer_data array. Then in line 65, we check if there is any data in the serial line, and if there is, we write the value we've received to the pin of the LED that indicates the DSP state. As with previous projects, it's better to control this LED from Pd to make sure that it will turn on only when the DSP is really on. Notice that reading data from the serial line and writing it to the DSP LED pin takes only one line of code. We'll be sending a single number from Pd to the Arduino, so we don't need to send it with the message "print", we'll just send that number as it is. This means that the number will arrive in the Arduino as a raw byte and not in its ASCII values. So if we send the number 1, the Arduino will receive the number 1, not 49. This enables us to write this value to the DSP LED pin as it is, without needing to subtract '0', which is the ASCII 0. Also, in the digitalWrite function in this line, we use the Serial.read() function as its second argument. This is perfectly legal and works as expected. Just make sure you close all the parenthesis in this line.

After that we continue the same way we did in Listing 8-5. There are three differences. The first is the pin number of the switch that controls the push buttons, which is now 3 instead of 2 since the DSP switch is now wired at pin 2. The second is the offset given to the digitalRead function in line 75, which is 4 instead of 3. The third is the offset given to the digitalWrite function in lines 89, 90, 94, and 95, which is 8, as with the setup function, since all digital pins have been shifted by one position because of the DSP switch on digital pin 2.

The comments in lines 5 to 7 not only explain how many bytes we need to send to the serial line, but also the order in which we receive these bytes in Pd. This means that we'll first receive the six analog values, and then the DSP state value, the carrier oscillator waveform value, and lastly, the modulator oscillator waveform value. Let's now look at the circuit on a breadboard. As you can imagine it's a combination of the two circuits we've already shown, along with one more switch and one more LED, for the DSP state control and display. It is shown in Figure 8-17.

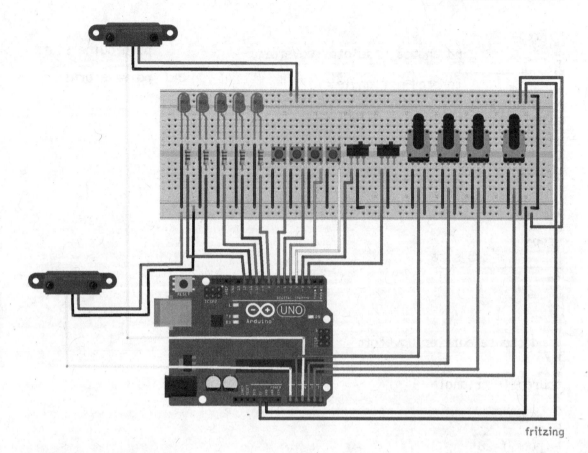

Figure 8-17. *The full test circuit*

Take good care of the wiring of both the analog and the digital pins. The first two switches should be wired to digital pins 2 and 3, after that the four push buttons should be wired, and lastly the five LEDs. The first proximity sensor should be wired to analog pin 0, then the first potentiometer is wired to analog pin 1, then the second proximity sensor goes to analog pin 2, and lastly the remaining three potentiometers are wired to the rest of the analog pins. If you make a different wiring you'll get strange results in Pd and the whole interface won't work. This is because of the way we've written the Arduino sketch and the Pd patch. We're not going to build a test patch to check the input of the Arduino in Pd, this is an addition of the patches we've built so far, which take input from the Arduino. Instead we're going to start building the final patch.

Building the Final Pd Patch

The final version of the Pd patch will contain a phase modulation with two table lookup oscillators, where the carrier will be built with [tabread4~] and the modulator with [tabosc4~]. We'll be able to choose the waveform for each oscillator independently, and we'll also have some other features, like controlling the sensitivity of the two proximity sensors, the overall brightness of the sound, and the tuning of the instrument. Figure 8-18 shows the full patch.

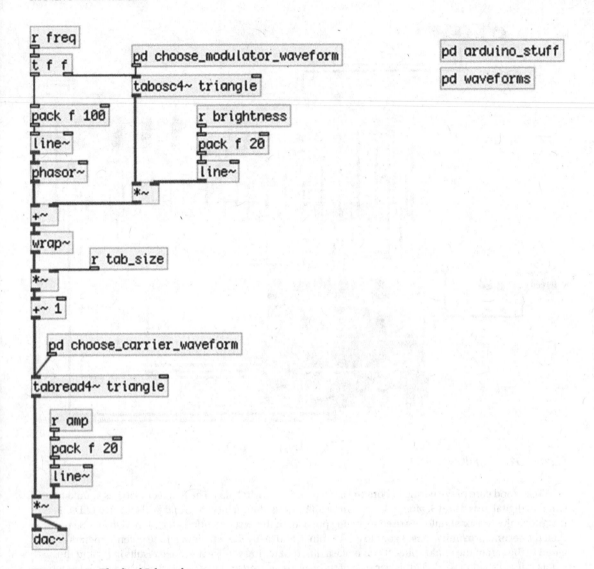

Figure 8-18. *The final Pd patch*

This patch is maybe the simplest patch of all the final versions we've built so far. Its main part has all its objects in the parent patch, instead of being encapsulated in a subpatch. There are only four subpatches that contain only a few things. The main part is the phase modulation of the [tabread4~]-based lookup oscillator by the [tabosc4~]-based lookup oscillator. Just as a reminder, since we don't need to have access to the phase of the modulator oscillator, we're using [tabosc4~], which is a nice package that reads waveforms stored in tables with sizes that are a power of 2 plus 3 (for the three guard points that are automatically being added with the sinesum feature of Pd). For the carrier oscillator, we can't use this object because we need to have access to its phase, so we've built it using [tabread4~] instead, which might make things a bit more complicated, but gives access to the phase of the oscillator. (Actually, [tabosc4~] is a combination of all the objects between [phasor~] and [tabread4~]—including these two— and excluding [+~], all written in the C programming language and compiled as a single object).

The Subpatches for Choosing Waveform

The two subpatches [pd choose_carrier_waveform] and [pd choose_modulator_waveform] are almost identical, and they are illustrated in Figures 8-19 and 8-20.

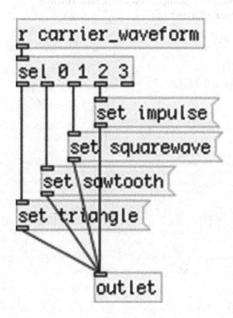

Figure 8-19. *The choose_carrier_waveform subpatch*

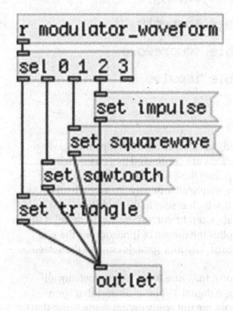

Figure 8-20. *The choose_modulator_waveform subpatch*

These two subpatches will receive input from the push buttons of the circuit, where pressing the first push button, according to the position of the switch that controls the buttons, the number 0 will be sent to either [r carrier_waveform], or [r modulator_waveform]. Pressing the second button the number 1 will be sent, and so forth. These numbers will bang the respective message that will set the table to read from for either [tabread4~], or [tabosc4~], meaning, the carrier or the modulator (triangle, sawtooth, squarewave, and impulse are the names of the four tables which store the four waveforms).

Waveforms Subpatch

The contents of the [pd waveforms] subpatch are illustrated in Figure 8-21.

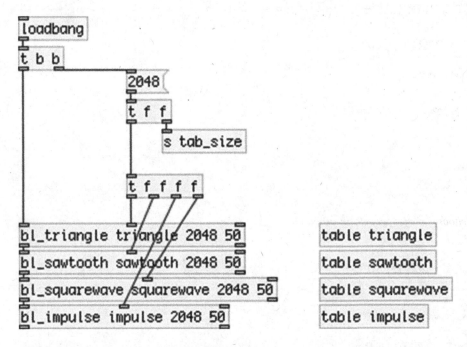

Figure 8-21. *Contents of the waveforms subpatch*

Here we're using the four abstractions from the "band_limited_waveforms" GitHub repository, where each abstraction creates one of the four waveforms. These abstractions are connected in a chain, because they all output a bang out their left outlet when they're done computing their waveform. This way we can bang the first one with [loadbang], and they will bang each other, creating all four waveforms on load of the patch. Before we bang the first abstraction, we bang the value 2048, which is sent to the middle inlet of each abstraction, which sets the size of the table, and is also sent to [s tab_size]. Its corresponding [receive] is in the parent patch, connected to the right inlet of [*~], which multiplies the output of [phasor~] by the size of the waveform tables, so that [tabread4~] reads the whole waveform. The four [table]s contain the stored waveforms, which are read by [tabread4~] and [tabosc4~].

Finally, the contents of the [pd arduino_stuff] subpatch is shown in Figure 8-22. We're receiving all the data in [serial_write], where we're using the arguments "analog 6 digital 3". This indicates that we're expecting 6 analog values and 3 digital ones. Well, the last two values are not really digital in the sense that they're not raw readings of digital pins, but they are values that result from the use of digital pins (the push buttons and their control switch), which are well within the range of one byte. (The argument "analog" tells the abstraction that these values are beyond the range of one byte, and they will arrive disassembled, so the abstraction will need to reassemble them.) This abstraction was initially designed to receive raw analog and digital readings, that's why it uses this syntax. We can nevertheless take advantage of the fact that *digital* actually means a value within one byte, and use it to transfer the values for the waveforms of the two oscillators.

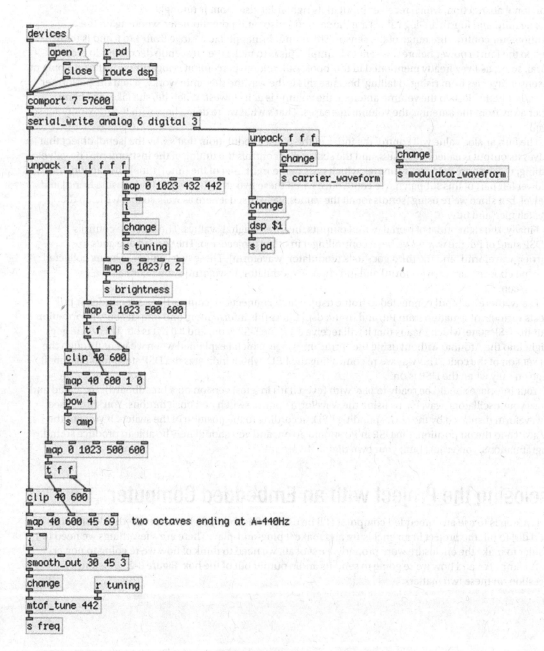

Figure 8-22. *Contents of the arduino_stuff subpatch*

The very first value of the "analog" group is the first proximity sensor, which is mapped and smoothed the way we've already shown. The second value comes from the first potentiometer, which controls the sensitivity of the sensor. When the potentiometer is all the way to its minimum value, the sensor will utilize its full range. As the potentiometer is turned toward it maximum value, the range of the sensor will shrink, with its sensitivity ending at a further distance. The result of this mapped and smoothed value goes to the [mtof_tune] abstraction, which receives input in its right inlet also, from [r tuning].

The third and fourth analog values are the second sensor and potentiometer, where again the potentiometer controls the range of the sensor. The output is mapped to a range from 1 to 0 and is then raised to the fourth power before it is sent to [s amp]. This is to make the crescendo decrescendo more natural, since, as I've already mentioned in this book, our volume perception is logarithmic. The mapping of this sensor inverses from rising to falling, because this is the way the theremin works. When the performer has his/her hand close to the volume antenna, the volume is at its lowest. When he/she takes the hand further away from the antenna, the volume increases. That's what we're doing here with the proximity sensor as well.

The fifth analog value will control the index of the phase modulation, that's why the [send] object that receives its output is called *brightness*, and the sixth value controls the tuning of the instrument. The values [s tuning] receives are going to [r tuning], which goes to the right inlet of the [mtof_tune] abstraction, in the lower left part of this subpatch. We could have wired these two straight away, without using [send] and [receive], but since we're using [send]s for all the values, we've used it here as well, so that we keep the subpatch nice and tidy.

Finally, the right outlet of [serial_write] outputs the three "digital" values. The first one controls the DSP state of Pd, the way we've been controlling it in other projects too. The second one goes to [s carrier_waveform], and the third goes to [s modulator_waveform]. These two values are being collected in the [pd choose_carrier_waveform] and [pd choose_modulator_waveform] subpatches, as we've already seen.

Lastly, there's a [r pd] connected to [route dsp], which connects to [comport]. As a reminder, [r pd] collects various information from Pd, and [route dsp] filters this information, and outputs only information about the DSP state, which means that it will receive a 1 if the DSP is on, and a 0 if it is off. This value goes straight into the Arduino without using the "print" message, which I explained when we went through the final version of the code. This way we're controlling the LED, which indicates the DSP state, making sure it will go on only when the DSP is on.

Your interface should be ready to play with (even if it's in a test version on a breadboard). Go ahead and choose your oscillators' waveforms using the waveform control switch and push buttons. You should see the waveform displayed by the corresponding LED, according to the position of the switch. If you put the DSP switch to the on position, the DSP in Pd should go on, and you should now be able to produce sound. Congratulations, you've just built your own digital theremin!

Enclosing the Project with an Embedded Computer

The last step is to use an embedded computer (I'll be using a Raspberry Pi, but any embedded computer would do) to put the project in an enclosure and make it plug-and-play. There are a few things we need to consider to make the enclosure work properly. First of all, we need to think of how we're going to power up the computer, and how we're going to send its audio output out of the box. Figure 8-23 illustrates a suggestion on these two matters.

Figure 8-23. *Bringing the power input and audio output of the Raspberry Pi outside of the enclosure*

I have used a female USB type B to a male micro-USB cable to power up the Pi. All I need to do is replace the cable of the Pi charger to a type B USB instead of a micro-USB one. I've also connected a stereo male 1/8-inch jack to the Pi's audio output, and connected its left channel and ground pins to the two pins of a mono female 1/4-inch jack, so that I can easily get the Pi's audio output from the box. It's not necessary to follow these exact steps, you could try another way with your own enclosure; this is only a suggestion. One more thing to consider is the space you have in the project box that you're going to use. For this project, I've used an Arduino Uno, which is not that small, so I made sure I have enough space inside the box for both the Pi and the Arduino with its circuit. It's better to make a project slightly bigger but more comfortable to build, rather than trying to enclose everything in a tiny space and have a hard time realizing your project.

I've also used a Proto Shield to build the circuit, as it minimizes the soldering you need to do to a great extent, since it provides easy access to the analog and digital pins of the Arduino. If you're using an Arduino Nano, or some other Arduino type or microcontroller, then you'll probably need to use a perforated board for that. As usual, I've daisy chained the voltage and ground pins of the circuit components so that I don't have to send a bunch of wires to the Proto Shield from the components mounted on the box.

Figure 8-24 shows the final version of the enclosure I've built for this project. You can see the two proximity sensors on two sides of the box's surface. The right one controls the pitch of the instrument, so it's wired to analog pin 0 on the Arduino. The left one controls the volume, so it's wired to analog pin 2. This is the way the theremin is designed, as it's aimed at right-handed people, since the "good" hand is more likely to control the pitch. If you're left-handed you can just switch the wiring of the two sensors.

Figure 8-24. *The final version of the enclosure*

On the bottom side, I've mounted the four potentiometers, keeping a good distance between them to be able to use them easily. If there're placed too close to each other, it won't be easy to twist them. On the top side I've placed the switches, the push buttons, and the LEDs. The bottom row has the two switches and the four push buttons. The left switch controls the DSP, so it's wired to digital pin 2 of the Arduino (the first digital pin we're using). The right switch controls which oscillator the push buttons will set the waveform for, so it's wired to digital pin 3. The four push buttons are wired sequentially to the next four digital pins. The top row includes the five LEDs, where the leftmost one is the DSP LED, so it's wired to digital pin 12 (remember the code we've written). The rest LEDs are wired to digital pins 8 to 11, from left to right, so the second from left is wired to digital pin 8; the next one on its right is wired to pin 9; and so forth.

Adding a Push Button for Switching off the Embedded Computer

Notice that in Figure 8-24 next to the LEDs there's another push button. We haven't included that in our test circuit, or the code we've written so far. This is placed there to shut the Pi down (or whichever embedded computer you're using). Since enclosing the computer inside the project box, we won't have access to its terminal, therefore we need some way to properly shut it down, instead of just powering it off. The circuit we've built so far leaves one digital pin free, so we can use that. That is pin 13, the last digital pin, so go ahead and wire it there. In order to use it we must modify both the Arduino code and the Pd patch a bit. We'll start with the Arduino code first. We're not going to rewrite it nor present the whole code here, since it's only a few additions. The first addition is shown in Listing 8-7.

Listing 8-7. First Addition to the Arduino code

```
1.   // number of bytes to send are the number of analog pins times 2
2.   // one byte for the waveform of the carrier, one for the waveform of the modulator
3.   // one byte for the DSP, one to shut the Pi down
4.   // and one for the beginning of the data stream
5.   const int num_of_data = (num_of_analog_pins * 2) + 5;
6.   // store value of last index for the shut down switch
7.   int last_index = num_of_data - 1;
```

These lines of code should replace the place where we compute the number of bytes we need to transfer from the Arduino to Pd, at the beginning of the code (not the very beginning, starting in line 5). We'll need this last_index so that we store the value of this additional push button to the correct position of the transfer_data array. This is because at the end of the code, we're using the ctl_switch variable to determine the position of the array, instead of using the post-increment technique (the index++ technique). If we try to use the post-increment technique for the last push button, it won't work. I'll explain this thoroughly in a bit, when we reach the spot where we'll store the push button value. The next addition is shown in the following line of code:

```
int shut_down_switch = 13;
```

Put this line right before the setup function. This is the pin number where we've wired the additional push button. The next thing we need to add is the following:

```
pinMode(shut_down_switch, INPUT_PULLUP);
```

As you can imagine, this should go in the setup function. It doesn't really matter where in the function, it should just be in the function. Listing 8-8 shows the last addition. The addition is only the last line of the listing, but the last part of the code is shown here to explain some things.

Listing 8-8. Last Addition to the Arduino Code

```
1.   if(button_pressed){
2.      digitalWrite((old_car_or_mod[!ctl_switch] + 8), LOW);
3.      digitalWrite((car_or_mod[!ctl_switch] + 8), HIGH);
4.      old_car_or_mod[!ctl_switch] = car_or_mod[!ctl_switch];
5.      button_pressed = false;
6.      index += !ctl_switch;
7.      transfer_data[index] = car_or_mod[!ctl_switch];
8.   }
9.
10.  transfer_data[last_index] = !digitalRead(shut_down_switch);
```

This is the part where we check if we're pressing a push button, so we can send the waveform value to the corresponding oscillator, depending on the position of the control switch, and we can turn off the previously lit LED and turn on the new one. The if test of the previous version is exactly the same as the one of Listing 8-8. If the control switch is at the on position, the ctl_switch variable will hold 0, but since we're reading its inverse, we'll be reading 1. In this case, line 6 would result in storing the value 16 to the variable index. If this happens, we could use the post-increment technique in line 7 and we could use the same variable in line 10 since it would hold the value 17, which is now the last position of the transfer_data array. In case the control switch is in the off position, then line 6 would result in storing the value 15 to the variable index, where the post-increment technique would be no good, since line 10 would store the value of the shutdown push button to the index 16 of the transfer_data array.

The preceding holds true only while we kept a push button pressed, because if we release it, the line 1 test won't be true and its code won't be executed, so the calculation in line 6 won't happen and line 10 results in storing the shutdown push button to position 15 instead of 17. For this reason, in the beginning of the code, we're storing the number of the last index of the array, and using that in line 10 instead of using the variable index with the post-increment technique. This ensures that the shutdown push button will always be stored in the last position of the transfer_data array.

Reading the Extra Push Button in Pd

These few additions to the Arduino code are what we need to make the extra push button work. Now let's look at the additions we need to make to the Pd patch so that the code and circuit work. Figure 8-25 shows the [pd arduino_stuff] subpatch with the additions made to it. We're now receiving four "digital" values from the Arduino, so the first thing we need to do is change the last argument of the [serial_write] abstraction from 3 to 4. Then we need to add another "f" argument to the [unpack] on the right side of the patch, since we'll be unpacking four values. The last of the four will be the value of the shutdown switch, which first goes through [change], so that it is being output only when it changes, and is then sent to [s shut_down]. We'll see in a bit where that goes.

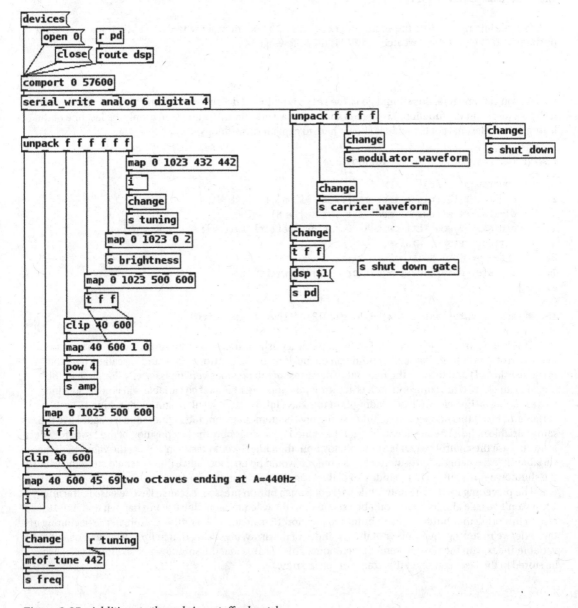

Figure 8-25. *Additions to the arduino_stuff subpatch*

358

The last addition is the [t f f] receiving the DSP state value, which is sent to [s shut_down_gate] too, along with [s pd]. We'll use that value to disable shutting down while the DSP is on. We could have used [r pd] for that, like we do to control the DSP LED on the Arduino circuit, but in case the DSP got stuck for some reason, we wouldn't be able to use the shutdown push button, so it's better to use the DSP switch value instead.

Notice also that the [smooth_out] abstraction is not used here. First of all, when making a circuit on a perforated board, you get less noise than what you get with a breadboard. Also, in my Raspberry Pi, which is a Pi 1, model B (so, not the latest Pi), I got lots of audio dropouts with this abstraction. A *dropout* is when the computer doesn't have enough time to do the audio calculations, and the sound card is asking for the next sample block before that has been calculated. This results in an audible click that's quite annoying. Since the circuit was stabilized quite well, I removed the abstraction altogether. The Raspberry Pi 2 has a quad code processor, so you might not be getting dropouts; test for yourself. Also, I tried this with the GUI, so using it without the GUI (launching Pd with the -nogui flag) might also improve the performance of the Pi.

Figure 8-26 shows the contents of the [pd shut_down] subpatch. We haven't made this subpatch in the previous version of the patch, so make it in the parent patch and put whatever you see in Figure 8-26 in it. In this subpatch we can see the two [receive] objects that receive input from the respective [send]s in the [pd arduino_stuff] subpatch, [r shut_down] and [r shut_down_gate]. As you can see, the value received in [r shut_down_gate] is being inverted, so when the DSP switch is turning the DSP in Pd on, [r shut_down_gate] will receive 0, closing the [spigot] below it. This will disable the rest of the patch from running, making it impossible to accidentally shut the Pi down while the DSP is running.

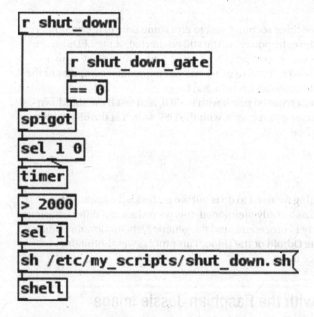

Figure 8-26. *Contents of the shut_down subpatch*

When the [spigot] is open, the value of the switch will go through when it changes. So when the push button is pressed, [r shut_down] will receive a 1 which will go to [sel 1 0]. [sel] will send a bang out its left outlet, since the value it received matches its first argument. This bang goes to the left inlet of [timer], which resets the timer. When the push button is released, [r shut_down] will receive a 0, and [sel] will output a bang out its middle outlet (the rightmost outlet outputs any value that doesn't match any of the arguments, as is). This bang goes to the right inlet of [timer], which will output the amount of milliseconds passed since it was reset. If this value is above 2000 (two seconds), [> 2000] will output a 1, and [sel] below it will bang the message "sh /etc/my_scripts/shut_down.sh", which goes to [shell]. This will result in running the script called shut_down.sh, which is in the directory /etc/my_scritps.

Writing the Script to Shut the Embedded Computer Down

Of course, we need to write the script for this subpatch to work, and store it in the right directory. As we did with the MIDI keyboard project in this book, we'll create a directory called my_scripts, in the /etc directory. Since /etc is not in the home directory, we need super-user privileges to do anything in there. To make this directory type the following:

```
sudo mkdir /etc/my_scripts
```

This will create the new directory. In that directory, we'll write the script to shut the Pi down. We've already done this, but we're writing it here too to refresh our memory. It is shown in Listing 8-9.

Listing 8-9. Script to Shut Down the Embedded Computer

```
sudo killall pd
sleep 3
sudo poweroff
```

This script will first quit Pd, then it will wait for three seconds (just to give some time to the computer to quit Pd properly), and then it will shut down. Since the power will be still connected, some LEDs on the Arduino might still be on, while the computer has shut down. Just give it some time (around one or two minutes should be enough) and then unplug the power. If you're using another embedded computer or the Raspbian Wheezy image, the last command should probably be sudo halt.

To make sure everything works as expected, open the Pd patch with its GUI, and test if the shutdown push button works. If after holding it down for at least two seconds, with the DSP switch at the off position, Pd quits and the Pi shuts down, everything works fine.

Loading the Pd Patch on Boot

If all the earlier steps function properly, the last thing we need to do is tell the embedded computer to launch Pd with the theremin patch when it boots. As already mentioned, this procedure can differ between embedded computers, so we'll cover the Raspberry Pi procedure, and the generic Debian/Ubuntu Linux procedure, for other embedded computers like the Odroid or the Udoo. This process was thoroughly explained in Chapter 5, but we'll go through it once more.

Launching Pd on the Raspberry Pi with the Raspbian Jessie Image

If you're using the Raspbian Jessie image on your Pi, then editing the /etc/rc.local script won't do. We'll have to use the crontab editor instead. First, we'll need to create a script that will launch Pd without the GUI and will open the theremin patch. In your home directory, make a new directory named my_scripts, and then run the following:

```
nano my_scripts/launch_pd.sh
```

The nano editor will open an empty file. In there, type the following lines:

```
#!/bin/sh
/usr/bin/pd -nogui -open /home/pi/pd_patches/theremin.pd &
```

This script first tells the shell which program to use to interpret the script, and then launches Pd without the GUI, opening the theremin patch. Once you've written the script, you need to tell the computer to run it on boot. In your home directory, as user "pi", run the following command:

```
crontab -e
```

This will open the crontab file of the user "pi". If it's the first time you run the crontab editor, you're asked which editor you want to use to edit crontab, while being prompted to use nano, because it's the simplest one. The choices will be numbered, and nano will probably be the second choice, so type **2**, and hit **Return**. The crontab file will open in nano and you'll be able to edit it.

If you haven't edited this file, it should only have a few lines with comments, telling you what this file does and how to use it (well, giving some minimal instructions). After all the comments add the following line:

```
@reboot sleep 20 ; sh /home/pi/my_scripts/launch_pd.sh
```

This tells the computer that when it reboots (or boots) it should first do nothing for 20 seconds, and then it should run the script we've just written. Take care to place the semicolon between the sleep 20 command and the sh command. You might want to experiment with the number 20, and try a smaller number. Give the Pi enough time to fully boot, but 20 might be too much. Try a few different values and see which is the lowest that works. You'll know if it works by turning the DSP switch to the on position. If the DSP LED goes on, it means that Pd has launched and everything should work properly. Test to see if you're getting sound. If you do, then congratulations, you have a new DIY digital theremin!

If you want to edit your patch further, or use your Pi for something else, as soon as you log in, type the following line:

```
sudo killall pd
```

So that Pd will quit and you'll be able to work further with your Pi.

Launching Pd on Other Embedded Computers or on the Raspberry Pi with the Raspbian Wheezy Image

If you're using another type of embedded computer, or the Raspbian Wheezy image, you can use the /etc/rc.local script to launch Pd on boot. You'll need super-user privileges to do this, so type the following:

```
sudo nano /etc/rc.local
```

The script will open in the nano editor, and you'll be able to edit it. Right before exit 0, type the following:

```
su -c '/usr/bin/pd -nogui -open /home/user/pd_patches/theremin.pd &' - user
```

Replace user with the user name. For example, on the Odroid Debian Jessie image, the user is *odroid*; on the Udoo Debian Jessie image, the user is *debian*; and on the Raspberry Pi Raspbian Wheezy image, the user is *pi*. The word *user* appears twice in the preceding command, so make sure you change it in both places.

This command will launch Pd without the GUI and open the theremin patch, as the default user of the computer (odroid, debian, pi, etc.), and not root. This is because [comport] doesn't like to be used by root, so we need to launch Pd this way.

Every embedded computer, or its operating system, has its own peculiarities; so even with these two ways, you might still not be able to make it work. It's always a good idea to sign up to forums or mailing lists and join the communities of the tools you're using, so you can always ask other members of the community how to solve some problems.

Once you have your on-boot script working, if you want to work further on your project or use your embedded computer for something else, as soon as you log in, type the following:

```
sudo killall pd
```

This will quit the Pd instance that runs on boot, so you can do anything else you want. Congratulations, you have made your own DIY digital theremin!

Conclusion

We have built our own digital theremin from scratch, using low-cost and open source tools. The theremin version we've built is not a straight representation of the actual instrument, but we've added our own things to make it work the way we like. You can, of course, change the Pd patch or even the Arduino sketch and circuit to make a different version of the theremin—one that better fits your needs. As with every chapter, this version is only a suggestion of how you can make something.

This chapter was the first to enclose an embedded computer in a box of our choice. The MIDI keyboard project also covered the enclosure, but in that project, we were restricted by the enclosure of the keyboard we had available. That made things more difficult. In this chapter, we could choose any project box, so we used a rather big box to easily fit all the instrument's components. Enclosing devices in boxes is a process in which you get better with time and practice. As you build more projects, you'll realize that you've become more flexible in what you can do with a given set of equipment, plus you'll make better decisions as to what kind of peripherals you'll use (USB cables, perforated boards, etc.). Just make sure not to start off with a box that barely fits the instrument's components; eventually, it will look nicer. It's better to build something not as nice as a start; but with time, you'll be able to master fairly complex and very nice projects.

CHAPTER 9

■ ■ ■

Making a Looper

In this chapter, we're going to build a celebrated mechanism, the *looper*. The looper is used widely by guitarists and other instrumentalists, even by people beatboxing. It enables you to gradually build complex sounds out of a single instrument or a voice. There are plenty of loopers on the market, even cheap ones, but this chapter will guide you through the process of building one. Once you understand how it works, then you can customize it to a great extent, making a looper that would either be very expensive as a commercial product, or one that includes features that are not present in the loopers that you can find on the market.

Even though a looper sounds like something fairly simple to build, I can assure you, it's rather complex. There are certain features that are nice to have (such as the looper knowing when to stop recording while overdubbing so that the overdub doesn't exceed the length of the existing loop, or like recording backward), but which are a bit difficult to accomplish. This mechanism is a great example of what programming really is and what it can do. We'll see in practice that the computer knows nothing by default, and that we need to provide it every single little piece of information that we might otherwise take for granted. On the other hand, we'll see that by providing all the necessary information, it is possible to achieve nearly anything you can think of. You might need to go through this chapter more than once to fully understand everything, but eventually, you'll learn how to manipulate recorder sounds in many ways.

Parts List

To build this project, we'll need all the components mentioned in Table 9-1.

Table 9-1. *Looper Parts List*

Project type	Foot-switches	LEDs	LED holders	Female 1/4 inch jacks	Project box	Tactile switches	Proto Shield	IDC connectors	Ribbon cable
Prototype	0	3	0	0	0	3	0	0	0
Final Project	3	3	3	3	1	0	1	2 × female 2 × male	1

As always, we'll first build the circuit on a breadboard, and then on a perforated board to make the final version. This project won't include an embedded computer because we'll need to have audio input and that requires an external sound card, which is beyond the scope of this book. I'll assume you can have audio input in your personal computer. The final version will include only the Arduino, which is meant to be used with your personal computer.

What Is a Looper?

Basically, a looper is a device with which you can record a short fragment of live sound. This fragment can be played over and over in a loop, hence its name. Different loopers do different things. Some can record in multiple channels, so the user can record multiple sounds and control them independently. Some can play the recorded sounds in various ways: forward, backward, change the playback speed or pitch, and so forth. Some others can utilize a tap tempo where the user can specify the playback speed by tapping on a switch, where the intervals of the tapping will control the playback speed.

The possibilities are endless. You can always come up with something new with this mechanism. Since this is the first looper that you're building, it's a good idea to restrict the capabilities that this looper will have. This doesn't mean that we'll make a simple looper that will just play one sound in a loop. The one we'll build will be able to do the following:

- Play back the recorded audio both forward and backward

- Record an arbitrary length of live audio

- Overdub an existing recording while it's playing back

- Synchronize end of recording with beginning of playback

- Overdub while playing backward

These five features are already enough to get us started. They might even become overwhelming once we realize how many things we need to take into account even for the simplest of these. Once you understand this project's procedures, you'll be able to add features of your own, which is another advantage of building your own projects—the ability to expand them.

Recording Our First Sounds in Pd

We'll go over the four features mentioned earlier in the order that they were covered, so we'll start by recording sound, and then we'll move on to record a sound of an arbitrary length. The first step to simply record some sound in Pd is pretty simple, and we've already seen it in this book. The object for this is [tabwrite~], which writes its audio input to a table ("tab" stands for *table*). A table is essentially an array, and that's where we'll store sounds in Pd. Actually, all digital recordings are stored in arrays, since sound is a stream of numbers. Figure 9-1 shows the simplest way to record audio to a table.

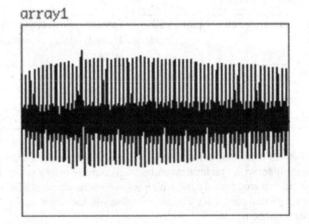
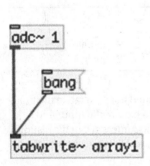

Figure 9-1. Recording sound in Pd

The patch in Figure 9-1 is very simple. As simple as the first patches shown in Chapter 1 of this book. It's pretty straightforward and doesn't need a lot of explanation. [tabwrite~] takes one argument that is the table to which it will write, in this case array1, the name of the array. When it receives a bang, it writes the audio it receives in its input to the specified table. Array1 in Figure 9-1 has 22,050 elements, which means that it's capable of recording half a second of sound, if the sample rate is set to 44,100, which is the most widely used sample rate. (A 44,100 sample rate means that the computer takes 44,100 amplitude samples of sound per second. 22,050 is exactly the half of it, meaning that it can hold half a second.)

Build the patch in Figure 9-1 and try it. Your computers built-in microphone, or a simple microphone you might be using for video calls will do. You'll probably notice that once you hit **bang**, the output of your recording will appear in the array half a second later. That is because [tabwrite~] will write as many samples as the size of the table it will write to, in this case 22,050, which will take exactly half a second.

There's one drawback for our project with this kind of recording, which is the set amount of time we have for a loop, half a second. When [tabwrite~] receives a bang it will write samples until it fills the specified table, unless it receives the message "stop". But sending the message "stop" brings another problem, which involves knowing the length of the actual recording. We'll deal with these issues later on.

Playing Back the Recorded Sound

Let's play back the sound we just recorded. Figure 9-2 shows the extension of the patch in Figure 9-1, which will play back the audio of the array. As you can see, the patch has already become quite more complex. There is some information we need to know to play the sound back at the proper speed, otherwise it will sound awkward. The information we need is the sampling rate and the length of the table we want to read, which we get with the objects [samplerate~] and [arraysize]. Diving the two provides the right frequency in hertz, at which the recorded sound will be played back properly.

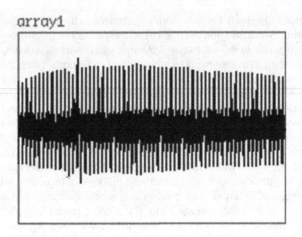

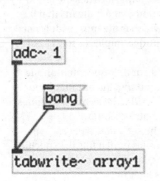

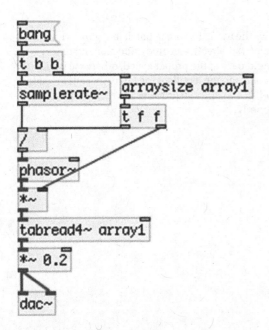

Figure 9-2. *Playing back the recorded sound*

In the patch in Figure 9-2, we're using [phasor~] to read the table. In Chapter 1 we briefly went through playing back recorded sounds using [tabplay~]. The advantage we have using [phasor~] and [tabread4~] instead is that we have access to the whole mechanism that reads the stored sound, enabling us to interfere in various ways (for example, play the sound backward). In this chapter we see another use of [tabread4~], which is playing any kind of stored audio, and not only oscillator waveforms.

Using [phasor~] means that we need to specify its range. [phasor~] is actually the phase of the looping mechanism, as it's a ramp that goes from 0 to 1, repeatedly, in the frequency specified either as an argument, or provided to its left inlet. The array in Figure 9-2 has 22,050 samples, so we need to multiply [phasor~]'s output by that number to provide the correct indexes to [tabread4~]. [tabread4~] takes indexes in its inlet and outputs the values of these indexes. By multiplying [phasor~] with the length of the table, we'll send the indexes from 0 to 22,049 to [tabread4~], which are all the indexes of Array1 ([phasor~] doesn't go until 1, but slightly below it). This way we can play back the recording that we just made at its normal speed.

Smoothing the Start and End of the Recording

One thing you'll probably notice is that you get clicks at the beginning of each loop. This is because the recorded sound is not silent at the beginning and the end. These clicks occur because when the recording starts over, the amplitude of the sound jumps from whichever value the amplitude of the sound is at the very end of the recording, to whichever value the amplitude is at the beginning. Even if a recording starts with complete silence, if it doesn't also finish with silence, the click will still be there. Also, it's impossible to start with complete silence unless you edit the sound, because there's always a bit of noise in an audio signal, even with very good microphones. This is where the smoothing window we've already used in some other projects of this book comes in. Reading the recording and at the same time the smoothing window, we'll make sure that both the beginning and the end of the playback will have zero amplitude, avoiding clicks. Figure 9-3 shows the patch.

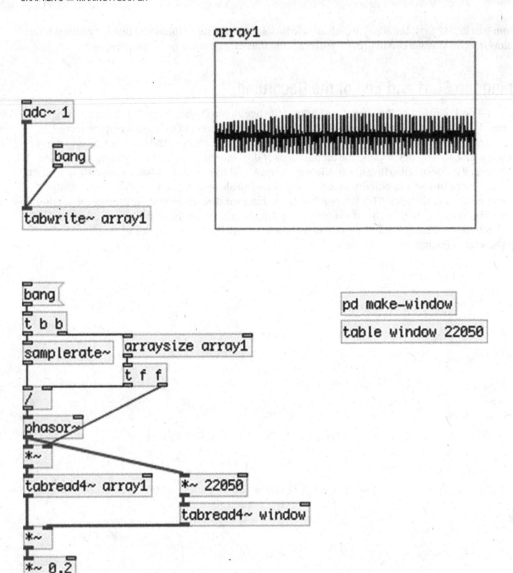

Figure 9-3. *Smoothing out the playback*

We can use [phasor~] to drive both the [tabread4~] that reads the stored audio and the [tabread4~] that reads the smoothing window. In the case in Figure 9-3, the sizes of the two arrays happen to be the same, which means that we could send the scaled output of [phasor~] to [tabread4~ window] as well. But this won't always be the case, so it's better to send the raw output of [phasor~] and scale it separately from the [tabread4~ array1] input signal. What this smoothing function does is to multiply the beginning and end of [tabread4~ array1] by 0, and quickly ramp up to 1 after the beginning, and quickly ramp down to 0 toward the end. Figure 9-4 shows the window, and Figure 9-5 shows the contents of the [pd make-window] subpatch, which creates the actual window. We've already seen this smoothing window in Chapter 7, but we're presenting it here as well to refresh your memory.

window

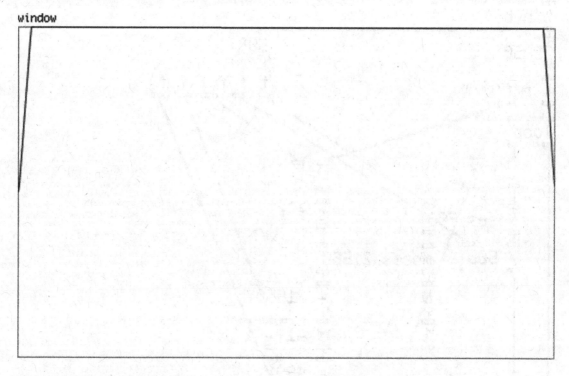

Figure 9-4. *The smoothing window*

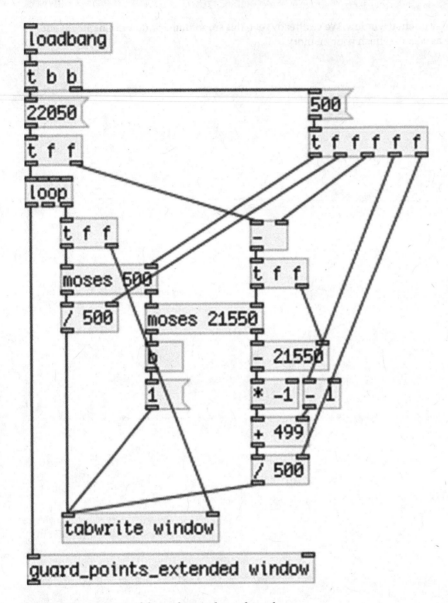

Figure 9-5. *Contents of the make-window subpatch*

Changing the Direction of the Playback

To play back the sound backward, all we need to do is supply the negative of the frequency resulting from dividing the sample rate by the size of the table to [phasor~]. A minor modification to the patch in Figure 9-3 can provide this simple feature. The patch is shown in Figure 9-6.

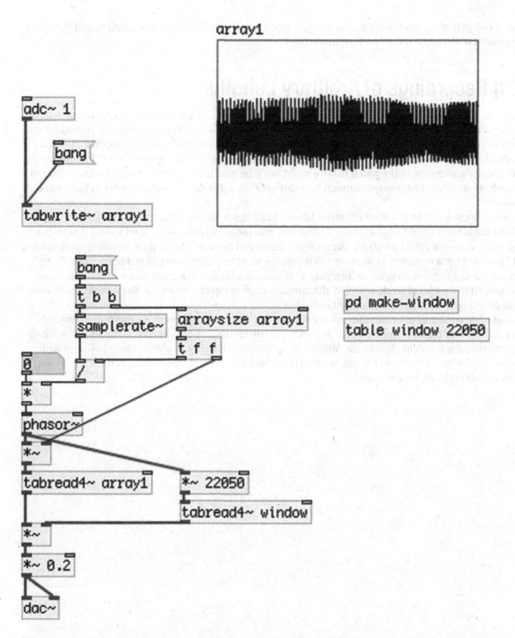

Figure 9-6. *Adding the direction reverse feature to the playback patch*

The only modification is that now the result of the division between the sampling rate and the table size is being stored in the right inlet of [*], and in its left inlet we can send either a 1 or a –1. A –1 will result in the negative of the frequency, resulting in reading the table backward. This happens because if we provide a negative frequency to an oscillator, it will reverse its period. So, providing a negative frequency to [phasor~], will result in a falling ramp rather than a rising one. Connect a [phasor~] to a [tabwrite~] and check for yourself. Any values in between will play the recording slower, and its pitch will also change. Any value above

1 or below –1 will play the sound faster and at a higher pitch. In this project, we're not going to add the speed control feature as it will make things pretty complicated.

Making Recordings of Arbitrary Lengths

The steps we've taken so far are pretty simple, but they also do something very simple, record and play back a sound of a predefined length. Of course, that's not what a looper is about. With a looper, you should be able to record sounds of "any" length. I put the word "any" in quotation marks because it can also mean a very long sound, which would last hours, for example. A looper is usually used to record short fragments of sound that last a few seconds. In the patch that we build, we'll be able to record rather lengthy loops—up to approximately 45 seconds. That's way too much for most uses of such a device, but it's better to have some space available.

To record a sound, the length of which is not known in advance, the best approach is to have enough space to hold the longest sound we'll ever need to record, and record as long as we want within that space. In Pd (as well as C, on which Pd is based) we can increase the size of an array in real time. Sending the message "resize $1" (where $1 is a number) to an array will resize the array to that number of elements. Even though that might sound as a solution to what we're trying to achieve, it's really not a good idea to use this feature. This will cause Pd to do a lot of work, trying to allocate new memory every time we resize a table. It will most likely cause dropouts, which are definitely not desirable.

The route that we'll take allocates enough memory in our computer by creating a pretty big array. We can record in that array for any length of time that we want within the size of the array. By using [tabwrite~], it's possible to stop the recording before the object fills up the table it writes to with samples, by sending it the message "stop". Figure 9-7 shows the patch we've been building since the beginning of this chapter, but modified to record arbitrary length sounds.

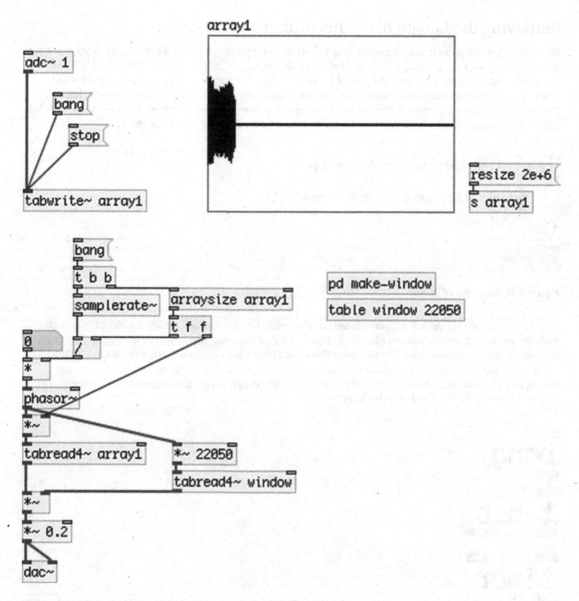

Figure 9-7. *Using a large array to record sounds of arbitrary length*

The message on the right side of the patch in Figure 9-7 resizes the array to a length of 2,000,000 elements ($2e + 06$ is called *scientific notation* and it means 2×10 to the positive sixth power, where e stands for 10). With a 44,100 sampling rate, this is enough to record 45 seconds of sound (to get the maximum length of a sound that fits in an array we need to divide the array length by the sampling rate, the opposite of what we do to obtain the right frequency to play the sound back. $2,000,000 \div 44,100$ yields ~45.35). This length should be enough for a looper. On the top-left part of the patch we're again triggering the recording by sending a bang to [tabwrite~], and when we want to stop, we send the message "stop". You can see in the figure that only a part of the array has been filled with samples, and the rest of it remains empty.

Retrieving the Length of the Recording

We can easily control the length of a recording, but what about playing it back? If you use the patch the way it is, it will play back the recording at its normal speed, but [phasor~] will go on and [tabread4~] will read the whole array, including the greatest part of it which is silent. To play back only the desired part of the array, we need to know how long the recording lasted. Thankfully, Pd provides an object for that, [timer]. This object calculates logical time since it was reset until it is banged in its right inlet. It is maybe the only Pd object where the right inlet is the hot one. Figure 9-8 shows a very simple use of this object.

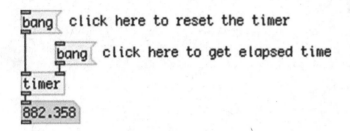

Figure 9-8. *Basic use of [timer]*

As you can see in Figure 9-8, [timer] is sub-milliseconds accurate, which makes the object very accurate and suitable for our needs. You can also see in Figure 9-8 that you must first bang the left outlet, and then the right to get the elapsed time. The design of this object is very logical, even though Pd has the right-to-left execution order. It is especially for this execution order that this object is designed like this, because by having its right inlet as the hot one, we can easily use it with [trigger] to get the elapsed time between successive bangs. Figure 9-9 shows this in practice.

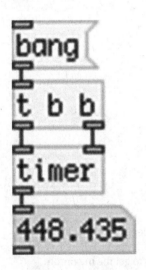

Figure 9-9. *Utilizing [timer]'s inverted hot and cold inlets*

Let's look at the simple patch in Figure 9-9. Whenever we bang [t b b], it first sends a bang out its right outlet, which bangs the right inlet of [timer]. We'll get the time elapsed since the last time it was reset, and then [t b b] will send a bang out its left inlet, which will reset [timer]. This way, the next time we bang [t b b], we'll get the time elapsed since the last time we banged it. This patch demonstrates the approach in the design of [timer].

Now let's import [timer]'s feature to the patch in Figure 9-7 so we can read only the length of the array where we've recorded sound. Figure 9-10 shows that patch. In newer versions of Pd, [timer] can calculate elapsed time in samples rather than milliseconds, which is very convenient for a looper, since we're recording in samples, and what we really need to know is the number of samples that we've recorded (that's what the size of the array was telling us when we were recording in the whole array). This is achieved by giving two arguments to the object, [timer 1 samp]. Unfortunately this is not really functional in Pd-extended, so we have to use [timer] with no arguments, get the elapsed time in milliseconds, and then convert it to samples with a small abstractions, [ms2samps], which is shown in Figure 9-11. If you're using Pd-vanilla 0-45 or newer, wherever you see [timer] connected to [ms2samps], use [timer 1 samp] instead.

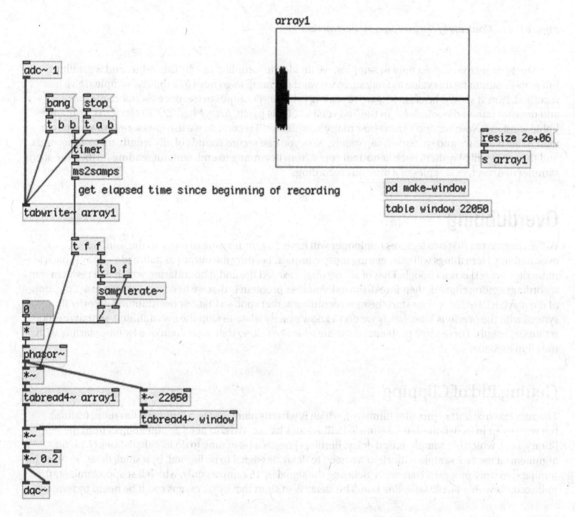

Figure 9-10. *Calculating elapsed time when recording sounds of arbitrary length*

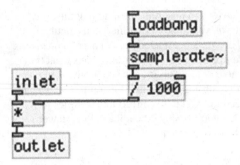

Figure 9-11. *Contents of the [ms2samps] abstraction*

Once we get the elapsed time in samples, we divide the sampling rate by this value, and we multiply [phasor~]'s output by this value as well, since we want the ramp to go from 0 to the last sample we've just recorded. Now it should be clear why we're scaling [phasor~]'s output twice: once for [tabread4~ array1] and once for [tabread4~ window]. In the first version of this patch, Array1 had 22,050 elements, as many as [table window]. Now, we don't know how many samples we'll record, so we use [phasor~]'s raw output to read [table window], and we scale it separately. Now you can record sounds of any length, up to 45 seconds, and the patch will play them back at normal speed from beginning to end, without reading further, the silent samples of Array1, or samples of a previous recording.

Overdubbing

We've covered the first two features our looper will have, so now it's time to move to the third one, overdubbing. Here things will start getting more complex. Feeding the output of [tabread4~ array1] back to [tabwrite~ array1] is not enough. First of all, we might exceed the audio boundaries, which will result in our recordings getting clipped, therefore distorted. Another problem is that we need to know the exact position of the playback so that we can start the new recording at that spot, so that our overdubbing is perfectly synced with the previous loop. Lastly we don't know exactly when to stop the overdubbing, so that the loop retains its length. These three problems are quite a lot of work, so we'll take them one by one, starting with the clipping issue.

Getting Rid of Clipping

The zexy external library provides [limiter~], which is what its name states, a limiter (it has more features, but we'll use it in its default state). Figure 9-12 illustrates its use. We also need another object from the zexy library, [z~], which is a sample-based delay. [limiter~] needs a bit of time to do its calculations (with no arguments it uses a 9 sample buffer), so we need to delay the signal to be limited, by a small delay, in order for [limiter~] to work properly. Here we're delaying the signal by 16 samples only, which last approximately 0.36 milliseconds with a 44,100 sampling rate. This delay is so short that by no means can it be heard by humans.

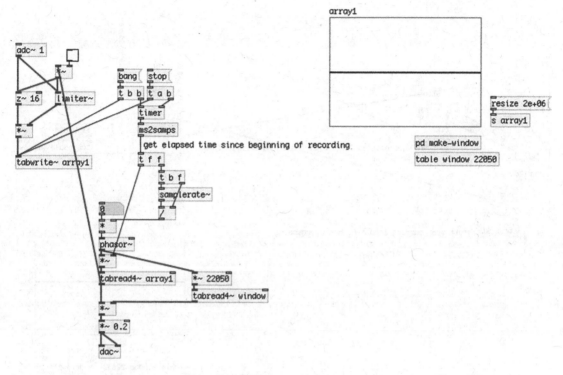

Figure 9-12. *Applying a limiter to avoid clipping when overdubbing*

In the patch in Figure 9-12 we are sending both signals from [adc~] and [tabread4~ array1] to [limiter~], which calculates the amplitude corrections, and also to [z~ 16], which delays their signals by 16 samples, before they are multiplied by the amplitude correction factor supplied by [limiter~]. Remember that when two signals are sent to a single audio inlet, they are being summed, which is what we want here. We've also included a toggle to control whether the loop from [tabread4~ array1] will go to [tabwrite~ array1] or not. This enables us to either overdub or record a new loop.

Getting the Start Position of the Overdubbing

The next step is to figure out the position of the loop while we're overdubbing. We'll need to get the position of [phasor~] scaled by the length of the recording as a float. [snapshot~] provides the first value of a sample block of a signal when banged. In digital audio, we have the restriction of the sample block, since all calculations are being made in blocks of samples, by default in blocks of 64 samples. This means that as soon as we bang [snapshot~], we'll get the first value of its input signal's next sample block, when this sample block arrives in its inlet. Still, 64 samples last approximately 1.45 milliseconds, which is accurate enough for our purpose, so using [snapshot~] is perfectly fine. Once we get the loop's position, we can use that value along with the message "start" to force [tabwrite~] to start writing at the given sample of the table. Figure 9-13 shows this. The objects of the patch have been repositioned, so that it is easier to "read" the patch. The bang that goes into [snapshot~] will trigger the overdubbing, starting at the position indicated by [snapshot~]. What is left is to calculate when the overdubbing will stop.

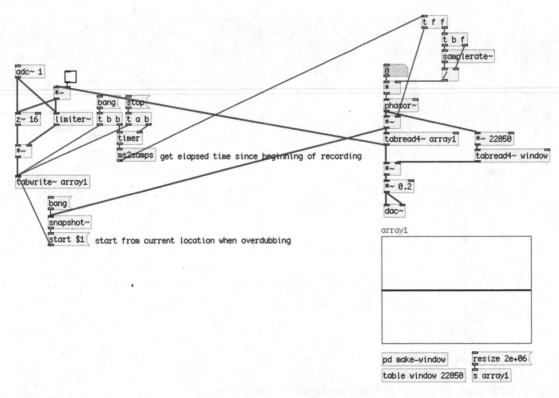

Figure 9-13. Getting the position of the loop for proper syncing of the overdubbing

Knowing When to Stop Overdubbing

Since we have both the position of the overdub start and the length of the initial recording, it is a matter of a simple subtraction to define when the overdubbing should stop, so that the loop retains its initial length. Subtracting the total number of samples, which we obtain with [timer] and [ms2samps] (or [timer 1 samp] on Pd-vanilla) from the value we get from [snapshot~] when we start overdubbing will yield the number of samples left for us to overdub. By converting this value to milliseconds, we can use it with [delay] to send the message "stop" to [tabwrite~] at the right moment. Figure 9-14 shows the patch. Once we make an initial recording, we store the amount of samples of the recording in [swap], and as soon as we start overdubbing, we subtract the position of the loop from the total number of samples and convert it to milliseconds so that [del] is used properly. If you're using Pd-vanilla you can replace the combination of [samps2ms] connected to [del] with [del 1 1 samp], which will apply a delay based on samples. Figure 9-15 shows the contents of the [samps2ms] abstraction. You might already have it in your abstractions folder, but it's shown here in case you don't.

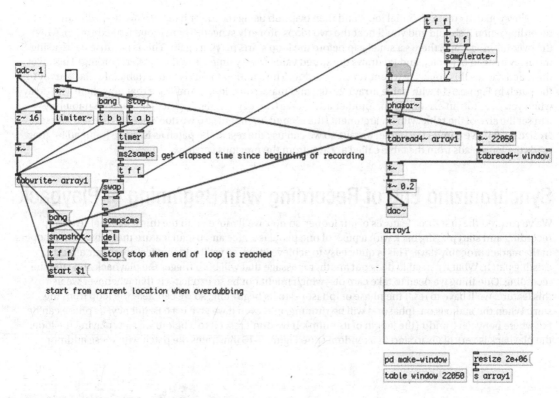

Figure 9-14. *Calculating the remaining time for overdubbing*

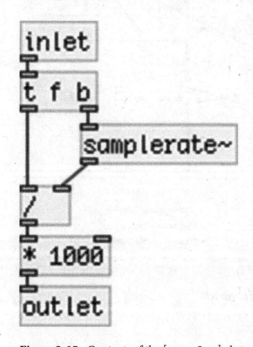

Figure 9-15. *Contents of the [samps2ms] abstraction*

Now you can create an initial loop, and then overdub using the lower bang. [tabwrite~] will stop recording when it should and you'll hear the two loops properly synced. There is one issue through, when the overdubbing stops, there's a small gap before the loop starts playing again. This is because we're using the array instead of [table], and Pd draws the stored values every time we make a new recording. This takes time, and causes this small lag whenever we overdub. The solution to this is to use [table]. Replace **array1** in the patch in Figure 9-14 with [table array1 2e+06] and make some loops. You'll see that now there's no delay when you overdub an existing loop. Another advantage (not of great significance) is that with [table] we can set the size of the table with an argument (the second argument), so we don't need to use the message "resize 2e+06". We named the [table] array1 so we can use the rest of the patch as is, since [tabread4~] that reads the loop reads from the table called array1, from the previous versions.

Synchronizing End of Recording with Beginning of Playback

We've covered the first three features of our looper, so now we'll move on to the third. Being able to stop recording and start playing back with a push of one pedal is a nice and useful feature that most of the loopers in the market probably have. This is quite easy to achieve in Pd, so we'll add that feature to our already existing patch. What we need to do is put another message that will both trigger the playback and stop the recording. One thing we need to take care of—which might not be so explicit—is that whenever we use this feature, we'll have to reset the phase of [phasor~] to its beginning, so we can hear the loop from the start. When the audio is on, [phasor~] will be running, and even if we give it a 0 frequency, its phase can be anywhere between 0 and 1 (the length of its ramp). If we don't reset it to 0 before we start playing the loop, the playback is very likely to start at a random spot. Figure 9-16 illustrates the patch with these additions.

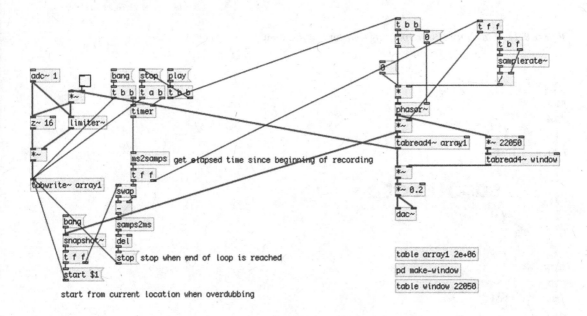

Figure 9-16. *Synchronizing end of recording with beginning of playback*

We've added the message "play", which we can hit instead of "stop", to both stop the recording and start the playback. Notice that the left outlet of [t b b], which the message "play" connects to, goes to another [t b b], which will first set [phasor~]'s phase to 0, and will then send 1 to [*], which will calculate the right frequency so that [phasor~] plays back the recording at normal speed. There's also the message "0", which enables us to stop the playback in case we want to make a new recording and erase the existing one. You can also see in Figure 9-15 that the array has been replaced by [table array1 2e+06], and that the message "resize 2e+06" has been removed.

Start Building the Interface

Since we've completed four out of the five features our looper will have, it's a good time to start building the Arduino code and circuit. This project's code adds some new features, so it might be a bit tricky to understand. Pay close attention to all explanations, and if needed, go through it once more. Eventually, you'll be introduced to some very flexible C++ features , which are very likely to be helpful in other projects you might build.

Writing the Arduino Code

Before we start diving into the new C++ concepts, let's first analyze what we want the controller for our looper to do. We've set five features we want the looper to include, and we've already completed four of them in Pd. The first basic thing is that we want to record loops of arbitrary lengths. We'll use a foot-switch (for the prototype, we'll use breadboard-friendly push buttons) to start and stop the recording. We'll need another foot-switch to start the forward playback, and one more for the backward playback. We'll also need one LED per foot-switch to indicate the state of the looper. Now let's take the following scenario. We start with recording the first loop, and to end it, we hit the forward playback foot-switch. This means that the LED of the recording foot-switch will be on for as long as we're recording, and as soon as we press the playback foot-switch, the recording LED should turn off, and the forward playback LED should turn on. As the loop is playing, we want to overdub, so we press the recording foot-switch. What should happen is that the forward playback LED should stay on, and the recording LED should light up, and stay lit until we reach the end of the loop, where it should turn off by itself. The same approach goes to the backward playback. What we see is that we'll need to control three LEDs in various ways, where using arithmetic like we've done so far (for example, i + 6) won't really help, as things are not so simple. We need another way to easily control these LEDs. This is where the first new C++ concept comes in—the enumeration. The enumeration in C++ enables us to define our own data types, but in this project, we'll use it in a simpler form. We'll use it to define self-explanatory names to variables that will hold the numbers of the pins of the LEDs we'll use, so we can easily control them in our program. Taken that the first LED will be wired to digital pin 5 (since we have three foot-switches and we start using digitals pins from 2 onward), we could define the following enumeration:

```
enum leds { recording_led = 5, playback_led, backward_led };
```

The first word is the keyword to define an enumeration. leds is the identifier of this enumeration, in which case it is also the new data type (we can define variables of the leds data type, once we define the enumeration, which can take any of the values between the curly brackets, but we won't deal with that). Inside the curly brackets, we set names for the LEDs, which are constants (we can't change them or their values throughout our code). We assign a value to the first one only, and the rest will take incrementing values (you can assign different values to the rest of the constants if you want, but if you don't assign any, they will get the incrementing values). The preceding line of code results in recording_led being equal to 5, playback_led being equal to 6, and backward_led being equal to 7. From this point on, we can use these three self-explanatory names to control the LEDs, and we don't need to care about their wiring.

Writing Functions for the Foot-Switches

The next thing we need to do is define a function for each foot-switch. This will make our code modular and eventually easy to understand and use. We'll write three functions in total, which we'll call from the main loop function.

The Recording Foot-Switch Function

The first function we'll start with is the recording foot-switch, which is shown in Listing 9-1.

Listing 9-1. Function for the Recording Foot-Switch

```
1.   void record(){
2.     if(!recording){
3.       if(playBack || backWard){
4.         Serial.print("overdub ");
5.         Serial.println("bang");
6.         digitalWrite(recording_led, HIGH);
7.         recording = !recording;
8.       }
9.       else{
10.        Serial.print("record ");
11.        Serial.println("rec");
12.        digitalWrite(recording_led, HIGH);
13.        recording = !recording;
14.      }
15.    }
16.    else{
17.      Serial.print("record ");
18.      Serial.println("stop");
19.      digitalWrite(recording_led, LOW);
20.      recording = !recording;
21.    }
22.  }
```

As you can see, apart from the recording_led, we're using another three variables, recording, playBack, and backWard (actually recording_led is a constant). These variables are of the type bool, and we'll define them in the beginning of the code, when we write the whole of it. What you need to know is that they must be initialized as false, since they denote the state of each function, and when the program starts, it's neither recording, nor playback, either forward or backward. In line 2, we check if the recording Boolean is false. We achieve succeeding in the if test of line 2 when the recording Boolean is false by using the exclamation mark, which inverts the variable's value. The value inside the test's parenthesis will be true when the actual value is false. If the test succeeds, it means that we're not recording, and we've pressed the first foot-switch to start a new recording instead of finishing it. If this test is true, in line 3, we check whether we're currently playing back a loop, either forward or backward. The || is the Boolean OR, which checks whether either of the two conditions is true. If any of the two, or both of the Booleans we check, playBack and backWard, is true, then this test will succeed. Of course, only one of the two can be true, as we can't play the loop both forward and backward at the same time, but this is how the Boolean OR functions; that's why I mention it.

If the test of line 3 succeeds, it means that we're overdubbing, so we'll send a bang to Pd with the tag "overdub ". Line 6 is where we use the enumerated constants for the first time. Since we're recording, regardless of whether it's a new recording or if we're overdubbing, we want to turn the recording LED on. In this line, we use the recording_led constant to control this LED. In the case of the recording foot-switch, the use of the recording_led enum is pretty simple, but in the other two functions we'll write, we'll see how this naming capability simplifies our code.

In line 7, we update the recording variable, because it will now be true, since we'll be recording. Simply using the exclamation mark reverses the state of the Boolean variable, so this line works perfectly for our needs.

In line 9, we have the else in case the test of line 3 fails (take care not to mix the two ifs of lines 2 and 3. The else in line 9 belongs to the if of line 3). This will happen if we want to record without overdubbing. If no playback occurs at the moment we press the recording foot-switch, the test in line 3 will fail, and the code of the else in line 9 will be executed. This code sends the string "rec" with the tag "record " to Pd, and turns the recording_led LED on. Finally, it updates the recording variable the same way with line 7. Mind that lines 7 and 13 will never be executed together, as we'll be either recording a new loop, in which case line 13 will be executed, or we'll be overdubbing, in which case line 7 will be executed.

In line 16, we have the else of the if of line 2, which will be executed if we press the recording foot-switch to stop the current recording. When we start recording, the recording Boolean will be true, so the next time we press the recording foot-switch, the test in line 2 will fail, and the code of the else of line 16 will be executed. In this code, we send the string "stop" to Pd with the tag "record ". Then we turn off the recording_led LED, and finally, we update the recording Boolean, which will be switched to false, since we're stopping the recording.

The Forward Playback Function

The other two functions are similar to the first one, and very similar between them, so it should be easy to follow. Listing 9-2 shows the function for the forward playback foot-switch.

Listing 9-2. Function for the Forward Playback Foot-Switch

```
1.   void playback(){
2.     if(!playBack){
3.       // if we're recording, quit recording first and then start playing back
4.       if(recording){
5.         Serial.print("record ");
6.         Serial.println("stop");
7.         digitalWrite(recording_led, LOW);
8.         recording = !recording;
9.       }
10.      Serial.print("playback ");
11.      Serial.println("1");
12.      digitalWrite(playback_led, HIGH);
13.      playBack = !playBack;
14.      if(backWard){
15.        digitalWrite(backward_led, LOW);
16.        backWard = !backWard;
17.      }
18.    }
```

```
19.    else{
20.      Serial.print("playback ");
21.      Serial.println("0");
22.      digitalWrite(playback_led, LOW);
23.      playBack = !playBack;
24.    }
25.  }
```

As with the recording function, the first thing we check is whether we're already playing back, and pressing the playback foot-switch should stop the playback (instead of start it). The switches that we'll use in this project are *momentary switches*, which are much like push buttons, so these tests are necessary, since we need to alternate between states with the same information we'll get from the switch. (A 0 when the switch is pressed, and a 1 as long as it's not pressed—due to the internal pull-up resistors). Therefore, line 2 checks if the playBack Boolean is false. If it is, in line 4 we're checking if we're pressing the playback foot-switch immediately after recording, which should stop the recording and start the playback at the same time. In line 3, there's a comment indicating that too. In line 4, we check if recording is true, which will mean that we're recording. If it is, we stop the recording by sending the string "stop" under the tag "record " to Pd, and then we turn off the recording_led and update recording. This last step is necessary, because if we omit it when we play back immediately after recording, the next time we try to record, pressing the recording foot-switch won't have any effect, as recording will still be true. (Remember that the initial tests in these functions use the exclamation mark, which means that they will succeed if the Boolean they test is false.)

Once the code of the if of line 4 has finished (if it is executed), we'll print the string "1" to Pd with the tag "playback ". This will trigger the forward playback of the loop. In line 12, we turn on the playback_led, and in line 13 we update the playBack Boolean. In line 14, we check if we were playing the loop backward, before we pressed the forward playback foot-switch, by checking if the backWard Boolean is true. If it is, we need to turn off the backward playback LED, and update the backWard Boolean, turning it to false. What we see in this function is that we might have to control another LED apart from the LED of this foot-switch, and this other LED might be any of the other two LEDs. It should be clear how handy the enumeration technique we used for naming the LEDs is; since we really don't need to care about pin numbers, all we need are the names of the LEDs. Of course, we could have defined three different variables with the corresponding pin numbers, but the enumeration is very handy because it happens in one line of code, and the number assignment is automatic, after the first constant.

In line 18, the if test of line 2 finishes, and in line 19 we have its else code, which will be executed when we press the forward playback foot-switch to stop the playback. Since we have updated the playBack Boolean the first time we pressed the forward playback foot-switch, the next time we'll press it, playBack will be true and the test in line 2 will fail, resulting the in execution of the code of else. In this chunk of code we just send the string "0" to Pd with the tag "playback ", we turn off the playback_led, and again we update the playBack Boolean. And this is the end of the forward playback function.

The Backward Playback Function

This function is almost identical to the function of the forward playback. Its code is shown in Listing 9-3.

Listing 9-3. Function for the Backward Playback Foot-Switch

```
1.  void backward(){
2.    if(!backWard){
3.      // if we're recording, quit recording first and then start playing back
4.      if(recording){
5.        Serial.print("record ");
6.        Serial.println("stop");
```

```
7.          digitalWrite(recording_led, LOW);
8.          recording = !recording;
9.        }
10.       Serial.print("playback ");
11.       Serial.println("-1");
12.       digitalWrite(backward_led, HIGH);
13.       if(playBack){
14.         digitalWrite(playback_led, LOW);
15.         playBack = !playBack;
16.       }
17.       backWard = !backWard;
18.     }
19.     else{
20.       Serial.print("playback ");
21.       Serial.println("0");
22.       digitalWrite(backward_led, LOW);
23.       backWard = !backWard;
24.     }
25.   }
```

This function is more or less the opposite of the forward playback function. Since we can stop recording by pressing the forward playback foot-switch, we should be able to do the same thing by pressing the backward playback foot-switch too. Compare this function to the previous one and you'll see that the playBack and backWard Booleans are reversed. There's no real need to explain this function, since I've explained its forward counterpart. Try to understand it on your own, although it shouldn't be difficult.

Calling the Foot-Switch Functions Dynamically

We've written the three functions for the foot-switches, but how are we going to call them? We want to read the digital pins of the foot switches, and when a switch is pressed, call the corresponding function. The switch / case control structure could be useful here, but there's another way to do this. What we actually need is an array of functions that we can call according to the digital pin that will read a switch press. We haven't seen something similar up to now. This technique requires another new C++ concept, that of the pointers. Before I explain how we can call these three functions dynamically, I must talk a bit about pointers.

Introducing Pointers in C++

A pointer is a somewhat advanced topic in C++ programming, but also a very useful one. A pointer is a special variable that points to a variable of one of the data types we already know. A pointer doesn't contain the value of the variable it points at, but its location in the memory of the computer. A parallelism to the concept of pointers is house addresses. A variable of some data type (an int, for example) could be paralleled to a house (to the contents of the house actually), and its pointer could be paralleled to the address of the house. When we use pointers, we tell the computer (or the Arduino) which place in the memory to look at. An extensive explanation of pointers is beyond the scope of this book, but we'll take a quick look at them.

Take the following example. We define a variable of type int, and assign to it the value 10. In C++ pseudo-code, the following chunk of code would result in printing the number 10 to the computer's monitor:

```
int var = 10;
print var;
```

Introducing pointers to this pseudo-code, we could do write the code in Listing 9-4.

Listing 9-4. C++ Pseudo-Code Using Pointers

```
1.   int var = 10;
2.   int *pnt;
3.   pnt = &var;
4.   print pnt;
5.   print *pnt;
6.   *pnt = 20;
7.   print var;
```

The first line defines the same variable we used in the previous pseudo-code. The second line defines a pointer that points to a variable of the data type int. To define a pointer we must use an asterisk, the same way used here (or with a whitespace between the asterisk and the identifier of the pointer, which would be this: int * pnt;). Defining a data type for a pointer doesn't mean that this pointer will acquire the defined amount of memory (an int in the Arduino language is 2 bytes, for example). Pointers are special variables, and they acquire the same amount of memory, no matter what data type they point to. Defining a data type for a pointer defines the number of bytes this pointer will jump, if it is told to point to the variable next to the one it is currently point at. This way pointers can be used much like arrays (array identifiers are actually constant pointers).

Line 3 assigns the address of the variable var to the pointer pnt. The use of the ampersand is called the *reference operator*, and it yields the address of a variable. Line 4 prints the address of the variable var, which has been stored in our pointer. This line will print a value in hexadecimal, like 0x7fff5df4c974, which is of no interest to us. This is just the address of the variable in the memory of the computer.

Line 5 might be a bit confusing. The use of the asterisk in this line is different than the use of it in line 2. Here it is called the *dereference operator*, and yields the value stored in the address the pointer points at. This line would print 10 to the monitor, because this is the value assigned to the variable var, which the pointer points at. Line 6 assigns another value to the address the pointer points at (the address of the variable var), again using the dereference operator, so line 7 will print 20, because this is the new value assigned to that memory location.

Pointers can be a bit daunting in the beginning, but they can be very useful and powerful when programming. In the Arduino they are rarely used, and even in the documentation of the Arduino's official web site there's no explanation about them, there's only a mention saying that it is a rather advanced topic that will most likely not be needed when programming the Arduino.

In our code, we're not going to use pointers the way they're used in Listing 9-4, but instead we'll use a pointer to functions. This is an even more advanced topic, but we'll make a moderate use of it, so it shouldn't be very difficult to follow. The earlier explanation of pointers was included only to give a hint as to what pointers are.

Using Pointers to Functions

To call the three functions for the foot-switches dynamically, we need to define a pointer to them. Let's first take a simple example of a pointer to a function. Listing 9-5 shows a small Arduino sketch that utilizes this feature.

Listing 9-5. Using a Pointer to a Function

```
1.   void one_func(int x){
2.     Serial.print("first function called with argument ");
3.     Serial.println(x);
4.   }
5.
6.   void another_func(int x){
7.     Serial.print("second function called with argument ");
8.     Serial.println(x);
9.   }
10.
11.  // define a pointer to a void function with one int argument
12.  void (*function)(int);
13.
14.  int count = 0;
15.
16.  void setup() {
17.    Serial.begin(9600);
18.  }
19.
20.  void loop() {
21.    if(!count){
22.      // call the first function through the pointer
23.      function = one_func;
24.      function(5);
25.      delay(1000);
26.
27.      // call the second function through the pointer
28.      function = another_func;
29.      function(10);
30.
31.      // increment the counter so if won't succeed again
32.      count++;
33.    }
34.  }
```

If you upload this code on your board and open the Arduino IDE serial monitor, you'll get the following (take care to set the baud rate of the serial monitor to 9600):

```
first function called with argument 5
second function called with argument 10
```

Let's start with line 12, where we define the function pointer. The declaration of a function pointer is a bit different than the declaration of a pointer to some variable, like we saw in Listing 9-4. The difference is that we must include the pointer name and the asterisk inside a parenthesis, and afterward place another parenthesis, which will include the arguments the function this pointer will point to is expecting. This

387

pointer will point to either the one_func function, or the another_func function, which have been defined in lines 1 and 6. Both these functions are of the data type void, and they take one argument of the data type int. The function pointer function is defined exactly this way. It is of the data type void and it takes one argument of the data type int. In the argument definition of the function pointer, we should not include the identifier of the argument (the name of the argument), like in the two functions of lines 1 and 6, where the argument they expect is defined as int x, but only the data type of the argument.

The way the function pointer and the two functions have been defined in Listing 9-5 makes it possible to use the same pointer to call any of the two functions. We can see that in lines 23 and 28, where we assign the function pointer to point first to the one_func function, and then to the another_func function. Even though we assign the function pointer to point to a function, we shouldn't include any parenthesis in this assignment (actually, the name of a function is a constant pointer itself!). In line 24, we call the one_func function through the function pointer and pass the argument 5 to it. In line 29, we call the another_func function through the function pointer and pass the argument 10. As we can see in the printed result, we get what we expected.

Defining a New Data Type for the Function Pointer

The example in Listing 9-6 is not very useful. For one, the two functions don't really do anything useful; they just print the value they receive (but that was intentional, since it is only an example to better understand function pointers). It's also not useful for our project, because in the code in Listing 9-6, we need to know which function we want to call and assign the function pointer to point to that function before we actually call it. We want to call any of the three functions for the foot-switches without knowing in advance which one we'll call. To do this, we need to create an array that will hold these three functions, and call them from there (they will be called from inside a for loop, using the loop's variable as the index of the function array). To achieve this we need to define a pointer to a function as a new data type, and then we'll use this new data type to create the array, which will contain the names of the three functions. Listing 9-6 shows the steps we need to take to create the function pointer as a new data type and the array of this data type.

Listing 9-6. Creating a Function Pointer and an Array of Functions

```
1.   // define a pointer to the functions
2.   typedef void (*function)();
3.   // set an array of the function pointer type with the names of the functions
4.   function my_functions[num_of_pedals] = { record, playback, backward };
```

We'll start with line 4 instead of line 1. This line creates an array of the data type function (we suppose that we have defined the constant num_of_pedals and assigned to it the value 3). This data type does not exist in C++, we have created it in line 2 by using the keyword typedef. The syntax of line 2 is almost the same as the function pointer definition in Listing 9-5. Here we use the keyword typedef, which stands for *type definition*, and it enables us to define our own data types. Therefore, line 2 creates a function pointer data type, which is not a C++ native data type. This way we're able to create an array of this data type, which will include function pointers (instead of ints, for example). The function pointers of the array are constant, just like the identifier of a function. This means that we can call these functions either by using their name or by using the array name with the appropriate index. After the declaration in Listing 9-6, if we have defined the three functions for the foot-switches as well, the following:

```
record();
```

will be equivalent to this:

```
my_functions[0]();
```

Both lines call the record function. This enables true dynamic function calls, since we can use a variable to control the function pointer array index inside a loop, like this:

```
for(int i = 0; i < num_of_pedals; i++){
  my_functions[i]();
}
```

This loop would call all three functions, one at every iteration. Of course, this is not what we want, but we can use tests inside the loop to check whether we should call a function or not.

The First Version of the Arduino Code

We have now covered all the new concepts this project introduces, so we can move on to write the first version of the code for the project. It is shown in Listing 9-7.

Listing 9-7. The Full Arduino Code

```
1.   const int num_of_pedals = 3;
2.   int old_val[num_of_pedals];
3.   const int first_led = num_of_pedals + 2;
4.
5.   // define a pointer to the functions
6.   typedef void (*function)();
7.
8.   // set an array of the function pointer type with the names of the functions
9.   function my_functions[num_of_pedals] = { record, playback, backward };
10.
11.  // booleans used in the functions to alternate states with every switch press
12.  bool recording = false, playBack = false, backward = false;
13.  // name the LED pins
14.  enum leds { recording_led = first_led, playback_led, backward_led };
15.
16.
17.  // define the functions
18.  void record(){
19.    if(!recording){
20.      if(playBack || backWard){
21.        Serial.print("overdub ");
22.        Serial.println("bang");
23.        digitalWrite(recording_led, HIGH);
24.        recording = !recording;
25.      }
26.      else{
27.        Serial.print("record ");
28.        Serial.println("rec");
29.        digitalWrite(recording_led, HIGH);
30.        recording = !recording;
31.      }
32.    }
```

```
33.    else{
34.      Serial.print("record ");
35.      Serial.println("stop");
36.      digitalWrite(recording_led, LOW);
37.      recording = !recording;
38.    }
39.  }
40.
41.  void playback(){
42.    if(!playBack){
43.      // if we're recording, quit recording first and then start playing back
44.      if(recording){
45.        Serial.print("record ");
46.        Serial.println("stop");
47.        digitalWrite(recording_led, LOW);
48.        recording = !recording;
49.      }
50.      Serial.print("playback ");
51.      Serial.println("1");
52.      digitalWrite(playback_led, HIGH);
53.      playBack = !playBack;
54.      if(backWard){
55.        digitalWrite(backward_led, LOW);
56.        backWard = !backWard;
57.      }
58.    }
59.    else{
60.      Serial.print("playback ");
61.      Serial.println("0");
62.      digitalWrite(playback_led, LOW);
63.      playBack = !playBack;
64.    }
65.  }
66.
67.  void backward(){
68.    if(!backWard){
69.      // if we're recording, quit recording first and then start playing back
70.      if(recording){
71.        Serial.print("record ");
72.        Serial.println("stop");
73.        digitalWrite(recording_led, LOW);
74.        recording = !recording;
75.      }
76.      Serial.print("playback ");
77.      Serial.println("-1");
78.      digitalWrite(backward_led, HIGH);
79.      if(playBack){
80.        digitalWrite(playback_led, LOW);
81.        playBack = !playBack;
82.      }
83.      backWard = !backWard;
84.    }
```

```
85.    else{
86.        Serial.print("playback ");
87.        Serial.println("0");
88.        digitalWrite(backward_led, LOW);
89.        backWard = !backWard;
90.    }
91. }
92.
93.
94. void setup() {
95.    for(int i = 0; i < num_of_pedals; i++){
96.        pinMode((i + 2), INPUT_PULLUP);
97.        pinMode((i + 5), OUTPUT);
98.        old_val[i] = 1;
99.    }
100.
101.    Serial.begin(115200);
102. }
103.
104. void loop() {
105.    if(Serial.available()){
106.        byte in_byte = Serial.read();
107.        // the only information we'll receive is when the recording has stopped
108.        // so we need to turn off the recording LED
109.        digitalWrite(recording_led, in_byte);
110.        // we also need to update the recording boolean
111.        recording = !recording;
112.    }
113.
114.    // read the foot_switches
115.    for(int i = 0; i < num_of_pedals; i++){
116.        int foot_switch_val = digitalRead(i + 2);
117.        // check if the state of a switch has changed
118.        if(foot_switch_val != old_val[i]){
119.            // if it has changed, check if the switch is pressed
120.            if(!foot_switch_val){
121.                my_functions[i]();
122.                old_val[i] = foot_switch_val; // update old value
123.            }
124.            old_val[i] = foot_switch_val; // update old value
125.        }
126.    }
127. }
```

Lines 1 to 14 have already been explained, we haven't seen only line 3, where we define a constant that holds the pin number of the first LED, which we use in the enumeration of line 14, and line 12 where we define the Booleans we use in the three functions for the foot-switches. Then in line 18, we start defining the functions. In line 105, we have an if test to see if there is data coming in the serial line. This is primarily there so we can turn off the recording_led when we're overdubbing. Since Pd will know when the overdub should stop, it will tell the Arduino to turn that LED off. We also update the recording Boolean, so that next time we hit the recording foot-switch, the record function will run as expected. We haven't implemented that in the Pd patch yet, but we'll do it when we get back to patching.

In line 115, we run the for loop that reads the digital pins of the foot-switches. This loop runs the same way we usually run it to read digital pins. When a switch press is detected, the corresponding function will be called using the function pointer array, controlled by the i variable of the loop. And this concludes the Arduino sketch.

Building the Circuit on a Breadboard

There's nothing special about the circuit for this project, it's just three momentary switches and three LEDs. Figure 9-17 illustrates it. Upload the code on your Arduino board and test if it works properly, by checking the LEDs (we can't use the Pd patch yet because we haven't finalized it). Pressing the first button should light up the first LED. Pressing it again, should turn it off. If the first LED is on and you press the second or third button, the first LED should go off and the LED of the button you just pressed should go on. The second and third LEDs should switch when one of them is on and you press the button of the other. Testing the overdub ending—where the recording LED should go off by itself—is not possible because that will happen by receiving data from Pd.

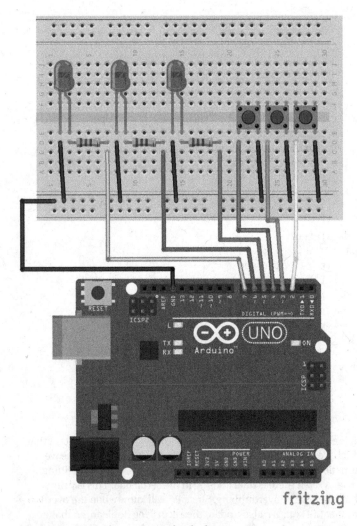

Figure 9-17. *Circuit prototype for the looper*

Working Further on the Pd Patch

Apart from not having built the last of the five features our looper will have, we have to also work on the communication with the Arduino, plus organize the patch a bit, as working further on it the way we've built it so far will most likely result in a mess. When building programs that tend to become rather complex, a good practice is to make them modular. A program that is modular is possibly easier to debug when something is not working properly, as each module does something different and it is easier to track down which part of it is malfunctioning. Another advantage of a modular program is that it is easier to expand it, since the addition will be another module. Lastly, making a program modular makes it easier to keep it tidy and easy to read.

What we already have is the first four features of the looper, which are playing back a loop in both directions, recoding loops of arbitrary length, overdubbing, and synchronizing the end of the recording with the beginning of the playback. We're going to first implement these features in a modular patch, and we'll apply the communication with the Arduino to control these features. Afterward, we'll add the last feature and we'll make some final touches to it.

The Recording Module

Let's start with the first module of the patch, which is the recording module, since without it we can't really have a looper. It is shown in Figure 9-18. On the top part we receive the messages with the "record" tag from the Arduino, where the message "rec" will send a bang to [tabwrite~] and to [timer], and the message "stop" will be sent as is to [tabwrite~] and will bang [timer] to get the elapsed time since the beginning of the recording. There are a few more things like [r current_tab] and [outlet set_tab], but we'll see a bit further down what they do. Create a subpatch and call it *rec*, and put all these objects in it.

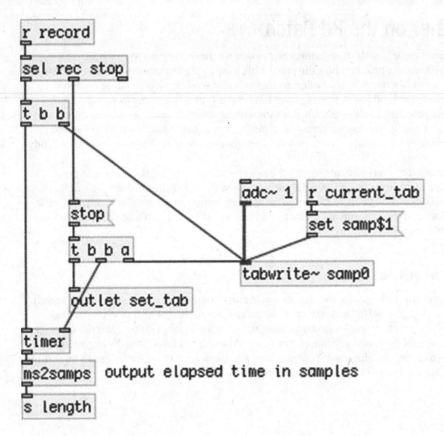

Figure 9-18. *The recording module*

The Phase Module

The next module is the phase module, which is the driving force of the looper. It is shown in Figure 9-19. This one is a bit more complicated. In the middle part of the patch, we receive the total length of the recording, which is sent from the recording module (you can see the corresponding [send] in Figure 9-18). We send this value to [swap] to divide the sampling rate by it and get the right frequency we need to send to [phasor~]. We also send it to [*~], which goes to [snapshot~], which is banged on every sample block and gives the current position of the loop in samples. Lastly the length value is sent to another [*~], which multiplies [phasor~]'s output to be sent to a [tabread4~].

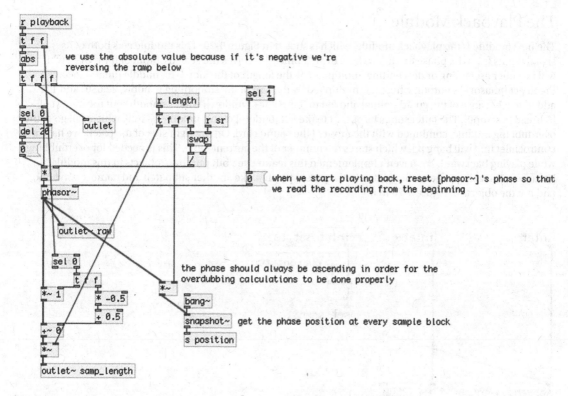

Figure 9-19. *The phase module*

On the top part of the patch, we receive the values with the "playback" tag from the Arduino. These values will be either a 1, a -1, or a 0. We need the phase to be always incrementing to get the right position of the loop when we want to overdub, even if we're overdubbing over a backward loop. For this reason we send the absolute of the "playback" value to [phasor~], so that it never inverts its phase. This value also goes to [sel 0], which excludes 0 and outputs the 1 and -1, scaled to give the right scale and offset to [phasor~] so we can playback the loop both forward and backward.

The absolute "playback" value first goes to [sel 1], which bangs a 0 and forces [phasor~] to reset its phase, so that we can play the loop from the beginning as soon as we press the forward or backward playback switch. It then goes out the third outlet of this module. This outlet will control the amplitude of the looper, where a 1 will make a 20 milliseconds ramp from 0 to 1 in the looper's output, so that we get a smooth start of the looping, and a 0 will make a ramp from 1 to 0, so we smoothly stop looping. For this reason this absolute value will go to [*] (which is multiplied by the right frequency [phasor~] needs to play back the loop at normal speed) through [sel 0]. A 1 will go to [phasor~] intact, but a 0 will be delayed by 20 milliseconds for the amplitude envelope to finish.

The output of [phasor~] goes to [snapshot~] so we can get the loop's phase when we want to overdub. It also goes to the second [outlet~] of this module, because it will control the smoothing window, which needs a different scaling than that for [tabread4~]. And it goes to the scaling and offset set by the "playback" value, which will go to [tabread4~], either as an ascending or a descending ramp. Notice that the output of [snapshot~] is being rounded with [i]. This way we get the integral part of the value ([snapshot~] is very likely to output decimal values) which is necessary for the calculations we'll need to make when we overdub while playing the loop backward. Make another subpatch in the parent patch and name it *phasor*. Put all the objects seen in Figure 9-19 in it.

The Playback Module

The next module is the playback module, which is shown in Figure 9-20. This module gets both of its signal inputs from the phase module, where the left [inlet~] receives the scaled output of [phasor~], which will be either ascending or descending, multiplied by the length of the loop. The middle [inlet~] receives the intact [phasor~]'s output, where it is multiplied by the size of the smoothing window, 22,050 (and 1 is added to it because of the guard points) and fed to [tabread4~ window], which smooths out the output of [tabread4~ samp0]. The latter sends its output to the left [outlet~] of this module as well, which will go to the overdubbing module, combined with the input of the sound card. On the right side of the patch, we have a control inlet that will bang [f], which stores the number of the current table. This is needed for overdubbing while playing backward. We haven't implemented this feature yet, but put these objects in this module because we're going to need them later. In the parent patch make another subpatch and name it *playback*, and put the objects of Figure 9-20 in it.

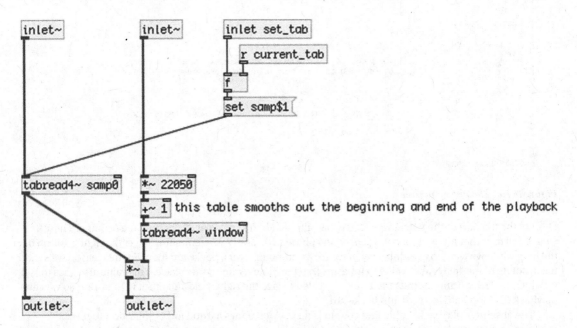

Figure 9-20. *The playback module*

The Overdub Module

The next module is the overdub module, which is shown in Figure 9-21. In this module we receive the output of [tabread4~ samp0] from the playback module, which outputs the loop, through the right [inlet~]. This signal is combined with the input of the sound card, where both signals are sent to [limiter~] to make sure we won't get any clipping. The left [inlet] receives messages for [tabwrite~ samp0]. These messages will be either "start $1", where $1 will be the loop's position at the point where we want to start overdubbing, or "stop" to stop overdubbing. The "stop" message will be sent automatically by Pd, we won't have to do anything about this. These message first go to [list split 1]. This is because we want to check whether we received a "start $1" message, but since we cannot know what value $1 will be, we split this message to "start" and "$1", where [list spit 1] will output "start" out its left outlet, and "$1" out its middle outlet. The left outlet of [list split 1] goes to [sel start], which outputs a bang if the message it receives is indeed "start". This bang will force a ramp from 0 to 1 in the input of the sound card. Since when we overdub, the [tabread4~] that reads the loop will play exactly the same sample that is already recorded, theoretically there won't be any change in the loop from there. The sound card's input is very likely to introduce a sudden jump to the new recording, especially if there are effects in the incoming signal (like an overdrive, for example). A ramp will probably smooth this sudden change avoiding a possible click. There's also a [r current_tab], like in the recording and playback modules, which sets the table we're currently using. Make another subpatch and name it *overdub*. Put the objects shown in Figure 9-21 in it.

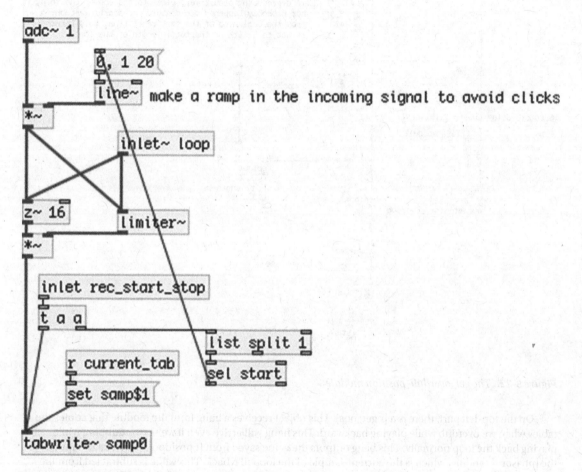

Figure 9-21. *The overdub module*

The Overdub Position Module

The next module is the module that gets the loop's position for overdubbing. It is shown in Figure 9-22. This is the most complex module so far. I'm introducing a feature that I haven't mentioned yet in this module. This feature is to make sure that if we want to overdub from the beginning of the loop, and we accidentally press the recording switch just before the end of the loop, we'll delay sending the "start $1" message to the overdub module until we begin the next iteration of the loop. This is implemented on the right side of the patch in Figure 9-22, but let's first talk about the left side first.

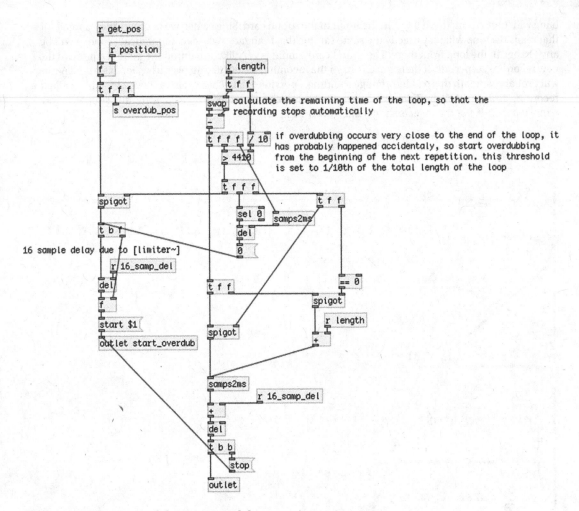

Figure 9-22. *The get_overdub_position module*

On the top-left part, there is a [r get_pos]. This object receives a bang from the module that copies the tables when we overdub while playing backward. This bang will arrive even if we're overdubbing while playing back the loop normally. This bang outputs the value saved from [r position], which takes input from the [phasor~] module, which is the current sample of the loop playback. This value is subtracted from the total length of the loop, at the top-right part of the patch. There is a test to check if this difference is more than one-tenth of the total length of the loop (it is set to a default of 4,410 samples, which is one-tenth of 44,100, which is one second of sound at a 44,100 sampling rate). This test is for the new feature mentioned earlier. If it is, it means that we've pressed the recording switch well before the end of the loop, and the [spigot] at the left side will open. If the [spigot] opens, the sample of the overdub beginning will go to [t b f], and will be sent to the "start $1" message with a 16 sample delay. This delay is applied because in the overdub module we're using [limiter~] with a 16 sample delay. The delay amount is calculated in a subpatch we haven't created yet, and received via [r 16_samp_del]. The [outlet start_overdub] sends the message "start $1" to the overdub module (we'll see all these connections when we put all the modules together in the parent patch).

Now let's go to the middle part of the patch. Underneath [t f f] there's another [spigot], which will open if the difference between the total length of the loop and the overdub beginning is more than one-tenth of the loop. This [spigot] will receive this difference value in samples. This value is being converted to milliseconds with the [samps2ms] abstraction; then the 16 sample delay time is added to it (again because of the 16 samples delay we use with [limiter~]) and it is sent to [del]. This object will delay a bang for the amount of time specified in its left inlet (or right, check its help patch). This amount of time is the remaining time of the loop since the beginning of the overdub (think about it, subtracting the beginning of the overdub from the length of the loop, yields the remaining time). Once the remaining time of the loop is over, [del] will send a bang to the message "stop", which will go to the overdubbing module, and then out its right outlet. This outlet will send this bang to a message "0", which will go to the Arduino module to turn off the recording LED.

There's little left to explain about this module. After we've calculated the difference between the length of the loop and the beginning of the overdub, if the test whether this difference is greater than one-tenth of the total length of the loop fails, the two [spigot]s we've already been through will close, and the third [spigot] on the right side of the patch will open. [sel 0] below this test will output a bang that will go to [del] in the middle of the patch. This [del] first receives the remaining time of the loop in its right inlet (converted to milliseconds), and then a bang in its left inlet. This bang will be fired when the loop starts again (since it's delayed by the remaining time of the loop), sending a 0 to the message "start $1". This will result in the overdub to occur at the beginning of the next iteration of the loop. This is a safety feature in case we press the recording switch very close to the end of the loop, while we want to record from the beginning of it. If this happens, the far right [spigot] will output the remaining time of the loop, to which the total length of the loop will be added, and then the 16 sample delay time (again due to the use of [limiter~]). This means that the recording LED will turn off when the next iteration of the loop ends.

Make another subpatch and name it *get_overdub_position*. Put the objects seen in Figure 9-22 in it.

The Arduino Module

The next module is the Arduino module, shown in Figure 9-23. There's nothing really to explain here; we've been seeing this almost throughout the whole book. As you might have already guessed, you should name this subpatch *arduino*.

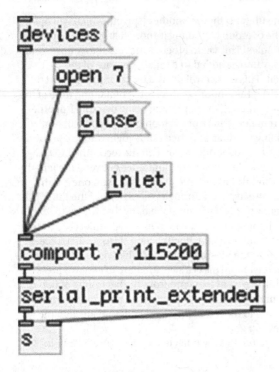

Figure 9-23. *The Arduino module*

The Table Copying Module

This is the last module that we'll make. It will implement the last feature of our looper, being able to overdub while playing the loop backward. This is also a rather complex module. Before I explain what's going on here, I need to explain what we need to do in case we overdub with a backward loop. When we want to grasp complex concepts, it's best to put the calculations in practice with small and round numbers. Let's say that we have a loop of 10 samples (of course, this is way too short to ever be able to realize, but it helps to understand what we need to do). The table of the loop will have stored values in its elements from 0 to 9. Suppose that we play this loop backward. Remember that in the phase module, even if we play the loop backward, we get the position of the loop's phase as if we were reading it forward. Let's say that we start overdubbing at the sixth sample (of the forward phase). In this case, we'll record four samples. If we try to record to the table we're reading from we'll make a mess, because we'll be reading backward, but [tabwrite~] only stores values in a forward fashion. For this reason, we need to store the backward playback and the overdub to another table. We'll call this table samp1 (since the table we're using so far is called samp0). The four samples we'll record are the samples from 3 to 0 of [table samp0], which will be stored at the indexes 6 to 9 of [table samp1]. For now, we'll suppose we write the loop as is, without giving any input to overdub, just to simplify the process. Figure 9-24 shows the copying of these four samples from one table to the other (it is illustrated with Arrays of 15 elements).

CHAPTER 9 ■ MAKING A LOOPER

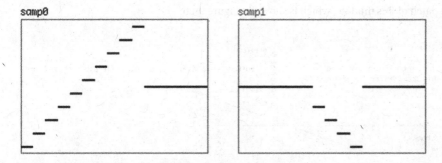

Figure 9-24. *Copying samples 3 to 0 of samp0 to the indexes 6 to 9 of samp1*

Once the overdubbing is done, we need to copy the remaining of the loop, which is the part of the loop before the overdub occurred. We started overdubbing at the sixth sample, which means that the first six samples, from 0 to 5, were played back without any overdubbing. Again, these sample numbers are of the forward phase. This means that the loop played backward without overdubbing the samples from 9 to 4. What we need to do is copy these six samples to the indexes from 0 to 5 in samp1. If we do this, we end up with the table shown in Figure 9-25.

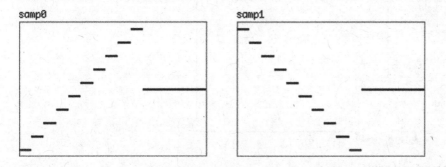

Figure 9-25. *Copying the remaining samples to samp1*

What we get is the exact inverse of samp0. There's a small detail here that we need to bear in mind. When the overdub ends automatically, the playback will still be backward. If we leave samp1 like this, reading it backward will have the same result as reading samp0 forward. To solve this we need to reverse samp1 (only the 10 samples we've recorded, of course). Again, we've supposed that we just copied one array to the other without adding any input, but the process is exactly the same.

Now let's check the patch of this module, which is shown in Figure 9-26.

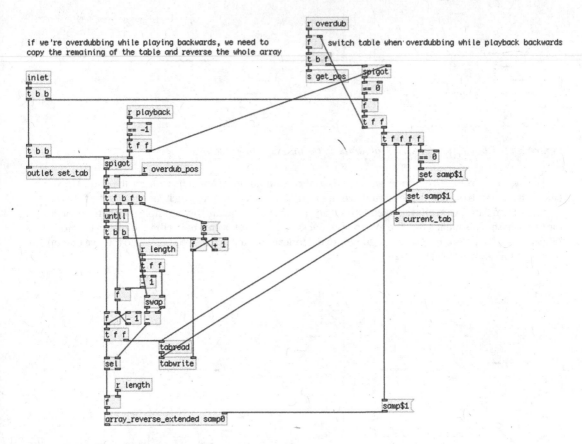

Figure 9-26. *The table copying module*

At the top middle part of the patch there's a [r playback] connected to [== –1]. This last object connects to two [spigot]s, which will open only when we're playing the loop backward. If this is the case, when the Arduino sends a bang with the "overdub" tag, [r playback] will bang [f] below it, which holds the number of the current table. If it's the first time that you're overdubbing while playing the loop backward, [f] will hold 0. This value will go through the [spigot] on the right side and then to [== 0], which will reverse this value to 1. This value connects to [t f f f f] and first goes to another [== 0], again being reversed to 0. This 0 goes to the message "set samp$1", which sets the table to read from for [tabread]. This means that [tabread] will read from [table samp0], which is the current table that holds the loop. Then [t f f f f] will send 1 unaltered to the same message (not the same message box) which this time will result to "set samp1", which is sent to [tabwrite]. This means that we'll be reading from [table samp0] and writing to [table samp1], like in the example in Figures 9-25 and 9-26. Then [t f f f f] sends this 1 to [s current_tab], so [tabwrite~] in the overdub and the recording modules will switch table (the playback module stores this value, but doesn't output it yet), and finally, it will send the message "samp1" to [array_reverse_extended]. This is an abstraction that reverses an array, which you can find in GitHub at https://github.com/alexdrymonitis/array_abstractions. After all this is done, the value 1 will be stored to [f], which is banged by [r playback], for the next overdubbing. Lastly, a bang is sent to [s get_pos], which is received in the module that gets the overdub position, triggering its process.

When the process of getting the position of the overdub is done, we'll record to the current table, which is [table samp1], in the overdubbing module, and a bang will be delay by the amount of time needed for the overdubbing to finish. This bang returns to this module, received in the [inlet] on the left side of the patch. This bang first goes to [f] that stores the current table number. We've already set this, but we need to bang it in case we're overdubbing while playing forward, since the [spigot] on the top-right side in Figure 9-26 will be closed, and the bang received by [r overdub] won't go through. After we bang [f] and set the proper tables, we bang the [spigot] on the left side of the patch, which will also be open. This [spigot] will let the bang through to [f], which stores the position where the overdubbing occurred. This value is sent from the overdub position module (check for the corresponding [s overdub_pos]) the moment we start to overdub. This value will trigger the rest of the process after we've recorded the overdubbed loop. This is the process that was explained with Figure 9-25, where we copy the remainder of the original table (from the end to the point the overdubbing occurred) to the new table (from its beginning to the point the overdubbing occurred). This process includes two counters: an ascending and a descending one, one for each table. Try to understand exactly how this process achieves what we want to do; it shouldn't be very difficult.

Once this process is done, the value of the total length of the loop is sent to [array_reverse_extended], which will reverse the recorded indexes of [table samp1] (this abstraction can reverse a part of a table, doesn't have to reverse all of it, check its help patch). After the table has been reversed, the lower [t b b] on the left side of the patch will send a bang to the module's outlet, which will go to the playback module. Recall the module from Figure 9-20, it has a [r current_tab] going to the right inlet of [f], and its right [inlet] (which receives this bang) goes to the left inlet of [f]. This means that only after the whole process of this module has finished, we'll set the new table to [tabread4~] that read the loop. This module enables us to overdub while the loop is playing backward. When the overdubbing finishes, the loop continues to play backward, but the stuff we wrote on top will play forward. If we play the loop forward, the previous recording will play forward and the new one will play backward.

If we're overdubbing while playing forward, the only things that will happen from this whole process is that [inlet] will bang [f], which stores the current table, and then it will send a bang to the outlet of the module, since the two [spigot]s will be closed. When we overdub forward we use the table we're reading from and we don't need to do any copying.

Putting It All Together

Now that we've built all the modules, what's left is to connect them the right way. Figure 9-27 shows the assembled parent patch. Take good care of the connections and try to understand the whole process of this patch. There are two subpatches that I haven't described: [pd init] and [pd tables], which are shown in Figures 9-28 and 9-29.

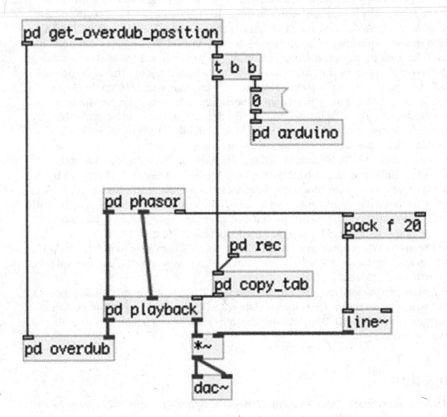

Figure 9-27. *The assembled patch*

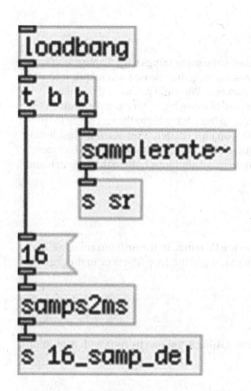

Figure 9-28. *Contents of the init subpatch*

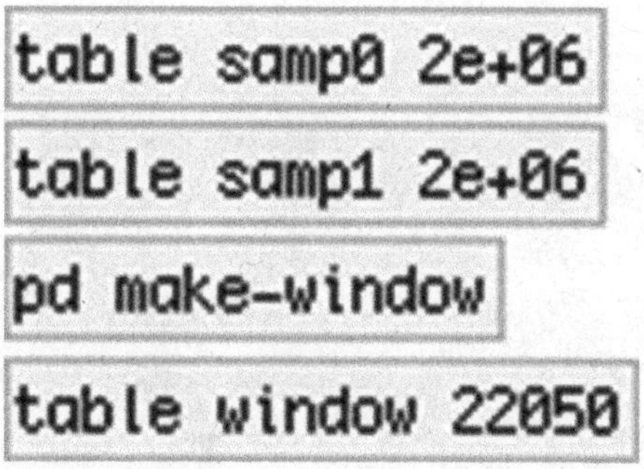

```
table samp0 2e+06
table samp1 2e+06
pd make-window
table window 22050
```

Figure 9-29. *Contents of the tables subpatch*

[pd init] initializes the sampling rate (which is extremely necessary when recording and playing back – it's actually extremely necessary in general) and the 16 samples delay in milliseconds, which we use due to the use of [limiter~], and [pd tables] holds the two tables we use for the looper and the smoothing table we use to eliminate possible clicks. The contents of [pd make-window] have already been shown in this chapter, so there're not shown here too. You can now test the Arduino code and circuit along with the patch. If all was done as described, you should have a fully functional looper!

You'll probably notice that when you overdub while playing the loop backward, there's a small delay before the overdubbed loop starts playing again. This probably happens because of the copying that happens in this case.

Making Some Final Improvements

The looper we've created so far is a pretty good one, which makes quite some things easy. There are a few more features that would be nice to have. For example, while overdubbing, maybe we want to quit recording before the end of the loop, because we want to overdub only a part of it. We could just as well not play for the portion of the loop we don't want to overdub, but in the case of noisy input, for example, with an overdriven electric guitar that has a loud hum, we might not want to introduce this to the rest of the loop. Another feature you might want to have is the ability to stop the loop immediately after overdubbing, letting the computer handle this instead of trying to synchronize it yourself (I don't think it's possible to be more synchronized than a computer). These features require only some small additions both to the Pd patch and the Arduino code.

Modifying the Arduino Code

We'll start with the Arduino code first and then we'll move on to the Pd patch. In the Arduino code, we'll need to modify the three foot-switch functions, as well as the beginning of the loop function, in the part where we handle data coming in from the serial line.

Modifying the Recording Function

We'll start with the first function, which is the recording function. Listing 9-8 shows its code with it additions.

Listing 9-8. Recording Function with Small Additions

```
1.   void record(){
2.     if(!recording){
3.       if(playBack || backWard){
4.         Serial.print("overdub ");
5.         Serial.println("bang");
6.         digitalWrite(recording_led, HIGH);
7.         recording = !recording;
8.       }
9.       else{
10.        Serial.print("record ");
11.        Serial.println("rec");
12.        digitalWrite(recording_led, HIGH);
13.        recording = !recording;
14.      }
15.    }
16.    else{
17.      // stop recording only if we'not overdubbing
18.      if(!(playBack || backWard)){
19.        Serial.print("record ");
20.        Serial.println("stop");
21.        digitalWrite(recording_led, LOW);
22.        recording = !recording;
23.      }
```

```
24.        // otherwise send a 0 through "overdub"
25.        else{
26.          Serial.print("overdub ");
27.          Serial.println("0");
28.          digitalWrite(recording_led, LOW);
29.          recording = !recording;
30.        }
31.    }
32. }
```

The additions to all three functions are in the code of else of the first if, which in this function occurs in line 16. Still, these listings show the whole function to avoid any mistakes (mainly syntactical) that might occur while making these additions. This chunk of code will be executed when we press the recording switch while we're recording. In this chunk, we're first checking whether we're currently playing back either forward or backward. This happens in line 18 where we use the same test with line 3, only this time with an exclamation mark to reverse the result of the test statement. If we're not playing back, then we'll send the message "stop" to [r record] in Pd. If we are playing back, we'll send a 0 to [r overdub]. In both cases we'll turn off the recording_led and we'll update the recording Boolean. Remember that [r overdub] was receiving only bangs up to now. This means that we'll have to modify the Pd patch to distinguish between bangs and floats. We'll see that when we modify the patch.

Modifying the Forward Playback Function

Let's check the forward playback function, which is shown in Listing 9-9.

Listing 9-9. Forward Playback Function with Small Additions

```
1.   void playback(){
2.     if(!playBack){
3.       // if we're recording, quit recording first and then start playing back
4.       if(recording){
5.         Serial.print("record ");
6.         Serial.println("stop");
7.         digitalWrite(recording_led, LOW);
8.         recording = !recording;
9.       }
10.      Serial.print("playback ");
11.      Serial.println("1");
12.      digitalWrite(playback_led, HIGH);
13.      playBack = !playBack;
14.      if(backWard){
15.        digitalWrite(backward_led, LOW);
16.        backWard = !backWard;
17.      }
18.    }
19.    else{
20.      Serial.print("playback ");
21.      Serial.println("0");
22.      // if we stop playing back while recording, it means we're overdubbing
23.      // in this case we'll wait for Pd to tell the Arduino when to turn the LED off
24.      if(recording) stop_playback = true;
```

```
25.       else{
26.          digitalWrite(playback_led, LOW);
27.          playBack = !playBack;
28.       }
29.    }
30. }
```

In this function, the addition has been made to the code of else of line 19. If we stop the playback, we'll send a 0 to [r playback] whether we're overdubbing or not. We'll take care not to stop playing back if we're overdubbing in the Pd patch. The addition is in line 24 where we use a new Boolean, stop_playback. This Boolean is initialized to false, and when we stop the playing, in case we're also recording (which means that we'll be overdubbing), we set this Boolean to true. We'll use it in the main loop function to turn off the forward playback LED when we receive data from Pd. If we're not recording when we want to stop the playback, like before we turn off the forward playback LED and we update the playBack Boolean.

Modifying the Backward Playback Function

The additions to this function are identical to the additions of the forward playback function, so there's no need to explain anything about it. The function is shown in Listing 9-10.

Listing 9-10. Backward Playback Function with Small Additions

```
1.  void backward(){
2.    if(!backWard){
3.      // if we're recording, quit recording first and then start playing back
4.      if(recording){
5.        Serial.print("record ");
6.        Serial.println("stop");
7.        digitalWrite(recording_led, LOW);
8.        recording = !recording;
9.      }
10.     Serial.print("playback ");
11.     Serial.println("-1");
12.     digitalWrite(backward_led, HIGH);
13.     if(playBack){
14.       digitalWrite(playback_led, LOW);
15.       playBack = !playBack;
16.     }
17.     backWard = !backWard;
18.   }
19.   else{
20.     Serial.print("playback ");
21.     Serial.println("0");
22.     if(recording) stop_backward = true;
23.     else{
24.       digitalWrite(backward_led, LOW);
25.       backWard = !backWard;
26.     }
27.   }
28. }
```

Modifying the Beginning of the loop Function

Last step we need to take is to modify the beginning of the loop function, where we receive data from the serial line. Listing 9-11 shows the code.

Listing 9-11. Modifications to the Beginning of the loop Function

```
1.   if(Serial.available()){
2.     byte in_byte = Serial.read();
3.     if(recording){
4.       digitalWrite(recording_led, in_byte);
5.       recording = !recording;
6.     }
7.     if(stop_playback){
8.       digitalWrite(playback_led, LOW);
9.       playBack = !playBack;
10.      stop_playback = false;
11.    }
12.    if(stop_backward){
13.      digitalWrite(backward_led, LOW);
14.      backWard = !backWard;
15.      stop_backward = false;
16.    }
17.  }
```

We've modified the if control structure at the beginning of the loop function, because now the data received from Pd might control any of the three LEDs of the circuit. We first store the data we receive (which is always a 0 anyway), and then we check all the LEDs to see if any of the is lit. If the recording Boolean is true, it means that we're recording and that the recording_led is on, so we turn it off. If the new Boolean, stop_playback is true, it means that we have stopped playing back while overdubbing, so when the overdubbing finishes and we receive a 0 from Pd, we must turn of the playback_led. The same applies to the backward_led. In all three cases, we update the Booleans of each function, and in the case of the playback, we set the new Booleans stop_playback and stop_backward to false.

At the beginning of the code, right after we set the Booleans for the three functions, include the two lines of the following code:

```
// booleans used in when receiving data from the serial port
bool stop_playback = false, stop_backward = false;
```

This concludes the modifications to the Arduino code.

Modifying the Pd Patch

In the Pd patch, we need to slightly modify three modules, the phase module, the overdub module, and the table copying module. Let's take them one by one.

Modifying the Phase Module

Figure 9-30 shows the top-left part of this module, where we've replaced [r playback] with a subpatch, which is shown in Figure 9-31.

409

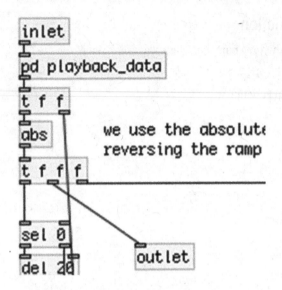

Figure 9-30. *Addition to the phase module*

if we're overdubbing and want to stop playback after the overdub has finished, hold the playback value until we receive a 0 after the necessary delays have been appied

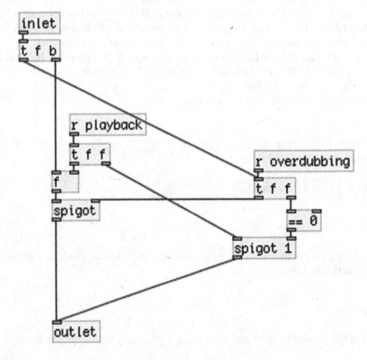

Figure 9-31. *Contents of the playback_data subpatch*

410

[r playback] is now placed in the [pd playback_data] subpatch. When [r overdub] will receive a bang, it will send a 1 to [s overdubbing], which will be received here. This will open the [spigot] on the left side and will close the [spigot] on the right side. If while we're overdubbing we press the forward or backward playback switch, this value will be stored in [f] and won't go through to the rest of the module. When the overdubbing finishes, this module will receive the 0 sent to the Arduino, which will first bang the playback value, which will be 0, and then it will reverse the [spigot]s. This will result in the looper to stop playing back as soon as the overdubbing finishes.

Modifying the Overdub Module

As with the phase module, we need to make only a minor addition to the overdub module. This addition is in the top part of the patch and it's shown in Figure 9-32.

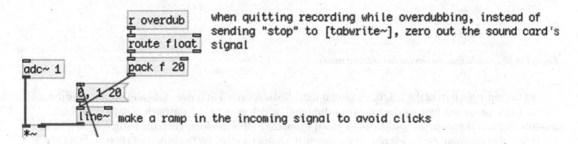

Figure 9-32. *Addition to the overdub module*

All we added here is a [r overdub] that will filter out the bangs and will output only the floats it receives. In case we want to stop the overdubbing before the loop is over, the Arduino will send a 0 to Pd. (This 0 might be considered an integer, but in Pd, all numbers are floats.) [route float] will let all floats out its left outlet, which will force [line~] to make a ramp to 0. Instead of sending the message "stop" to [tabwrite~], it's simpler to just eliminate the audio of the sound card's input signal and write the rest of the loop to [tabwrite~], because we've already built a mechanism to send a "stop" to [tabwrite~] at the end of the loop. This way we don't need to further interfere with our patch, since it's working nicely.

Modifying the Table Copying Module

This is the last module we need to modify. Figure 9-33 shows the top part of the patch, which has been modified.

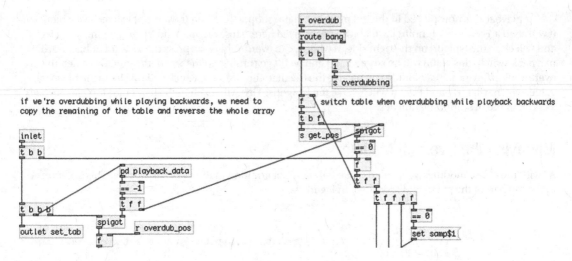

Figure 9-33. *Additions to the table copying module*

At the top-right part of the patch, we filter out the floats received in [r overdub] and output only the bangs. These bangs will first send a 1 to [s overdubbing], which goes to [pd playback_data] in the phase module, but it also goes to the [pd playback_data] subpatch in this module. Then the bang goes to [f] as it did before this addition. At the left part of the patch, the bang received by the [inlet] will first go to [f] like before, then to the [spigot] at the left side, then to [pd playback_data], and lastly to the [outlet].

Figure 9-34 shows the contents of the [pd playback_data] subpatch. This is very similar to the subpatch with the same name of the phase module. Again, we're holding the playback value while we're overdubbing until the loop has finished. Here, though, since the [inlet] will receive a bang, we send a 0 to [t f f], which controls the two [spigot]s to open the right one and close the left one.

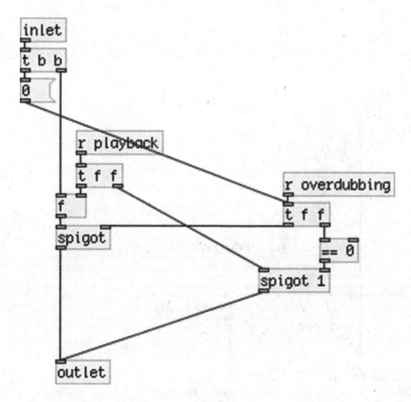

Figure 9-34. *Contents of the playback_data subpatch*

Since we've made these additions, we need to see the parent patch again, because there's a small addition to its connections. It is shown in Figure 9-35. The only addition is that the 0 sent to [pd arduino] also goes to [pd phasor], because we need to receive it in its [pd playback_data] subpatch. This concludes all the additions to the software in this project.

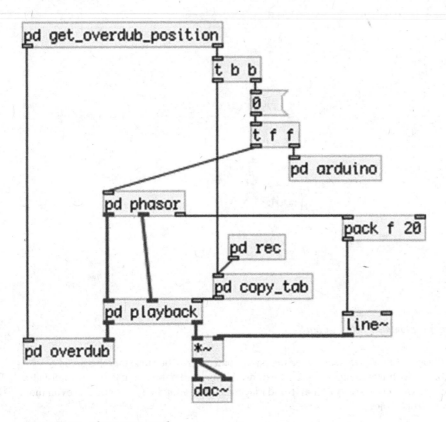

Figure 9-35. *The parent patch*

Enclosing the Circuit in a Box

Making an enclosure for this project is very similar to the same process in Chapter 7, the interactive drum set. This circuit is even simpler as it has one foot-switch and one LED less, and no other sensors. Since we're not covering the enclosure of the computer as well, it should be pretty simple to make an enclosure for this project. Figure 9-36 shows the front side of a box I made for this circuit.

Figure 9-36. *Front side of the enclosure box*

We can use IDC connectors to connect this box with an Arduino. We can also use a Proto Shield, which makes the building of the circuit a lot easier. Figure 9-37 shows the back side of the enclosure, where you can see the male IDC connector.

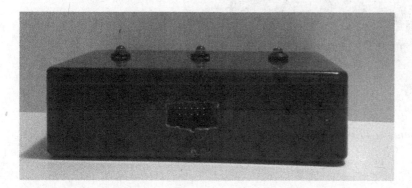

Figure 9-37. *The back side of the enclosure box*

Finally, Figure 9-38 shows the enclosure connected to an Arduino Uno with a ribbon cable through the IDC connectors. This way you don't "sacrifice" your Arduino for this project, but you can attach it to the project's circuit any time.

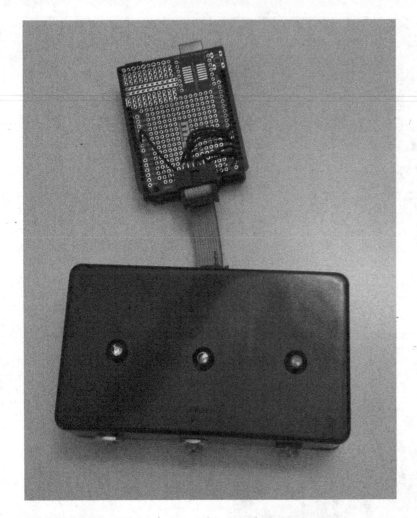

Figure 9-38. *The enclosure connected to an Arduino Uno*

Conclusion

We have finished our looper, which doesn't have many features, but it is very user-friendly, something that is very important when it comes to performing. This looper outputs only the recorded sound and not the audio signal of the sound card. This is because a looper usually bypasses its incoming signal and mixes it with the loop it outputs. I suppose you have a splitter or a small mixing desk that you can use with this interface, since it might be not such a good idea to digitize your signal only to let it pass through. If you want to have your signal pass through so that you can hear your real-time playing and the looping from your computer, then connect [adc~ 1] to [dac~]. But make sure that you use a [limiter~] or that you multiply both signals by 0.5 before you send them to [dac~].

This was a rather complex project to accomplish, but hopefully you understood the process and gained some insight on approaching the various issues that you might have while building a project. You can customize your looper to do pretty much anything you want. You might want some features that were not included in this chapter. That is one of the great things about programming—being able to hack things.

CHAPTER 10

■ ■ ■

A Patch-Bay Matrix Synthesizer

In this last chapter, we're going to build a patch-bay synthesizer, much like the analog synthesizers. We'll simulate the interface, but instead of an audio signal, we'll pass connection data through the patch cords. The data will be sent to the computer so that the corresponding connections can be done in the program. We'll use external chips (we'll call them *ICs* from now on, which stands for *integrated circuit*) to extend both the analog and the digital pins of the Arduino for more available input and output. Finally, we're going to use oscillators, filters, and envelopes to build the various parts of the program, which we have already used. We'll be able to connect these parts between them through the patch-bay interface that we'll build.

Parts List

Table 10-1 lists the components we'll need for this project. The 74HC595, 74HC165, and CD74HC4067 are ICs that you can find at your local electronics store or online. SparkFun has made a breakout board for each of these ICs, so you can choose those if you like. The 74HC595 and CD74HC4067 circuits will be built using these breakouts, as well as the ICs, since Fritzing (the program with which all the circuits in this book are built) provides parts for these breakouts. But the 74HC165 will be built with only the IC, because there's no part for the breakout in Fritzing. The pin names and numbers of the breakouts and the ICs will be mentioned in detail. If you have the data sheet for each IC, you'll be able to build the circuits by following the instructions in this chapter. (I've used the data sheets provided by SparkFun.) The breakouts have the pin names printed on the circuit board. If you use the breakouts, make sure that you solder male headers on the pins of the board; otherwise, you won't get proper connections when you use them with a breadboard.

The banana male terminals will be used to create the patch cords. To have five patch cords, we'll need 10 male bananas. The more you have, the better, since you'll be able to make more connections. You can make a few in the beginning and then make more at some later point.

Table 10-1. Parts List for the Patch-Bay Synthesizer

Arduino	74 HC 595	74 HC 165	CD74 HC 4067	Potentio meters	Banana female terminals	Banana male terminals	Resistors	Capa- citors	Diodes	Cable
1 (Nano preferred)	2	2	1	16	16	10 (at least)	18 × 10KΩ 2 × 220Ω	2 × 0.1uF	5 (half as many as the banana male terminals)	1 × 2 meters

417

What We Want to Achieve in this Chapter

Before we start, let's first visualize what we really want to achieve. Figure 10-1 shows a Pd patch with a few components that are not connected between them. We have two oscillators (a sine wave oscillator, which can have its phase modulated, and a sawtooth oscillator), a lowpass filter, an envelope, and the [dac~]. You might want to play around a bit with this patch and try different connections between these elements. Connect the output of the sawtooth oscillator to the lowpass filter and then to the phase of the sine wave oscillator. Connect the envelope to the amplitude of the sine wave oscillator (you'll need to add a [*~] for that). Connect the envelope to the output of the sawtooth and then to the phase of the sine wave. There can be quite a few connections, which can yield pretty interesting results. By the end of this chapter, we'll have built an instrument in which these connections were made using real cables, without the clicks we get when we dynamically change a Pd patch with the DAC on.

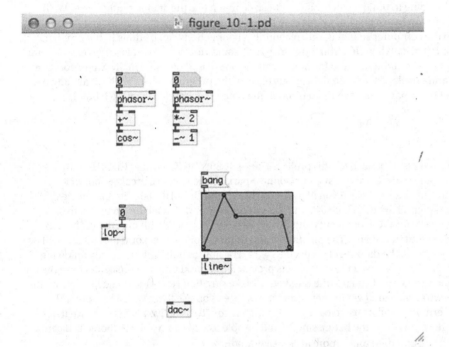

Figure 10-1. *A Pd patch without predefined connections*

This chapter is rather long and probably the most complex of all the chapters in this book. What follows are certain techniques that deal with expanding the analog and digital pins of the Arduino, and making a patch-bay matrix. After that, we'll go through the implementation in Pd. Read carefully through the chapter and be very careful when you're building the circuits. If something doesn't work for you, read again and retry the circuits until it works.

Extending the Arduino Analog Pins

The Arduino Uno has only six analog pins, which are quite few. Even the Nano, which has eight, is rather limited for some projects. To solve this problem we can use multiplexers. These are chips that can take up to a certain number of analog inputs (depending on the number of channels they have) and route them sequentially to one analog pin of the Arduino. The CD74HC4067 multiplexer has 16 channels, which means that with one of them we can have 16 analog inputs that will occupy only one analog pin of the Arduino. Of course, this multiplexer doesn't come so cheap, as it needs four digital pins to control the sequence of its inputs. To better visualize how this chip works, let's see it wired in Figures 10-2 and 10-3.

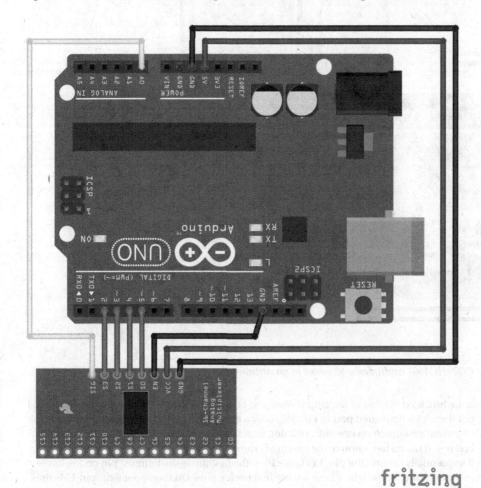

Figure 10-2. *The CD74HC4067 multiplexer breakout board wired to an Arduino Uno*

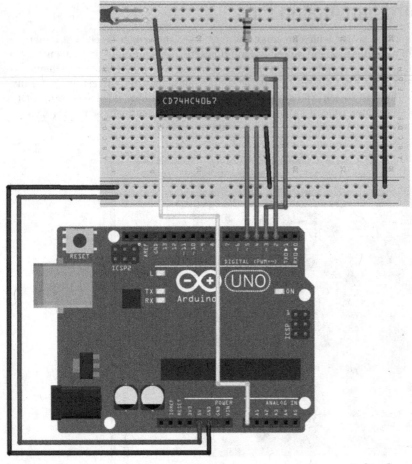

Figure 10-3. *The CD74HC4067 multiplexer IC wired to an Arduino Uno*

In Figure 10-2, the breakout version of the chip is used and Figure 10-3 uses the IC (mind that the actual IC is a bit thicker, but there's no dedicated part in Fritzing, only a generic IC part). Just a reminder on pin numbering, all ICs have either a notch on one side, or a dot, which indicates the first pin. The dot is usually on the side of the first pin. If the IC has a notch, the first pin is the one on the left side of the notch. Then pins are numbered sequentially around the pin. In Figure 10-3, the first pin is the leftmost pin on the lower side, and the successive pins are on its right. These are the first twelve pins. On the upper side, pin 13 is the rightmost pin, and the leftmost pin is the last one, pin 24.

There are no potentiometers wired to the breakout or IC for now, just to keep the circuit a bit clearer. On the bottom side of the board we have the 16 input pins, C0 to C15 (on the IC these are pins 9 to 2 and 23 to 16 labeled I0 to I15). On the top side, we have the SIG pin (signal pin, on the IC this is pin 1 labeled Common Input/Output), the S0 to S3 pins (on the IC they are pins 10, 11, 14; 13), the EN pin (enable pin, on the breakout this is pin 15 labeled E), and the VCC and GND pins (voltage and ground, on the IC these are pins 24 and 12). If you're using the IC instead of the breakout, you might want to connect pin 15 of the chip (E) to ground via a 10KΩ resistor, and use a 0.1μF capacitor between the voltage and the ground, close to pin 24 (VCC). The resistor and capacitor are included in the breakout board used here.

The pins C0 to C15 (or I0 to I15) take analog input. We'll connect the potentiometers to these pins. The SIG pin connects to one of the Arduino's analog pin. This pin will transfer the signals of the pins C0 to C15 to the Arduino. The four control pins (S0 to S3) are connected to four digital pins of the Arduino, where S0 connects to digital pin 5, S1 connects to pin 4 and so forth. These pins will determine which of the input pins (C0 to C15) will be connected to the SIG pin. The EN pin enables or disables the device, where with a low signal (zero volts or ground) it is enabled (this is referred to as *enable low*), and with a high voltage it is disabled. We won't deal with disabling the device, so we'll always have this pin wired to ground (with a 10KΩ resistor for the IC circuit).

How a Multiplexer Works

To understand how the S0 to S3 control pins work, I need to talk a bit about binary numbers. Listing 10-1 shows the first 16 numbers (including 0) in binary form.

Listing 10-1. The First 16 Numbers in Binary Form

```
0000
0001
0010
0011
0100
0101
0110
0111
1000
1001
1010
1011
1100
1101
1110
1111
```

To make the connection between these numbers and the control pins of the multiplexer, think of 0s as low voltage, and 1s as high voltage, meaning that with a 0, a digital pin will be LOW, and with a 1, it will be HIGH. The four control pins of the multiplexer, S0 to S3, and the four digital pins of the Arduino used in Figures 10-2 and 10-3 (pins 2 to 5) are paralleled with the digits of the numbers in Listing 10-1. What I mean by that is that the rightmost digit of a value represents the state of the pin S0 of the multiplexer and digital pin 5 of the Arduino, the digit on it left represents the state of the pin S1 and digital pin 4 and so forth.

This means that to connect pin C0 of the multiplexer to its SIG pin, which connects to analog pin 0 of the Arduino, we need to set all four of the Arduino digital pins that control the chip LOW. If we want to connect C1 to the SIG pin of the chip, we need to set digital pin 5 of the Arduino HIGH and the other three LOW. If we want to connect C2 to the SIG pin, we need to set digital pin 4 HIGH and digital pins 5, 3, and 2 LOW. Hopefully you understand the mechanism behind this. Since we're using digital signals to control the multiplexer, and since the multiplexer has 16 channels, it makes sense as to why we need four control pins, since 2 to the fourth power yields 16, which means that with four digital pins we can have 16 different combinations. If the multiplexer had eight channels, we would need three control pins, since 2 to the third power yields 8.

Writing Code to Control a Multiplexer

With some explanation, the mechanism of a multiplexer might seem rather simple and clear. But how can we implement this in Arduino code? One way would be to write the binary values in Listing 10-1 to a two-dimensional array and call each digit of each value to set the control pins of the Arduino appropriately. This doesn't look very elegant though, and the coding we would need to do would be a bit more than necessary. We want to read the values of the pins C0 to C15 sequentially and store them in an array, which we'll then transfer to Pd. To do this, we'll run a for loop. We'll use the loop's incrementing variable to set the control pins of the Arduino. Listing 10-2 shows the code.

Listing 10-2. Code to Read Analog Values from a Multiplexer

```
1.    // define a log base 2 function
2.    #define LOG2(x) (log(x) / log(2))
3.
4.    // change this number if using more chips
5.    const int num_of_chips = 1;
6.    // set the number of channels of the multiplexer
7.    const int num_of_channels = 16;
8.    // get the number of control pins according to the number of channels
9.    const int num_ctl_pins = LOG2(num_of_channels);
10.   // set the total number of bytes to be transferred over serial
11.   // * 2 to get 10bit resolution
12.   // + 1 for start character
13.   const int num_of_data = ((num_of_chips * num_of_channels) * 2) + 1;
14.   // buffer to hold data to be transferred over serial
15.   // with the start character set
16.   byte transfer_array[num_of_data] = { 192 };
17.
18.   // create an array for the control pins
19.   int control_pins[num_ctl_pins];
20.   // create an array to hold the masking value to create binary numbers
21.   // these values will be set in the setup function
22.   int control_values[num_ctl_pins];
23.
24.   void setup() {
25.     // set the value masks, the control pins and their modes
26.     for(int i = 0; i < num_ctl_pins; i ++){
27.       control_values[i] = pow(2,(i+1)) - 1;
28.       control_pins[i] = (num_ctl_pins - i) + 1;
29.       pinMode(control_pins[i], OUTPUT);
30.     }
31.
32.     Serial.begin(115200);
33.   }
34.
35.   void loop() {
36.     int index = 1;
37.     // loop to read all chips
38.     for(int i = 0; i < num_of_chips; i++){
```

```
39.        // this loop creates a 4bit binary number that goes through the multiplexer control pins
40.        for (int j = 0; j < num_of_channels; j++){
41.          for(int k = 0; k < num_ctl_pins; k++){
42.            // this line translates the decimal j to binary
43.            // and sets the corresponding control pin
44.            digitalWrite(control_pins[k], (j & control_values[k]) >> k);
45.          }
46.          int val = analogRead(i);
47.          transfer_array[index++] = val & 0x007f;
48.          transfer_array[index++] = val >> 7;
49.        }
50.      }
51.    Serial.write(transfer_array, num_of_data);
52.  }
```

This code is written in such a way that will work with any number of multiplexers, and the multiplexers can have any number of channels (usually it's eight or sixteen), as long as all multiplexers used have the same number of channels. (You could write code for multiplexers with different number of channels, but it would be slightly more complex, and we shouldn't really care about this right now, since we'll use one multiplexer.)

Defining Preprocessor Directives

Already from line 2, we meet new things. This is a preprocessor directive. It is not regular C++ code, but a directive for the preprocessor. This means that this line will be handled before the compilation of the code. We've used the #include preprocessor directive in Chapters 3 and 5, where we used the SofwareSerial library. Here we use the #define preprocessor directive to define what is called a *macro*. The #define directive is used to define constants. What it actually does is to replace text. Therefore, this line sets that wherever we use LOG2(x) in our code, it will be replaced by (log(x) / log(2)). This last definition uses the log built-in function of the Arduino language to create a function that returns the base 2 logarithm of a given value. The log function returns the natural logarithm (base e), so to get the base 2 logarithm of a value, we must divide the natural logarithm of this value, by the natural logarithm of 2. Since this function is a one-liner, we can define it as a macro, instead of the way we've been defining functions up until now.

We need the base 2 logarithm to get the number of control pins we need for the multiplexer. If your math is rusty, all you need to know is that the logarithm of a certain value yields the number we need to raise the base to get this value. I've already mentioned that we need four digital pins to control the multiplexer because it has 16 channels, and since 2 to the fourth power yields 16, the base 2 logarithm of the number of channels (which is 16) will yield 4 (which is the number of control pins we need). If the multiplexer we use had 8 channels, then the base 2 logarithm of 8 would yield 3, which would be the number of control pins we would need in this case. This way, we can limit the number of parameters we need to change manually in case we use multiplexers with a different number of channels.

If you're using an older version of the Arduino IDE, this will probably not work. In this case, don't include it in your code, and in line 9, you'll have to hard code the value of the num_ctl_pins constant to 4, instead of using this custom log base 2 function.

Explaining the Code Further

In lines 5 and 7, we define the number of multiplexers we'll use and the number of channels these multiplexers have. Line 9 is where we use the macro function we've defined in line 2. This line uses the constant that holds the number of channels of the multiplexer and yields the number of control pins this multiplexer has. If we hadn't defined the macro function, we'd have to hard-code this line, meaning that if we used this code with a different multiplexer, which had another number of channels, we would have to change this line manually.

In line 13, we define the number of bytes we'll be transferring to the serial line. (The earlier comments explain why we need a particular number of bytes. We'll use the Serial.write function, so we need to break the analog values into two.) In line 16, we define the array that will be transferred over serial. Line 19 defines an array that will hold the numbers of the control pins, but since we don't want to hard-code things, we'll set these numbers in the setup function using a for loop. In line 22, we define an array that will hold values, which will be used in the main for loop that reads the values of the multiplexer. The values of this array will be used to translate the decimal value of the incrementing variable of the for loop to binary. We'll see how this works in detail further on.

Setting the Control Pin Numbers and the Binary Masks

In the setup function we have two things, a for loop and the initialization of the serial communication. In the for loop, we set the values of the last two arrays we created in lines 19 and 22. I've stated that the control_values array will be used to translate decimal values to binary. To do this we'll need to mask the decimal values by the values of the array using the bitwise AND (&). After this loop has finished, the array will have the following values stored: 1, 3, 7, 15. Think how this is achieved (hint, the pow function raises its first argument to the power of its second argument). We still don't see the workings of this value translation, because it happens in the main loop function. The control_pins array will store these values: 5, 4, 3, 2, which are the numbers of the control pins. Again, we'll see further on why they are stored from the greatest number to the smallest.

Reading the 16 Values of the Multiplexer

In the main loop function, we have three nested for loops. The first one, in line 38, goes through all the multiplexers we're using (in this case, it's only one). Note its incrementing variable, i, as we'll need to keep it in mind to understand how these nested loops work. The second loop, which runs inside the first, in line 40, goes through all the channels of the current multiplexer. Its incrementing variable is j. The third loop, in line 41, goes through the four control pins, and its incrementing variable is k. The important lines we need to understand are 44 and 46. In line 44, we read the following:

```
digitalWrite(control_pins[k], (j & control_values[k]) >> k);
```

In this line we have two bitwise operations, the bitwise AND and the bitshift right. Just as a reminder, bitwise ANDing two values results in comparing the bits of each value and acting on them independently of the rest of the bits in a Boolean fashion. When both bits are 1, the resulting bit will be 1. When one of the bits or both of them are 0, the resulting bit will be 0. For example, bitwise ANDing the following two values:

```
00101101
10011011
```

will result in this value:

```
00001001
```

Shifting the bits of a value by a certain number of places results in just that: the bits of the value being shifted. If we shift the bits of the following value:

```
00101101
```

by one place to the right, we'll get this:

```
00010110
```

What happens is that the bits don't wrap around the end and beginning, but the new bits that are introduced (in this example, it's the leftmost bit) are 0.

All these operations are possible in all kinds of numeric representations, be it binary, decimal, hexadecimal, or octal. Regardless of the numeral system, these operations act on the bits of values (since we're dealing with computers, all values are eventually binary).

Now let's get back to line 44. In the first iteration of the first two loops, all iterations of the third loop will result in the second argument to the `digitalWrite` function being 0. This is because j will be 0, and no matter what value we bitwise AND it with and how many places we shift its bits, it will always result in 0. The loop of line 41 will then set digital pins 5, 4, 3, and 2 LOW (these four numbers are the values of the `control_pins` array).

This is where we need to keep track of the values of the incrementing variables of the three loops. At the first iteration of the loop of line 40, in all iterations of the loop of line 41 j will be 0, but k will take the values 0 to 3. In the second iteration of the loop of line 40 j will be 1. In the first iteration of the loop of line 41, the second argument to the `digitalWrite` function will be 1. This is because we're bitwise ANDing 1 with 1, which results in 1, and we're shifting its bit by 0 places, leaving the value 1 intact. So line 44 will set digital pin 5 HIGH. The other three iterations of the loop of line 41 will result in 0, because we'll be masking 1 with the values 3, 7, and 15, and we're shifting their bits by 1, 2, and 3 places. To visualize it, the numbers 1, 3, 7, and 15 are shown in listing 10-3 in both binary and decimal.

Listing 10-3. Values 1, 3, 5, and 7 in Binary and Decimal

```
0001 = 1
0011 = 3
0111 = 7
1111 = 15
```

Don't forget the bit shifting after we bitwise AND the two values. If j is 1, bitwise ANDing it with 3, 7, or 15 will yield 1. But shifting their bits to the right will again result in 0. Take some time to think and understand how these nested loops result in setting the control pins to the values we want. You might need to write the values down and compare them. A binary to decimal converter (and the other way round) will also be helpful here.

Go to line 46. This is where we're reading the Arduino analog pin. This happens at all 16 iterations of the loop of line 40. But instead of using the j variable to control the analog pin that we'll be reading, we're using i, which is 0 (because the loop of line 38 runs only once in our example), since all 16 inputs of the multiplexer end up in analog pin 0 of the Arduino. After we read the Arduino analog pin, we'll store the value we get every time to a successive index of the transfer_array.

In our example, we're using only one multiplexer, so i will always be 0. If we were using more multiplexers, in the second iteration of the loop of line 38, i would be 1, so in the loop of line 40, we would connect all the inputs of the second multiplexer to analog pin 1 of the Arduino. We would read these values and again store them in successive indexes of the transfer_array. This way, we could use the same four Arduino digital control pins for as many multiplexers that we're using. When reading the inputs of one multiplexer we don't care about the input from other multiplexers because we aren't reading the Arduino analog pins the other multiplexers are wired to.

Wiring 16 Potentiometers to the Multiplexer

We've already seen how to wire the multiplexer to the Arduino, but I'll show the circuit again, this time with the 16 potentiometers wired to the multiplexer (only with the breakout this time, but it should be easy to wire the potentiometers to the IC as well). It is shown in Figure 10-4. It might look a bit scary, but the logic is simple. Each potentiometer's wiper (the middle leg) is wired to one input pin of the multiplexer, and the side legs are wired to the voltage and ground provided by the Arduino. In Figure 10-4, the rightmost potentiometer is the first one, and this is only because of the orientation of the multiplexer breakout. It is oriented like this because it was easier to parallel its control pins with the binary values in Listing 10-1.

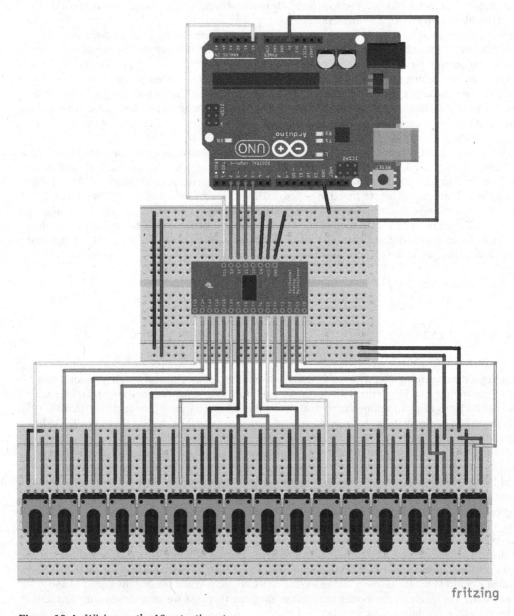

Figure 10-4. *Wiring up the 16 potentiometers*

Mind that if you're using a long breadboard (which you most likely are), the ground and voltage lines on both sides of the board are usually cut in the middle of the board. If you provide voltage and ground on one side of the board, this will go up to the middle of it and the other side won't get it. Figure 10-4 shows how you should connect both sides of voltage and ground of the board. This is done on the lower side of the board where no part is actually wired (all potentiometers get voltage and ground from the upper side of the board). This is done this way because it's a bit clearer to show how to wire a long breadboard as the upper side of it is filled with jumper wires.

Reading the 16 Potentiometers in Pd

Since we've used the Serial.write function in the Arduino code, it's easy to guess that in Pd we'll use the [serial_write] abstraction to receive the potentiometer values. The patch is shown in Figure 10-5. There's not much to say about this patch, we've already received analog input from the Arduino using this abstraction. We need to set only two arguments to [serial_write] "analog 16" and it will work.

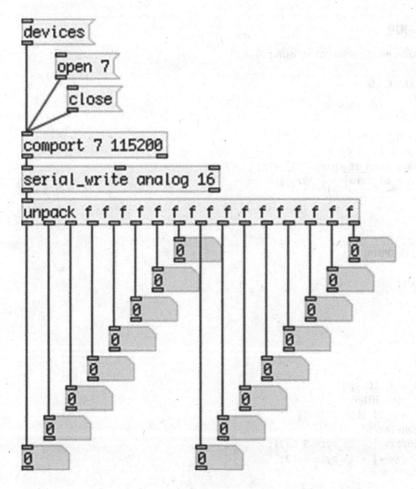

Figure 10-5. *Pd patch that reads the 16 potentiometers of the multiplexer*

Extending the Arduino Digital Pins

There are two kinds of ICs that enable extending Arduino digital pins: one for input and one for output. These ICs are called *shift registers*, because they shift their bits in the register. You don't need to know what this exactly means and how it is done, it's just mentioned to explain their name. One nice feature that we do care about is that these ICs are being daisy chained. This means that if we want to use more than one, we don't need to occupy more pins on the Arduino, because they talk to one another, passing data through the chain.

Using the Input Shift Register

We'll start with the input IC, which is the 74HC165 shift register. This chip has eight inputs, so we'll use it with eight push buttons first, and then we'll add another chip and another eight push buttons to see how we can daisy chain them and control them this way.

Writing the Arduino Code

We'll start with the Arduino code, which is shown in listing 10-4.

Listing 10-4. Input Shift Register Code

```
1.   #include <SPI.h>
2.
3.   const int latch_pin = 9;
4.   const int num_of_chips = 1;
5.   const int num_of_data = (num_of_chips * 2) + 1;
6.   byte transfer_array[num_of_data] = { 192 };
7.
8.   void setup() {
9.     SPI.begin();
10.
11.    pinMode(latch_pin, OUTPUT);
12.
13.    Serial.begin(115200);
14.  }
15.
16.  void loop() {
17.    int index = 1;
18.    byte in_byte;
19.
20.    digitalWrite(latch_pin, LOW);
21.    digitalWrite(latch_pin, HIGH);
22.    for(int i = 0; i < num_of_chips; i++){
23.      in_byte = SPI.transfer(0);
24.      transfer_array[index++] = in_byte & 0x7f;
25.      transfer_array[index++] = in_byte >> 7;
26.    }
27.
28.    Serial.write(transfer_array, num_of_data);
29.  }
```

Including the SPI Library

In line 1, we use the #include preprocessor directive to include the SPI library. SPI stands for Serial Peripheral Interface. It is a communication interface used for short distance communication. This is a standard library in the Arduino language, but we still need to include it this way. We'll use a class from this library to send bytes to the shift register, as it facilitates the control of the chip to a great extent. We'll also use this library to control the output shift registers, almost the same way.

The Arduino provides a built-in function to achieve what we want, shiftIn(), but it is slower than using the SPI library, and not as intuitive. Its advantage over the SPI library is that with this function we can use any pin of the Arduino to control the shift register, whereas with the SPI we need to use specific pins, as the former is a software implementation and the latter is a hardware implementation. For the project we want to build, using the specified pin of this library is not a problem as we don't need them for something else, so it servers our purposes just fine.

Defining the Global Variables and Setting up Our Program

In lines 3 to 6, we define our global variables. The first variable we define is the latch_pin. We've named this variable this way because of the pin it connects to on the shift register. This pin is called the *latch pin* because it is the pin that will trigger output from the shift register into the Arduino when it goes HIGH. Then we define the number of chips, which is the only variable we'll change when we use two chips, and depending on this value, we define the number of bytes we'll be transferring to Pd over serial. There are two different ways that we can send the data of the shift register to Pd. This chip sends in one byte to the Arduino, which represents the states of its eight input pins (a byte consists of eight bits). We can either break this byte into its bits in the Arduino, using the bitRead() function, or we can send the whole byte to Pd as is, and break it down in there. We'll go for the second approach because it's similar to the technique we'll use in our patch-bay synthesizer for detecting connections. Finally, we define the array that will be transferred to Pd over serial.

The first thing we do in the setup function is to begin the communication of the SPI library. We do this by calling the begin function of the SPI class. Notice that this is similar to beginning the serial communication of the Serial class, only for the SPI class we don't define a baud rate, but we leave the parenthesis of the begin function empty. Then we set the mode of the latch_pin, which must be OUTPUT, and finally we begin the serial communication.

The Main loop of the Program

In the main loop function, apart from defining the index offset we use when we transfer data with Serial.write, we also define a byte that will store the byte received from the shift register. What we then do is drop the latch_pin LOW and then immediately bring it HIGH. This will trigger output from the shift register into the Arduino, as stated earlier. Once we do that, we run a loop to read data from all the chips in our circuit (which is only one for now). Inside this loop, we store to the in_byte variable the byte returned by the SPI.transfer function. We need to provide an argument to this function call, but since we don't need to write to a device (as we'll do when we'll use the output shift registers), we provide a 0 as the argument. This function will then read the data received in the digital pin 12 of the Arduino, which is the MISO pin (MISO stands for *Master In Slave Out*). This pin receives the bits sent by the shift register and returns a byte for every eight bits. This byte is stored in the in_byte variable and then we use the same technique with the analog values to store it in the transfer_array. Notice that we mask in_byte with 0x7f instead of 0x007f because we're now dealing with bytes and not ints. An int is two bytes long, and two digits of a hexadecimal number cover the range of one byte. When we mask an int with a hexadecimal value, we must provide four digits to this value, but with a byte, we only need two digits.

Even though we're reading switches, which are digital, we send the states of eight switches in one byte. This means that this byte can possibly take the value of the start character, which is 192. To avoid this we use the technique of breaking up the byte to two bytes with the eighth bit always low (refer to Chapter 2 if you don't remember how this works). In this sketch, we're using one shift register only, so we could have just sent this byte as is to Pd. Later on, we'll use two shift registers and sending bytes without a start character won't be possible. Also, we want to make the code flexible so we can use the same sketch for any number of chips, only by changing the num_of_chips constant in line 4, so not using a start character would make this impossible. This means that we'll receive the byte of the shift register in Pd as if it was an analog value.

Making the Input Shift Register Circuit

The circuit for this IC is shown in Figure 10-6. Each time that we used push buttons in this book, we used the integrated pull-up resistors of the Arduino. With the shift register, we don't have this capability so we need to use external resistors. This time we'll use pull-down resistors as you can see in Figure 10-6. One leg of each push button is connected to a 10KΩ resistor, which is connected to ground. The other leg is connected to the 5V. There's one more 10KΩ resistor connecting the pin 15 (CE) of the chip to ground. There's also a 0.1μF capacitor between ground and 5V. You should place this capacitor as close to the 5V pin of the chip as possible. If you're using the breakout, the 10KΩ resistor of the pin 15 and the capacitor are on the circuit board (you don't even have a label for pin 15, it's not broken out).

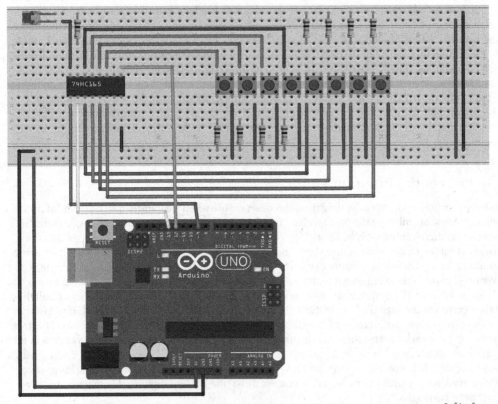

Figure 10-6. *Input shift register circuit*

Pin 1 of the chip (PL on the chip's data sheet, SH/LD on the breakout board) connects to pin 9 of the Arduino. This is the latch pin. Pin 2 of the chip (CP on the chip, CLK on the breakout) connects to Arduino's pin 13. This is the Arduino's SCK pin (Serial Clock). Quoting from the SPI library page of the Arduino web site, this pin sends "the clock pulses, which synchronize data transmission generated by the master." Pin 9 of the chip (Q7 on the chip, SER_OUT on the breakout) connects to Arduino's pin 12, which is the MISO pin, as I've already mentioned. Pin 8 is the ground pin and pin 16 is the voltage pin.

Connect the leg of each push button that connects to the pull-down resistor to the chip's pins 11, 12, 13, 14, 3, 4, 5, and 6. These are the pins D0 to D7 (on the breakout, they're labeled A to H).

The Pd Patch That Reads the Eight Switches

Figure 10-7 shows the Pd patch that breaks the incoming byte to its bits. As I've already said, we're using the [serial_write] abstraction and we're receiving the byte of the eight switches as if it was an analog value, since we're breaking it up to two bytes. Every time we receive a new byte, we run a loop that iterates eight times (as many as the pins of the chip). This loop raises 2 to an incrementing power, starting from 0 and ending at 7 (mind that [swap 2] has both its outlets connected to both inlets of [pow]). The result of this yields the values 1, 2, 4, 8, 16, and so forth, until 128. We bitwise AND these values with the incoming byte. If the result is not equal to 0 (which is checked by [!=] - this means *not equal to* in C), it means that the corresponding push button is being pressed, otherwise it is released. We're using the same incrementing value of [loop] to route each value to the appropriate toggle on the bottom side of the patch.

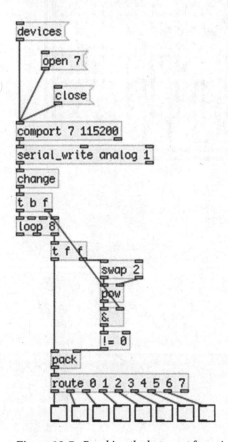

Figure 10-7. *Breaking the byte sent from Arduino to its bits*

Upload the sketch on your Arduino board and test it with this patch. You should see the corresponding toggle of each push button you're pressing "going on." Of course, you can press more than one push button at a time, you can even press them all together. They are independent of each other. We can see that with only three digital pins of the Arduino we can have eight digital inputs. Actually, we can have much more than that, and this is the next thing we're going to do— add another input shift register to our circuit.

Daisy Chaining the Input Shift Registers

A very nice feature of these shift registers is that they can be daisy chained. This way, we can extend the digital inputs of the Arduino even more, while using the same three digital pins of it. We don't need to show the code here, all you need to do is change line 4 in Listing 10-4 and set the num_of_chips constant to 2. This enough for the code to work. We need to check the circuit. This is shown in Figure 10-8. I've used curved wires for the second chip so you can easily distinguish its connections from the first.

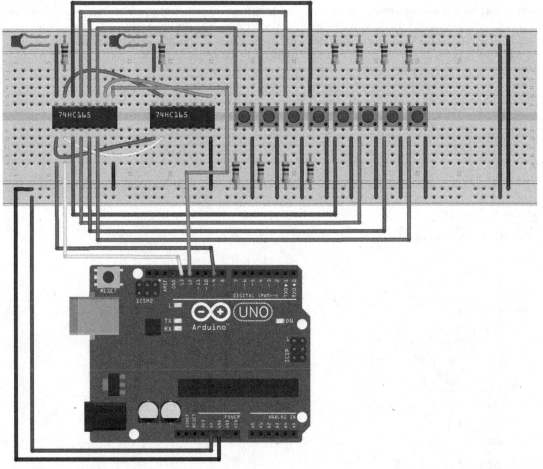

Figure 10-8. *Daisy chaining two input shift registers*

The push buttons of the second shift register are not included because it would look quite messy. Pins 1 and 2 (PL and CP on the chip, SH/LD and CLK on the breakout) of the second chip are connected to the same ones on the first chip. Pin 9 of the second chip (Q7 on the chip and SER_OUT on the breakout) connects to pin 10 of the first chip (DS on the chip and SER_IN on the breakout). Pin 15 of the second chip (CE on the chip, CLK on the breakout) connects to ground via a 10K resistor, but must also connect to pin 15 of the first chip (the resistor is included in the breakout).

The labels of the breakout board are more intuitive here. SER_OUT stands for *serial out*. This pin sends the byte of the chip serially out. SER_IN stands for *serial in*. This pin receives bytes serially from other chips and passes them through its own SER_OUT pin. This way, we can use many chips daisy chained where the byte of the last chip will pass through all the other chips until it reaches the Arduino. The rest of the wiring for the second chip is the same as with the first (including another eight push buttons and all the resistors and the capacitor – remember that if you're using the breakout board, the capacitor and the resistor that connects to the CE pin are included in the board).

Extending the Pd Patch to Read 16 Push Buttons

As with the circuit and code, the additions we need to make to the Pd patch are just a few. The patch is shown in Figure 10-9. This time we'll be receiving two bytes (actually, four bytes broken into two bytes each, which are being assembled inside the [serial_write] abstraction), so we need to change the second argument to [serial_write] from 1 to 2. Then we [unpack] these two bytes and we go through the same process for each.

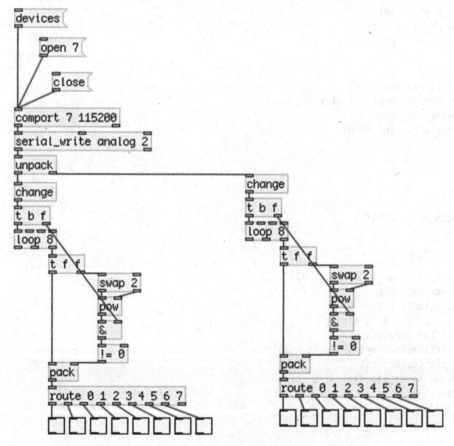

Figure 10-9. *Reading 16 push buttons from two shift registers*

Now on the right side we have the second shift register and on the left side we have the first one. With these two chips and only three Arduino digital pins, we already have 16 digital inputs, more than the digital inputs of the Arduino itself! Probably you can see how much potential this chip and technique give us. Getting digital input doesn't limit us to just using push buttons or switches, further on we'll see how we can use these chips to accomplish the patch-bay matrix for our synthesizer.

Using the Output Shift Register

Let's move on to see how the output shift register works. This is the 74HC595 chip, and like with the others, there's a breakout board by SparkFun for it. It is very similar to the input one. Again, we'll use only three pins of the Arduino for as many shift registers we use. We'll again use the SPI library to control these devices in a similar way with the input shift registers.

Writing the Arduino Code

Listing 10-5 shows the code for a basic control of one output shift register.

Listing 10-5. Simple Output Shift Register Code

```
1.   #include <SPI.h>
2.
3.   const int latch_pin = 10;
4.   const int num_of_chips = 1;
5.   byte out_bytes[num_of_chips] = { 0 };
6.
7.   void set_output(){
8.     digitalWrite(latch_pin, LOW);
9.     for(int i = num_of_chips - 1; i >= 0; i--)
10.      SPI.transfer(out_bytes[i]);
11.    digitalWrite(latch_pin, HIGH);
12.  }
13.
14.  void setup() {
15.    SPI.begin();
16.
17.    pinMode(latch_pin, OUTPUT);
18.  }
19.
20.  void loop() {
21.    for(int i = 0; i < num_of_chips; i++){
22.      // zero previous chip
23.      if(i) out_bytes[i - 1] = 0;
24.      for(int j = 0; j < 8; j++){
25.        // clear the previous bit
26.        if(j) bitClear(out_bytes[i], (j - 1));
27.        // set the current bit
28.        bitSet(out_bytes[i], j);
29.        set_output();
30.        delay(200);
31.      }
32.    }
33.
```

```
34.      for(int i = num_of_chips - 1; i >= 0; i--){
35.        // zero previous chip
36.        if(i < num_of_chips - 1) out_bytes[i + 1] = 0;
37.        for(int j = 7; j >= 0; j--){
38.          // clear previous bit
39.          if(j < 7) bitClear(out_bytes[i], (j + 1));
40.          // set current bit
41.          bitSet(out_bytes[i], j);
42.          set_output();
43.          delay(200);
44.        }
45.      }
46.    }
```

In this code, we're not controlling the chip through the computer so there's no serial communication involved here. In line 5, we define an array of the data type byte to hold the bytes to be transferred to the chips (for now it's only one). What we want to do is light up eight LEDs one after the other, back and forth. This means that we need to run two different loops, one for lighting up the LEDs in an ascending order, and one in a descending order. To avoid writing the same code twice, we define a function that will write the byte to the shift register in line 7, set_output.

This function is very similar to the way we read bytes from the input shift registers. There are only three differences. First, we set the latch_pin LOW, then we write the bytes, and then we set it HIGH. In the case of the input shift registers, we would set this pin LOW, then HIGH to trigger the output of the chips, and then we would read the bytes in the Arduino. Now we need to write the bytes to the chip and then trigger its output (which will go to the LEDs, not the Arduino). The other difference is that we're not receiving any value from the SPI.transfer function, but we only write a value through it, so we're not assigning whatever this function returns to any variable, as we did with the input shift registers. A last minor difference is that in the loop that writes the bytes to the chips, we set the loop's variable i to its maximum value and we decrement it, instead of starting it from 0 and increment it. This way, the bytes we'll write will go to the chips we expect. If we run this loop the other way round, the first byte would go to the last chip of the chain (in case we use more than one chip).

The bitSet and bitClear Functions

We encounter two new built-in functions of the Arduino language: bitSet and bitClear. These functions operate on the bits of a given value. bitSet writes a 1 to a specified bit of a given value, and bitClear writes a 0. They take two arguments: the value to write to its bit and the bit of that value to write to. These functions are being used in the main loop function where we run two nested loops where the first one iterates through all the chips of our circuit and the second one iterates through all the pins of each chip.

In line 23, we make sure that we're not at the first chip, so that we can set a 0 to the byte of the previous chip. (If we are at the first chip, there's no previous chip and this line would try to access a byte beyond the bounds of the out_bytes array, causing our program to either not run properly, or even crash). Then we iterate through the pins of each chip. In line 26 we check if we're not at the first pin (much like the way we check if we're not at the first chip), and if we're not, we write a zero to the previous pin, so we turn that LED off. In line 28, we write a 1 to the current bit of the current byte, causing the LED to go on and then we call the set_output function. Only when this function is called will the values we've set to the out_bytes array take effect on the LEDs.

In line 34, we run a loop the other way round, which starts from the last chip and goes down to the first. Now, instead of checking if we're not at the first chip, we check if we're not at the last one. The same goes for the pins. This way, we can erase the previous pin of the loop (which is actually the next pin of the chip). In both loops, there's a small delay of 200 milliseconds to make the whole process visible. If we don't include this delay, all LEDs will look like they're constantly lit, but they will be dimmed (it will work like doing PWM to an LED). This doesn't mean that we can't light up more than one LED at a time. On the contrary, shift registers provide independent control of each of their pins.

Making the Output Shift Register Circuit

Fritzing provides a component for the breakout board of the output shift register, so we'll use both this and the IC here. The circuit with the breakout is shown in Figure 10-10 and with the IC in Figure 10-11. A cool feature of the breakout board (apart from saving us from some wiring) is that it enables daisy chaining by breaking out the pins of the chain to the left and right of the board. The way the board is oriented in Figure 10-10, the first chip would take input from its right side, and it would then give output to the next chip from its left side. This way, it is extremely easy to daisy chain these devices.

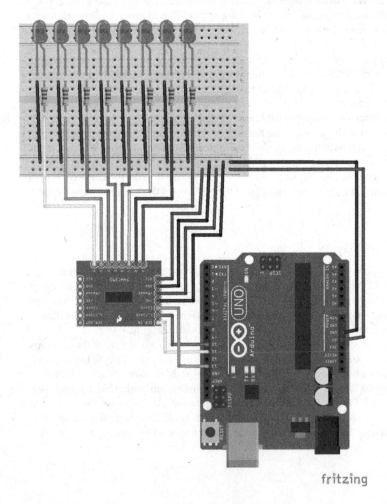

Figure 10-10. *One output shift register breakout circuit*

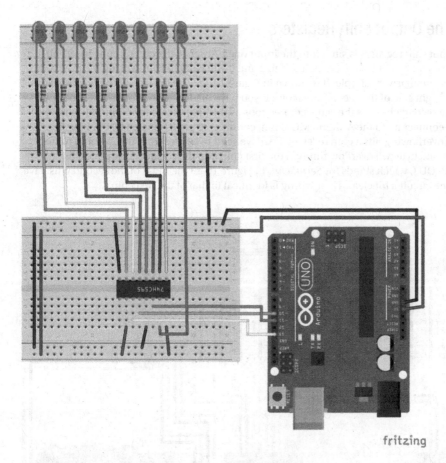

Figure 10-11. *The output shift register IC circuit*

On the left side of the board, starting from top we have the voltage pin (VCC, pin 16 on the IC), which connects to the voltage. Then we have the ground pin (GND, pin 8 on the IC) which connects to ground. The next pin is the /Reset pin (pin 10 on the IC labeled SRLCR), which connects to the voltage. The pin after that is the /OE (this stands for *Output Enable* and it is pin 13 on the IC) which connects to ground. Then we have the Clock pin (pin 11 on the IC labeled SRCLK). This pin connects to digital pin 13 of the Arduino. This is the same pin as pin 2 on the input shift register, which is marked as CLK on the breakout and CP on the chip. This pin receives the clock pulses from the Arduino to be synced. After the Clock pin we have the L_Clock pin (pin 12 on the IC marked RCLK), which connects to digital pin 10 of the Arduino. This latch pin triggers the output on the chip. Last pin is the SER_IN, which stands for *serial in* (pin 14 on the IC labeled SER). This pin connects to digital pin 11 of the Arduino, which is the pin called MOSI, which stands for *Master Out Slave In*. It's the opposite of digital pin 12, which is the MISO that we've used with the input shift register. The MOSI pin sends out bits to the devices, where the latter forms a byte from these bits and output it when their latch pin goes HIGH. We don't see the workings of the MISO and MOSI pins because it's being taken care of by the SPI library. This makes the whole process of using such devices very easy and intuitive (as well as fast). Before you test the program, note that the resistors used in this circuit are 220Ω. Run the sketch in Listing 10-5 and you should see the LEDs go on one by one.

Daisy Chaining The Output Shift Registers

Daisy chaining the output shift registers is similar to the input ones. As far as the code is concerned, all we need to do is set the num_of_chips constant in line 4 to 2 and it will work. If you're using the breakout for the shift register, the circuit is very simple. It is shown in Figure 10-12. The left side of the first breakout connects straight to the right side of the second breakout. If you have the ICs instead of the breakouts, check Figure 10-13. The second chip has the same connections as the first, except from pin 14 (marked SER) which, instead of connecting to the Arduino MOSI pin, connects to pin 9 (labeled QH') of the first chip. Pin 9 outputs the overflowing bits a chip receives. So, if we send two bytes from the Arduino to the chips, the bits of the second byte will overflow through the first chip into the second. This pin on the breakout is labeled SER_OUT, which stands for *Serial Out*. In Figure 10-13 the LEDs of the second chis have been omitted to keep the circuit a bit clear. Their wiring is identical to that of the first chip.

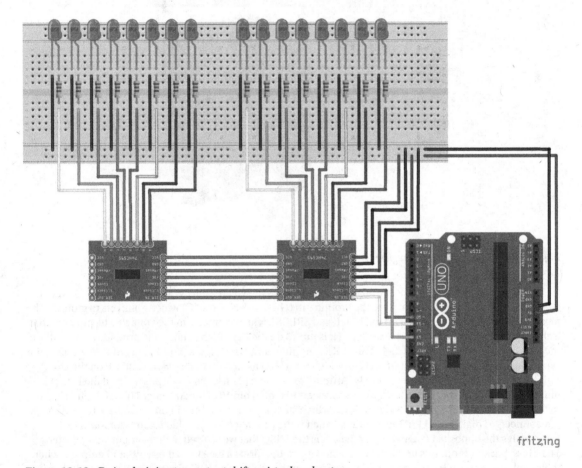

Figure 10-12. *Daisy chaining two output shift register breakouts*

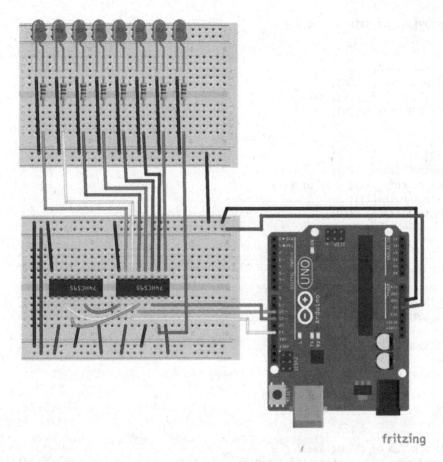

Figure 10-13. *Daisy chaining two output shift register ICs*

If you're using the breakouts, you can use angled headers to connect the two boards. This feature makes it extremely easy to use as many shift registers as you want. The same applies to the breakouts of the input shift register breakouts.

Combining the Input and Output Shift Registers

It should be fairly easy to combine the two kinds of shift registers. Listing 10-6 shows the code.

Listing 10-6. Combination of Input and Output Shift Registers

```
1.   #include <SPI.h>
2.
3.   const int input_latch = 9;
4.   const int output_latch = 10;
5.   const int num_of_chips = 2;
6.   byte in_bytes[num_of_chips];
7.   byte out_bytes[num_of_chips];
8.   const int num_of_data = (num_of_chips * 2) + 1;
```

```
9.   byte transfer_array[num_of_data] = { 192 };
10.
11.  void get_input(){
12.    digitalWrite(input_latch, LOW);
13.    digitalWrite(input_latch, HIGH);
14.    for(int i = 0; i < num_of_chips; i++)
15.      in_bytes[i] = SPI.transfer(0);
16.  }
17.
18.  void set_output(){
19.    digitalWrite(output_latch, LOW);
20.    for(int i = num_of_chips - 1; i >= 0; i--)
21.      SPI.transfer(out_bytes[i]);
22.    digitalWrite(output_latch, HIGH);
23.  }
24.
25.  void setup() {
26.    SPI.begin();
27.
28.    pinMode(input_latch, OUTPUT);
29.    pinMode(output_latch, OUTPUT);
30.
31.    Serial.begin(115200);
32.  }
33.
34.  void loop() {
35.    int index = 1;
36.
37.    get_input();
38.    for(int i = 0; i < num_of_chips; i++){
39.      transfer_array[index++] = in_bytes[i] & 0x7f;
40.      transfer_array[index++] = in_bytes[i] >> 7;
41.      out_bytes[i] = in_bytes[i];
42.    }
43.    set_output();
44.
45.    Serial.write(transfer_array, num_of_data);
46.  }
```

Now we define a function for each type of shift register, so that we don't include their code in the main loop function. These functions are defined in lines 11 and 18. Take a minute to compare them and to compare the get_input function to the part of the code in Listing 10-4, where we were reading the incoming bytes of the input shift registers.

We also define one array per shift register type, in lines 6 and 7, as well as an array for transferring data. The latter has a different size than the first two. In this sketch, we suppose we use the same number of chips for both input and output, because we'll control the LEDs of the output with the push buttons of the input, so we need one push button per LED.

In the main loop function, we first call the get_input function to get the input from the push buttons. Then we run a loop to copy the bytes that were written in the in_bytes array to the transfer_array. We do that because we'll need these bytes intact to control the LEDs. If we used the transfer_array straight in the get_input function, we wouldn't be able to control the LEDs properly, as we would have split the bytes to two to transfer them over the serial line. In this same loop, we copy the bytes of the in_bytes array to the out_bytes array. This step could have been avoided and we could have used the in_bytes array in the set_output function. This way, we used names that are more self-explanatory and we make the code a bit clearer. Once we do all the copying we call the set_output function to control the LEDs, and finally we write the transfer_array to the serial line.

There's no need to explain any circuitry because we can use exactly the same circuits we've already used. Notice that digital pin 13 of the Arduino, called the *Serial Clock*, connects to both the input and the output shift registers. This pin synchronizes the data transmission generated by the master, so that the communication between the chips and the Arduino is correct. The rest of the pins are different for the input and the output chips. We can also use the same Pd patch to receive the push buttons bytes. We have now extended the Arduino digital pins to a great extent. With only five Arduino digital pins, we have 16 digital inputs and 16 digital outputs!

Making a Patch-Bay Matrix

We have now been introduced to the tools we'll use for this project, so we can move on and start building the actual interface. The first thing we need to do is to understand how a patch-bay works and how we can implement that with the Arduino and Pd. A patch-bay matrix synthesizer is usually an analog synthesizer where the user can connect certain parts of it with between them using patch cords. These patch cords pass the signal of a part of the synthesizer to another part of it. For example, you can connect an oscillator of the synthesizer to a filter. The patch cord will pass the oscillator's signal to the filter and this way the oscillator will be filtered.

Implementing a Patch-Bay Matrix with the Arduino

It can be a bit tricky to implement a patch-bay with the Arduino. First, we must think about how this can be achieved. We're not going to use the analog pins (as you may have possibly thought) because these pins are input only. The Arduino Uno, Nano, or Pro Mini, doesn't have a DAC, so we can't produce analog output. We can only use the PWM pins, but that won't do for our purpose. Anyway, we don't want to implement the whole synthesizer with the Arduino, so we're not going to output any analog signal from it, since all the DSP will be done in Pd. We'll use the Arduino digital pins— more precisely the input and output shift registers, for this task.

Digital pins send what is called a *digital signal*. This signal is either high or low in voltage. We have to utilize this two-state voltage to detect connections between pins. Let's take an example of an 8-input and 8-output patch-bay. We'll set all the outputs low and we'll start bringing them high one by one, like we did with the LEDs and the output shift registers. When the first output pin is high, we'll read all the inputs. If this output is connected to one of the inputs, that input will read high. We'll take advantage of this synchronization, so when the first output is high, which ever input reads high means that it is connected to the first output. Then we'll bring the first output low and we'll bring high the second one. Again, if we detect any input that reads high, it means that it is connected to the second output. In general, this is the process of detecting connections between input and output pins. Listing 10-7 shows the code.

Listing 10-7. Patch-Bay Matrix Sketch

```
1.    #include <SPI.h>
2.
3.    const int input_latch_pin = 9;
4.    const int output_latch_pin = 10;
5.    const int num_of_input_chips = 1;
6.    const int num_of_output_chips = 1;
```

```
7.    byte input_bytes[num_of_input_chips] = { 0 };
8.    byte output_bytes[num_of_output_chips] = { 0 };
9.    // we need two bytes for the connection byte
10.   // one for the input chip that is connected
11.   // one for the output pin that is connected
12.   // and one for the start character
13.   const int num_of_data = 5;
14.   byte transfer_array[num_of_data] = { 192 };
15.
16.   const int num_of_output_pins = num_of_output_chips * 8;
17.   byte connection_matrix[num_of_output_pins][num_of_input_chips] = { 0 };
18.   bool connection_detected = false;
19.   int connected_chip, connected_pin;
20.
21.   void get_input(){
22.     digitalWrite(input_latch_pin, LOW);
23.     digitalWrite(input_latch_pin, HIGH);
24.     for(int i = 0; i < num_of_input_chips; i++)
25.       input_bytes[i] = SPI.transfer(0);
26.   }
27.
28.   void set_output(){
29.     digitalWrite(output_latch_pin, LOW);
30.     for(int i = num_of_output_chips - 1; i >= 0; i--)
31.       SPI.transfer(output_bytes[i]);
32.     digitalWrite(output_latch_pin, HIGH);
33.   }
34.
35.   byte check_connections(int pin){
36.     byte detected_connection;
37.     for(int i = 0; i < num_of_input_chips; i++){
38.       if(input_bytes[i] != connection_matrix[pin][i]){
39.         detected_connection = input_bytes[i];
40.         connected_chip = i;
41.         connection_detected = true;
42.         connection_matrix[pin][i] = input_bytes[i];
43.         break;
44.       }
45.     }
46.     return detected_connection;
47.   }
48.
49.   void setup() {
50.     SPI.begin();
51.
52.     pinMode(input_latch_pin, OUTPUT);
53.     pinMode(output_latch_pin, OUTPUT);
54.
55.     set_output();
56.
```

```
57.    Serial.begin(115200);
58.  }
59.
60.  void loop() {
61.    int index = 1;
62.    byte detected_connection;
63.
64.    for(int i = 0; i < num_of_output_chips; i++){
65.      for(int j = 0; j < 8; j++){
66.        int pin = j + (i * 8);
67.        bitSet(output_bytes[i], j);
68.        set_output();
69.        delayMicroseconds(1);
70.        get_input();
71.        detected_connection = check_connections(pin);
72.        bitClear(output_bytes[i], j);
73.        if(connection_detected){
74.          connected_pin = pin;
75.          break;
76.        }
77.      }
78.      if(connection_detected) break;
79.    }
80.
81.    if(connection_detected){
82.      transfer_array[index++] = detected_connection & 0x7f;
83.      transfer_array[index++] = detected_connection >> 7;
84.      transfer_array[index++] = connected_chip;
85.      transfer_array[index++] = connected_pin;
86.      connection_detected = false;
87.
88.      Serial.write(transfer_array, num_of_data);
89.    }
90.  }
```

This sketch might be a bit complicated so we'll take it step by step. The first 14 lines define some constants and arrays which we should already be familiar with. Only line 13 that sets the number of bytes to be transferred over serial might seem not so obvious. The comments above it explain why we need this number of bytes, but leave it for now, as it will become clearer when I explain the rest of the code.

In lines 21 and 28, we define the same functions for the shift registers as we did with Listing 10-6. In line 35, we define a new function, check_connections, which does what its name states, it checks if two pins are connected. We'll come back to it when it's time to explain how it works. I should mention that in line 16, we set the number of output pins, because we need it to define the rows of a two-dimensional array that we define in line 17. Plus, in line 18, we define a bool, and in line 19, we define two ints. The two-dimensional array and the bool will be used in the check_connections function, whereas the two ints will be used in the main loop function.

The Main Mechanism of the Code

In the main `loop` function, we first set the index offset for the `transfer_array`, and then we define the variable `detected_connection` of the type `byte`. This variable will hold the byte that will define a new connection between two pins. Then we go through the pins of each chip of our circuit (only one for now). In this loop, we first set the current output pin we'll be manipulating, in line 65. This pin number is the number of the current chip times eight, plus the incrementing variable of the nested loop. We'll need that value to detect a possible connection. The next thing we do is to set the current pin `HIGH` using the `bitSet` function. Once we do that, we call the `set_output` function to trigger the output on the output shift registers. Then we give a tiny bit of time to the Arduino so that the chips can do their job by calling the `delayMicroseconds` function, which pauses the program for the number of microseconds specified as an argument. Here it's only one, so it is really just a tiny bit of time. Once the delay is over, we call the `get_input` function to receive input from the input shift registers. We'll use that input the way that I described in the general overview of the patch-bay technique to detect a connection. Notice that this time we're first writing data to the output shift register and then we read the input shift register, in contrary to listing 10-6 where we used the input to control the output. The input is pulses sent from the output chips, which are detected from the input chips. Therefore, if we don't provide output first, we can't get any input to detect whether there is a connection between two pins or not.

The check_connections Function

In line 70, we call the `check_connections` function and we assign the byte it returns to the `detected_connection` variable. Now it's time to see how this function works. Let's go back to line 35. This function checks whether a new connection or disconnection has been made. It takes one argument, which is the current output pin we're manipulating. What we do here is to run through all the bytes of the `input_bytes` array and compare them with the bytes of the two-dimensional array `connection_matrix` of the row set by the argument of the function. This two-dimensional array holds as many columns as the number of the input chips, and the rows are as many as the output pins. Since the connections are actually input from the shift registers, they are being expressed in bytes, where one byte expresses all the connections of one chip. Storing one byte per input chip for each output pin, we can keep a log for all possible connections. So line 38 actually checks if a stored connection state has changed. If a byte of the row we're currently checking changes, we store it to the `detected_connection` variable. After that, we store to the `connected_chip` variable the current value of the loops incrementing variable, which is the number of the input chip where a new connection (or disconnection) has been detected. Then we set the `connection_detected` bool to true. Then we update the current byte of the current row of the two-dimensional array, and finally we call the `break` control structure to exit the loop. Finally we return the `detected_connection` variable. Mind that this variable is local to the `check_connections` function, and is totally independent of the `detected_connection` variable of the `loop` function.

Continuing with the loop Function

Now back to the `loop` function. In line 70, we assign to the `detected_connection` variable that is local to the `loop` function the byte returned by the `check_connections` function. Then we set the current output pin `LOW` by calling the `bitClear` function and we check if a connection has been detected. If it has, we assign to the `connected_pin` variable the value of the output pin that has just been connected or disconnected and we exit the loop. We need to call the `break` control structure twice, since we run two nested loops.

Finally, if a connection has been detected we must write some values to the `transfer_array`. These values are the input byte of the input chip that has its connection state changed, the number of this chip, and the output pin that has its connection state changed. The first byte is broken in two bytes to avoid collisions with the start character of the data stream. Then we set the `connection_detected` Boolean to false, so we can check for new connections. Lastly, we `write` these bytes to the serial line.

Making the Pd Patch That Reads the Connections

The Arduino code alone is not enough for the patch-bay matrix to work. Pd has a central role in this mechanism too. The patch that reads the connection data is illustrated in Figure 10-14. First of all, in this patch we're using an external object from the iemmatrix library, [mtx]. This stands for *matrix*, which is a two-dimensional matrix. It's a very useful object that is necessary for the operations that we need to do. To use it we must import the library using [import iemmatrix]. The matrix we must create should have the number of inlets and outlets of the shift registers passed as arguments with this order. For now, it's 8 for both.

In the Arduino sketch, we first sent the connection byte split in two and then the two bytes, which are the input chip number and the output pin number. This means that we'll be receiving one value as *analog* and two values as *digital*, and we set these arguments to the [serial_write] abstraction.

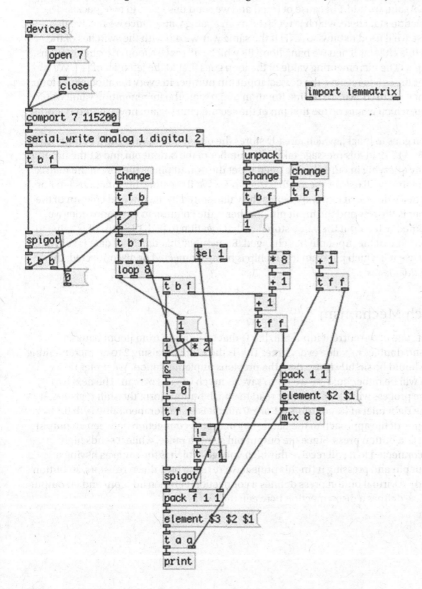

Figure 10-14. *Pd patch that reads connection data from the Arduino*

445

Explaining the Pd Patch

In the top part of the patch we make sure the whole process will be executed even if only one of the three bytes changes. I'll start explaining the last two bytes we're receiving from the Arduino, since they are being output first (it's the right to left execution order philosophy of Pd). The input chip is the first of the two values. Since we receive the input as a whole byte, we cannot know which pin in this chip has been connected or disconnected yet. We multiply this value by 8, because the shift registers have eight inputs, and we add 1, because [mtx] starts counting from 1, not 0. Then we store this value to [+], which we'll use later. We initialize the argument of [+] to 1, because if we connect to a pin of the first input shift register (which has index 0), the first of these two bytes won't go through due to [change]. The second value is the output pin. In the Arduino code, we're iterating through each pin individually, so we do know the exact pin that has changed its connection state. Again, we add 1 because of [mtx] and we send this value to two [pack]s.

Now let's go to the connection state byte, which is received as an analog value. Once we receive this byte, we store it in [&] because we'll need to bitwise AND it the same way we did with the switches of the input shift registers earlier in this chapter. Then we bang [loop 8], which will make a loop of eight iterations, since the pins of a chip are eight. The incrementing value of the loop goes first to the left inlet of [+]. We're adding the input chip number times 8 to retrieve the correct input pin number in every iteration of the loop. For now it's only one chip, but if we used two, at the first iteration of the loop, the incrementing value would be added to 9, so it would be 9, which would be the first pin of the second chip (again, now we start counting at 1, not 0).

The result of this addition goes to [pack], which already stores the output pin number and outputs this list to the message "element $2 $1". $2 in this message takes the number of the output pin and $1 the number of the input pin we're currently querying in our loop. We need to set the output pin as the row of the matrix because of the way an audio matrix we'll use for the synthesizer works. We'll see how this functions later on. The message "element" with two values sent to [mtx] queries the value stored in the row and column of the matrix. The output pin number is the row and the input pin number is the column, in the same order we provide arguments to [mtx]. [mtx] will output the value stored at that location to [!=]. We'll use this value to check if the stored connection state of the input pin has changed. Remember that the input data is sent as a whole byte, and only here can we tell which pin of an input chip is being connected or disconnected, that's why we need to make this inequality test.

The Heart of the Patch Mechanism

After we store the connection state of the current input pin, [t b f] that takes input from [loop] bangs a counter which starts from 1 and doubles its value every time. This is the same as raising 2 to an incrementing power starting from 0, but it should be slightly faster than the previous implementation. We prefer this because in the end our patch will be rather busy, so we try to save as much CPU as we can. The next two objects apply exactly the same process with the patch that read the push buttons form the shift registers. It checks if the byte stored in the right inlet of [&] outputs a 1 or a 0 after it is tested for inequality with 0.

This is the main philosophy of the approach to the patch-bay matrix. A connection between an output and an input pin is perceived as a button press. Since the output pin sends a pulse, while it sends high voltage, the input pin that is connected to it will receive this high voltage, which is the same as having a push button wired to the input pin and pressing it (in this project we're using pull-down resistors, so button presses are not being inverted). A virtual button press defines a connection between an input and an output pin, and a virtual button release defines a disconnection between the two pins.

Detecting Connections

The connection state that is being detected by [!= 0] is being sent to [!=] that has stored the previous connection state of the current input pin provided by [mtx] in its right inlet. If the two states are not equal, it means that the connection state has changed and [!=] will output a 1. This value first goes to [sel 1], which bangs the rightmost inlet of [loop] and stops the loop, and then to the right inlet of [spigot] and opens it. Afterward, the output of [!= 0] goes to the left inlet of [spigot], and since the [spigot] is open, it will go through to [pack f 1 1] (this is also initialized with its second and third arguments to 1). If it's 0 (meaning that the connection state has not changed) the loop doesn't stop and [spigot] remains closed.

[pack f 1 1] has stored the output pin number in its right-most inlet and the input pin number in its middle inlet, both initialized to 1. The value it receives in its leftmost inlet is the state of the connection between the two pins it stores in its other two inlets. This connection state will be a 1 if the two pins are being connected and a 0 if they are being disconnected. The output of [pack f 1 1] goes to the message "element $3 $2 $1" which is very similar to the message "element $2 $1", only it doesn't query the value of the row and column of the matrix but it sets the last value ($1) to that location. This means that once a connection state has been detected it will be set in [mtx], so when this state changes again it will be detectable. Finally the message "element $3 $2 $1" is being printed to Pd's console so we can monitor the functionality of the interface. Notice the arguments to [trigger], they are "a a", which stands for *anything*. We're neither passing floats nor symbols, but a whole message. For this reason, we must use [t a a].

Making the Patch-Bay Matrix Circuit

The circuit for the patch-bay matrix is very similar to the combination of the input and output shift register circuits. All pins of the shift registers are wired the same way as before, except from their input and output pins, which were wired to push buttons and LEDs. Now we're not using buttons or LEDs so these pins are left unwired. We'll use jumper wires to connect pins of the input and output shift registers in a patch-bay matrix fashion (connections between an output and an input pin are correct, you can't connect an output to an output or an input to an input). The circuit is shown in Figure 10-15, where the breakout is used for the output and the IC for the input. In this figure, there's a connection between the output pin 2 (labeled B on the breakout and QB—pin 15—on the IC) and the input pin 5, starting from 1, not 0 (labeled E on the breakout and D4—pin 3—on the IC). This is just to show how the inputs and outputs of the matrix connect to each other. If you make this connection, in Pd, you should get the message "element 2 5 1" printed to the console. Notice that the Clock pin of the breakout board and pin 2 of the IC (the CP pin) are both wired to digital pin 13 of the Arduino, the Serial Clock pin, which synchronizes the transmission of data.

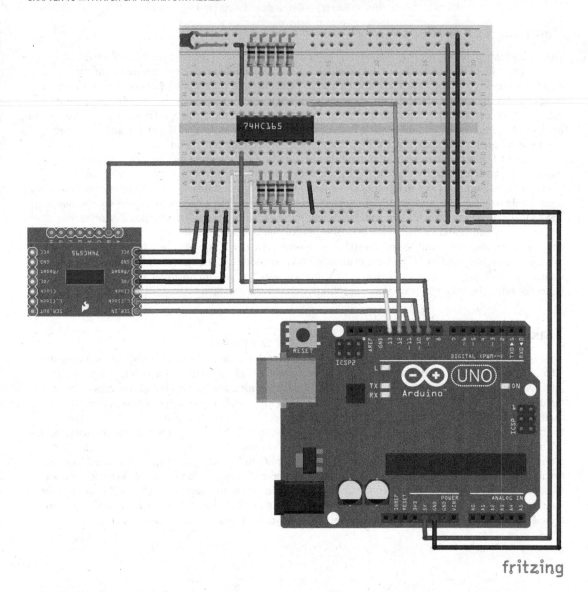

Figure 10-15. *A one input and one output shift register patch-bay matrix circuit*

This concludes the explanation of the patch-bay matrix. Making such a matrix is no easy or simple task, and we saw that in both the Arduino sketch and the Pd patch. We need to take into account several parameters and we need to keep a log of certain data for the mechanism to work. It might be a bit confusing as to how exactly this patch-bay works. You might need to go through both the Arduino sketch and the Pd patch more than once to really understand their workings. Once you completely understand how they work, you see that the whole approach is based on simple logic blocks that form a complex structure.

To use more inputs and outputs, the only thing we need to do is change the num_of_input_chips and num_of_output_chips constants to the number of chips that we want to use. Then daisy chain the shift registers in the circuit and change the arguments to the [mtx] object, where the first argument is the number of output pins and the second is the number of input pins. These few additions are enough for our matrix to work with any pin configuration. The number of chips used doesn't have to be the same for input and output, they are independent of each other. The only thing that could bother us is whether we have enough current in our circuit, in case we use too many chips, but we're not going to reach that limit in this project.

Start Building the Audio Part of the Pd Patch

We can now start building the audio Pd patch for our synthesizer. This will consist of a few different audio modules, which we'll be able to inter-connect in various ways. Even though the audio software will consist of modules, this is not a modular synthesizer since the hardware will be fixed, and not modular. As with all projects, it's best to make a rather simple interface that will be easier to achieve. This synthesizer will consist of two oscillators where each oscillator will output all four standard waveforms, one filter with various filter types, and two envelopes which we'll be able to use either for amplitude, frequency, or any other parameter that can be controlled by a signal.

A Signal Matrix in Pd

The heart of the patch is a matrix. The patch that reads the matrix message, which we've already built, is going to be part of our main patch, but that is not the actual matrix that will connect the various audio modules. The iemmatrix library provides another object (among many others in this library) which is a signal matrix. This object is called "mtx_*~". Figure 10-16 shows a very basic use of it. This object connects its signal inlets with its outlets in a [line~] fashion, making smooth transitions from one state to another. In all matrix messages in this library, the first argument is the number of rows and the second is the number of columns. In [mtx_*~] the rows are translated to outlets and the columns to inlets. The arguments for this object are the number of outlets, the number of inlets, and the ramp time. When trying to create this object you might get an error saying that Pd can't create it. In this case, type **mtx_mul~** instead.

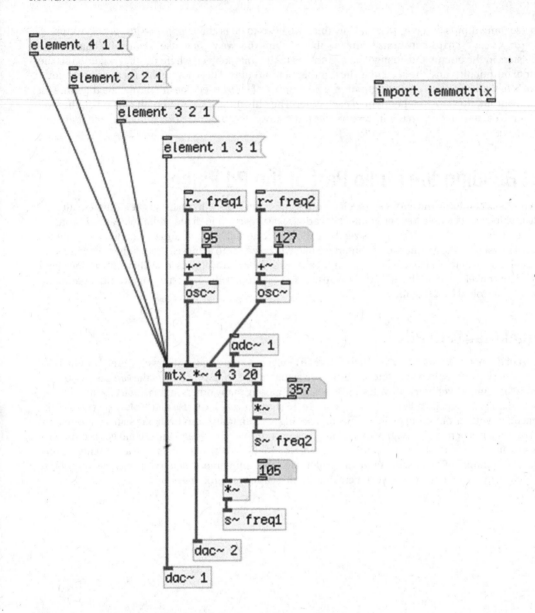

Figure 10-16. *A basic use of [mtx_*~]*

In Figure 10-16, the second argument is 3, but we get five inlets. This is because the first inlet takes only matrix messages, and the last one (which is a control inlet) overrides the last argument, the ramp time. A matrix message of the form "element 3 2 1" means, "multiply the input of inlet 2 by 1 and connect it to outlet 3." Again, the first number of the message is the row/outlet number of the matrix, the second is the column/inlet of the matrix, and the third is the value to write to that position of the matrix. In [mtx_*~] a value at a certain position means a multiplication by that value of the column/inlet, connected to the row/outlet.

450

In Figure 10-16, we're connecting the output of the first oscillator to the frequency input of the second (with the message "element 4 1 1"). This is followed by the output of the second oscillator to the right speaker of the computer, and then the output of the second oscillator to the frequency input of the first oscillator, and lastly the input of the computer's microphone to the left speaker. Using the messages generated by the patch that reads the patch-bay matrix from the Arduino, we can make connections dynamically according to the physical connections of the patch-bay matrix.

Here we are presented to a basic concept of the patch-bay matrix, which is a sort of a caveat. The outlets of the oscillators are connected to inlets of the matrix, and the outlets of the matrix sometimes end up in inlets of the oscillators. This might seem a bit obvious, but what we really need to get used to is that the inlets of the hardware interface will actually be outlets for the software, and vice versa. So from now on, you should bear in mind that the output shift registers will play the role of inputs and the input shift registers will play the role of outputs.

Building the Modules for the Synthesizer

Since the audio software in this project is modular, we can build these modules as abstractions and then load them in the main patch. We'll first build the two oscillator modules, where one will be a non-band limited and the other will be a band-limited oscillator (a table lookup oscillator with waveforms built with the sinesum feature of Pd). We'll start with the non-band limited one.

The First Module, a Non-Band Limited Oscillator

Figure 10-17 shows the first module abstraction, which is called "all_osc~", because it outputs all four standard oscillator waveforms. It is not a very complex abstraction, but it's not very simple either. Let's start looking at its subpatches.

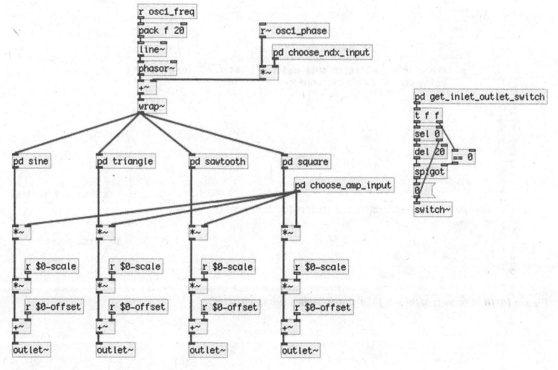

Figure 10-17. *Contents of the all_osc~ abstraction*

The get_inlet_outlet_switch Subpatch

The first subpatch we should look at is the [pd get_inlet_outlet_switch], which is shown in Figure 10-18. This subpatch deploys a basic concept in this project, which is taking advantage of connections of modules to control their DSP. On the left part of the patch we receive a message with [r mtx_msg], which stands for *matrix message*. This message is always of the type **element $1 $2 $3**. What we care about is whether an outlet in this module is connected to the matrix. If it is, it means that we are using this module so we should turn its DSP on. Remember that a module's outlet is a matrix's inlet.

Since this is an abstraction, we'll provide two arguments, the first inlet of the matrix it connects to, and the first outlet of the matrix it receives input from. These arguments are being passed to the abstraction from the bottom part of the subpatch in Figure 10-18. At the left part of the subpatch, we split the matrix message to two, where the second half will be the matrix inlet number and the connection state. We first store the connection state to [f], and then we check if the inlet of the matrix message is one of the outlets of this abstraction, which are four in total. If it is, we pass the connection state through the [spigot] and we scale it and give it an offset so that a 1 will be 1 and a 0 will be –1. This way, we can accumulate these values, because we might connect more than one of the module's outlets to the matrix, so we need to know when all its outlets will be disconnected, so we can turn its DSP off. We use [clip 0 100] to make sure we won't go below 0 (we'll never reach 100 connections, it's there just to give some headroom for the clipping).

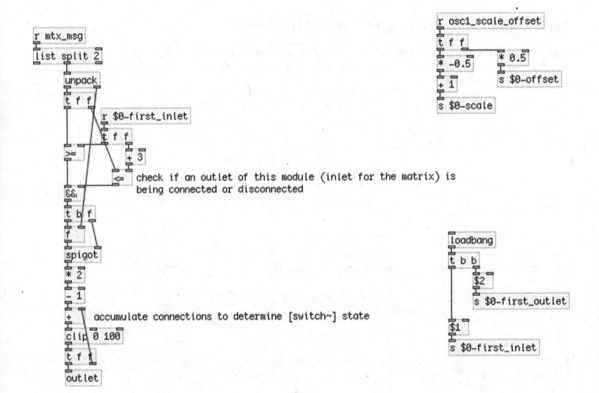

Figure 10-18. *Contents of the get_inlet_outlet_switch subpatch*

In the parent patch of the abstraction, shown in Figure 10-17, we can see that when the subpatch in Figure 10-18 outputs a 0, we delay it by 20 milliseconds, because this is the ramp time of [mtx_*~]. This way, we will avoid clicks when we turn the DSP of the subpatch off. We reverse the value to control a [spigot] because as we connect or disconnect a patch cord, the connection will wobble a few times (this will take a couple of milliseconds, if not microseconds). This wobbling might result in a sequence of 1s and 0s when we connect the patch cord, ending in 1. A 0 will delay its output by 20 milliseconds, so the last 0 before the last 1 will give its output after the last 1, resulting in turning the DSP of the abstraction off. Using a [spigot] this way prevents this from happening.

At the top-right side of the patch we will be receiving the values of a switch which we'll use to scale and offset the output of all four waveforms, so that we can either have the oscillator's full amplitude, or reduce it to a range from 0 to 1. We'll add this switch when we build the final circuit for this project because we may want to control the amplitude of the other oscillator, or its index, and we don't want negative values. In the parent patch of the abstraction we can see the corresponding [receive]s of these [send]s, at the output of each waveform.

The choose_ndx_input Subpatch

The next subpatch we'll look at is [pd choose_ndx_input], which is shown in Figure 10-19. In this subpatch, we set the control of the phase modulation index of the oscillator. Again, we're using the matrix messages to determine whether we connect a signal to the specific inlet of the module, so this time we query the outlet of the matrix message, that's why the implementation is slightly different than that in Figure 10-18. We add 1 to the outlet number because the first inlet of the module (outlet of the matrix) will be the signal that will modulate the phase of this oscillator.

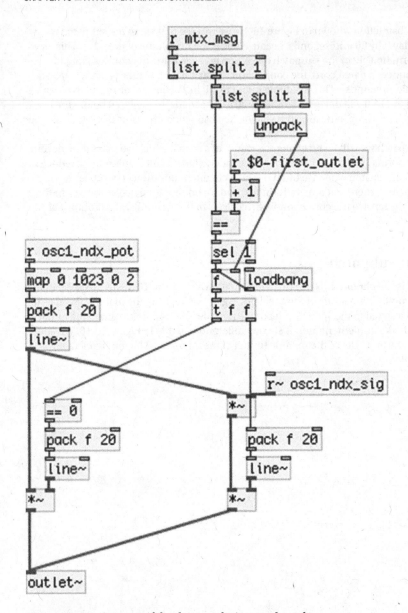

Figure 10-19. *Contents of the choose_ndx_input subpatch*

Because of the right-to-left execution order in Pd, the connection state will be output first out of [unpack] and the rest of the values afterward. What we do here is store the state value and query the matrix outlet number. If this number fits the matrix outlet number connected to [s~ osc1_ndx_sig], we output the connection state to determine whether a signal is connected to that outlet or not. We need to bang the connection state only when the outlet number of the matrix message fits the outlet we want to query; otherwise, we would be controlling this subpatch with any connection we made in the patch-bay matrix, and we don't want that.

Note that we don't really need to accumulate connections since we'll most likely have only one signal connected to this inlet. If there is a signal connected to this inlet, we'll choose that, if there's no signal connected, we'll choose the values we'll be receiving from a potentiometer, received with [r osc1_ndx_pot]. The value of this potentiometer will also control the amplitude of the input signal, because we might want to have a different range for the index than that of the input signal.

Since we need to determine which signal will control the index from the moment we'll open the patch, we bang [f] on load, which will hold 0 by default. This means that by default we choose the potentiometer values to control the index, and only if we connect a signal to the matrix outlet connected to [s~ osc1_ndx_sig] will we set that signal to control it.

The choose_amp_input Subpatch

This subpatch is almost identical to [pd choose_ndx_input]. The only things that change are the names of the [receive]s that receive the potentiometer value and signal, the mapping of the potentiometer values, and the number of the matrix outlet. It is shown in Figure 10-20. There's no real need to explain it, because I've already explained [pd choose_ndx_input]. Only note that we're adding 2 to [r $0-first_outlet], since this signal will arrive at the third inlet of the module.

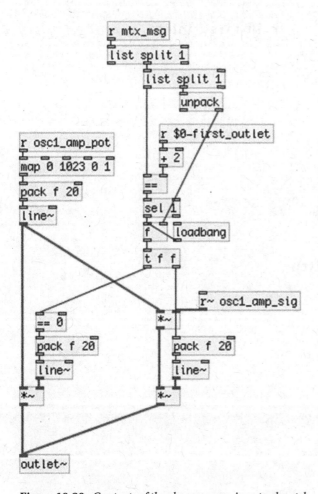

Figure 10-20. *Contents of the choose_amp_input subpatch*

The waveforms Subpatches

The last thing we need to look at is the subpatches of the four waveforms. We have already seen these waveforms a few times in this book, but here we add the connection feature that controls the DSP of a patch. They are shown in Figure 10-21 through 10-24. This time we care about whether the outlet of each subpatch is connected to the matrix, so we filter the matrix message to get the matrix inlet number and the connection state. Again, if the subpatch gets disconnected from the matrix we delay zeroing the subpatch by 20 milliseconds to avoid clicks because of the 20 millisecond ramp of [mtx_*~]. Each subpatch is controlled by a different outlet of the module, where the sine wave subpatch is controlled by the first outlet, and the next ones are controlled by the successive outlets. Notice that [pd square] receives a value in [r duty_cycle], which will be sent by a potentiometer in the final circuit.

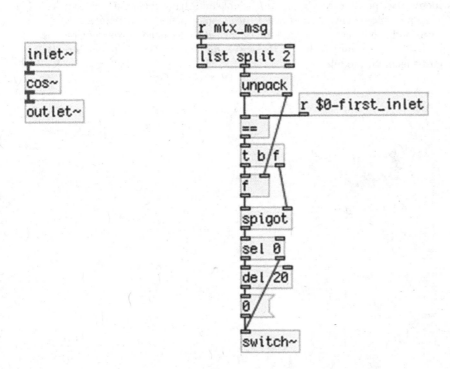

Figure 10-21. *Contents of the sine subpatch*

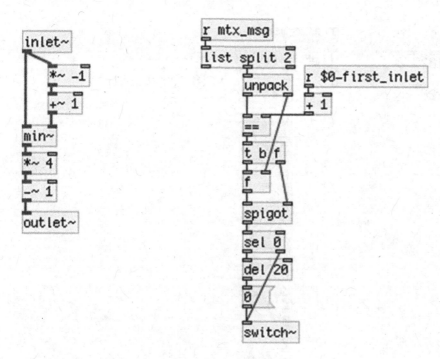

Figure 10-22. *Contents of the triangle subpatch*

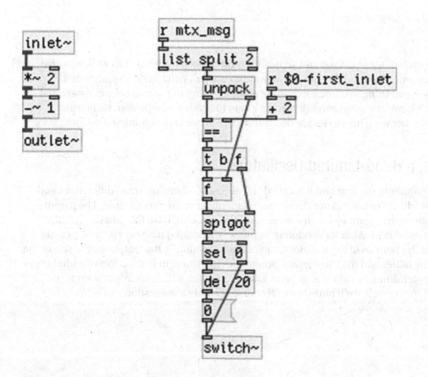

Figure 10-23. *Contents of the sawtooth subpatch*

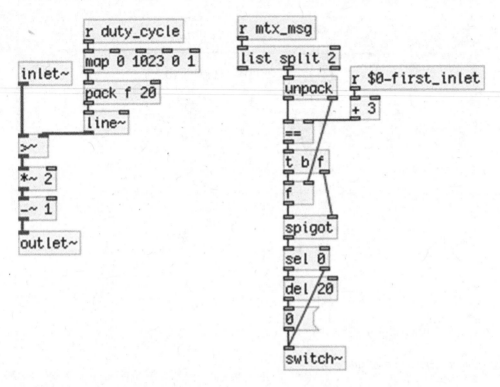

Figure 10-24. *Contents of the square subpatch*

Controlling the DSP of each of these subpatches separately, and of the parent patch as well, will save some CPU when this module won't be used. Eventually we want to use this patch with an embedded computer and CPU is always something to consider, as these computers are not as powerful as a personal computer. This project won't have that many modules, and it's very likely that all will often be playing simultaneously. Nevertheless, knowing how to utilize this feature will prove very helpful as a project grows.

The Second Module, a Band-Limited Oscillator

The second module is very similar to the first and it is called "bl_all_osc~". There are small differences and these are the ones we'll deal with. Whatever is not shown here is identical to the first module. The parent patch of the abstraction is shown in Figure 10-25. In the center of the patch we have the [phasor~], which drives the four table lookup oscillators. After we modulate its phase, we [wrap~] it around 0 and 1 to make sure we'll always stay within the bounds of the waveform tables and we multiply this output by the size of the tables and add 1, because the tables add the three guard-points with the sinesum feature. We've added some mapping to the frequency potentiometer values to get very low frequencies, which we might want to use in combination with the envelope module we'll build later. Of course, it's only a suggestion.

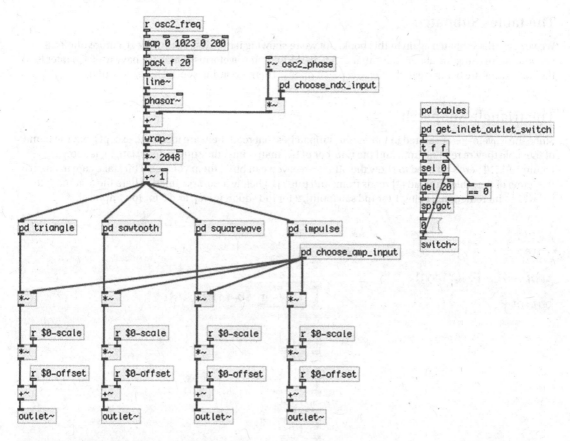

Figure 10-25. *The bl_all_osc~ abstraction*

Apart from that, the main difference is that all [receive]s have their names changed to *osc2*. So, wherever you have a [receive] make sure you apply this change (the rest of the name stays the same, as with [r~ osc2_freq]). These [receive]s are in [pd choose_ndx_input], [pd choose_amp_input], in [pd get_inlet_outlet_switch] and in the four subpatches that read the stored waveforms. We also have an additional subpatch that creates the waveform tables, which is shown in Figure 10-26.

```
loadbang
|
bl_triangle $0-triangle 2048 40          table $0-triangle
|
bl_sawtooth $0-sawtooth 2048 40          table $0-sawtooth
|
bl_squarewave $0-squarewave 2048 40      table $0-squarewave
|
bl_impulse $0-impulse 2048 40            table $0-impulse
```

Figure 10-26. *Contents of the tables subpatch*

The tables Subpatch

We've seen this subpatch again in this book, but we're showing it here too, because it is a more minimal version. We're using the abstractions that create band-limited waveforms, which we have used Chapter 8, so the waveforms are being created when the patch opens. There's nothing special here, only that.

The triangle Subpatch

Since the waveforms are stored in tables, the subpatches that read them are identical, except from the name of the table they're reading from and the number of the matrix inlet they query. [pd triangle] is shown in Figure 10-27. There's no need to show the other three; you can build them yourself. Only take care to change the name of the table [tabread4~] reads from, and put a [+] below [r $0-first_inlet], where the argument to [+] will be incrementing, being 1 for [pd sawtooth], 2 for [pd squarewave], and 3 for [pd impulse].

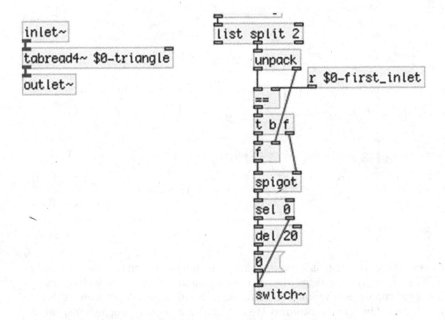

Figure 10-27. *Contents of the triangle subpatch*

This module doesn't have a sine waveform since there's no difference between a sine created by a [phasor~] and [cos~] and a sine created with sinesum. If you check the source code of [cos~] you'll see that it reads from a table where a cosine waveform is stored, which is essentially the same thing a [tabread4~] would do with a sine waveform created with sinesum. We've replaced the sine wave with an impulse waveform.

Third Module, a Multiple Type Filter

Our third module is simpler than the first two. We'll use a filter abstraction I've made with which we can have multiple filter types at the same time. This abstraction is [multiFilter_abs~] and you can find it on GitHub, in the same repository with [omniFilter_abs~], used in Chapter 5 (most likely you've already downloaded it if you built the project in Chapter 5). The module is shown in Figure 10-28. Name it **multi_filter~**. [multiFilter_abs~] gives eight outputs (for now, they might grow in the future) of which we'll use three. This is because we need to bear in mind how many inputs and outputs our synthesizer will have.

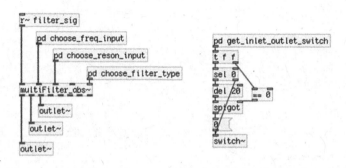

Figure 10-28. *The multi_filter~ module abstraction*

We have already built two modules both of which have three inputs and four outputs. Since the inputs and outputs are inverted because of the matrix, we are already occupying eight pins of the inputs shift registers (that makes an entire chip) and six pin of the output shift registers. The filter module takes another three inputs and three outputs, making a total of eleven pin of the input shift registers and nine pins of the output ones. We'll be using two chips of each shift registers, so we'll have 16 pins of each available. What' we're left with is five inputs and seven outputs. We'll need another two outputs for the [dac~], and two inputs and two outputs for the last module, which will be two separate envelopes. We also need to have three switches (one for each oscillator module and one for the DAC) and if possible, as many LEDs. If you think about it, we have already occupied all pins of all chips, so when building such an instrument we need to be careful as to what each module will do.

We're not going to show the contents of [pd choose_freq_input] and [pd choose_reson_input], because they are almost identical to the similar ones of the previous modules. Just make sure you change the names of the [receive] objects to [r filt_freq_pot] and [r~ filt_freq_sig] for [pd choose_freq_input] and [r filt_reson_pot] and [r~ filt_reson_sig] for the [pd choose_reson_input] ("reson" stands for *resonance*). Also, in [pd choose_freq_input] map the value of [r filt_freq_pot] to a range from 20 to 10,000 and put have a [+ 1] under [r $0-first_outlet] and in [pd choose_reson_input] map the value to a range from 0 to 1 and put a [+ 2].

The choose_filter_type Subpatch

Figure 10-29 shows the contents of [pd choose_filter_type]. The data we need is the connection state of the [outlet~]s of the abstraction, so we must use the inlet number of the matrix message along with the connection state. We're filtering the matrix message to isolate these two values. Then we subtract the number of the first matrix inlet this abstraction will be connected to. This way, the matrix inlet numbers of this abstraction will start from 0. We send the offset matrix inlet number along with the connection state to [route 0 1 2], which will set the [switch~] state of each filter type accordingly. [multiFilter_abs~] takes messages of the type **lowpass 1** or **lowpass 0** to turn the DSP of the lowpass filter on or off. Check its help patch for more information.

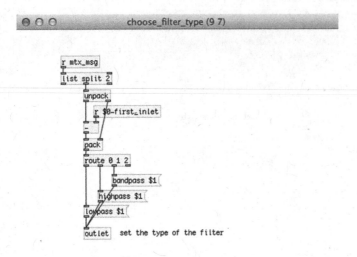

Figure 10-29. *Contents of the choose_filter_type subpatch*

The way that this subpatch works, as soon as we connect one of the abstraction's [outlet~]s to the matrix, [multiFilter_abs~] will receive the appropriate message and will turn on the DSP of the corresponding filter inside it. When we disconnect the [outlet~], it will receive the appropriate message to turn that DSP off. This way, we can save a lot of CPU, since filters can be CPU-intensive.

The get_inlet_outlet_switch Subpatch

This subpatch is nearly the same as the other two modules we've built, that's why we're showing it here. It is shown in Figure 10-30. The differences are that we don't have a switch to scale and offset the output signal of the module, and that we have three [outlet~]s in total. Thus we have a [+ 2] below [r $0-first_inlet], so we can check that the matrix inlets we get from the matrix messages correspond to the [outlet~]s in this module. Apart from that, it is the same as the corresponding subpatches in the previous two modules that we built.

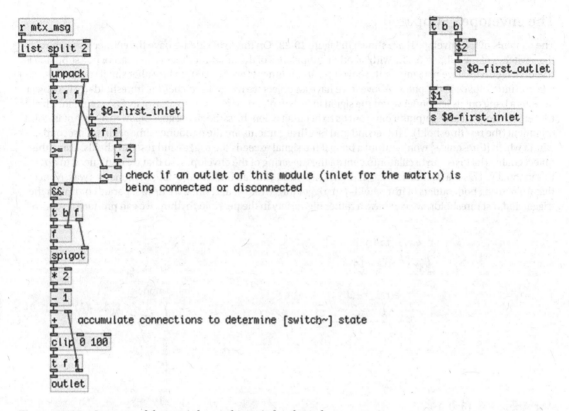

Figure 10-30. *Contents of the get_inlet_outlet_switch subpatch*

The Fourth Module, Two Envelopes

The last module that we'll build for our synthesizer is an envelope module, which we can use to control various parameters of the other modules. Its name is "envelope~" and you can see it in Figure 10-31. Its parent patch is very simple, but we'll need to take a look at its subpatches.

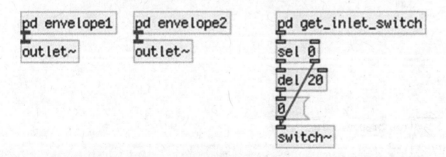

Figure 10-31. *The envelope~ module abstraction*

The envelope1 Subpatch

The contents of [pd envelope1] are shown in Figure 10-32. On the right side we have the connection control for the [switch~] of the subpatch, like with all other subpatches of the other modules. The heart of this subpatch is [ggee/envge], which we're using for the envelope. Its arguments are "100 60" to be rather small so that it doesn't take too much space in the patch. Above it we have an object we haven't seen before, [threshold~]. This object outputs a bang out its left outlet when the signal in its left inlet rises above the value of its first argument (the trigger threshold). It also outputs a bang out its right outlet when the same signal goes below the value of its third argument (the rest threshold). The second and fourth arguments are the debounce time set in milliseconds, within which [threshold~] won't output a bang if the signal exceeds the trigger and rest thresholds. Using this object enables us to use an oscillator to control the triggering of the envelopes, so that we don't need to do this manually. This is a desired feature in many cases, and you can build interesting rhythms this way. Notice that we're using both outlets of [threshold~] to trigger the envelope and since we have separate control on the trigger and rest thresholds, we can have a rather big variety in the possible rhythms we can produce.

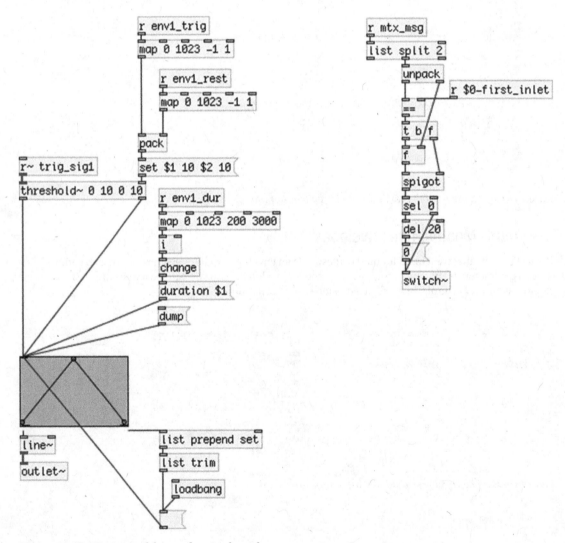

Figure 10-32. *Contents of the envelope1 subpatch*

In this subpatch, we receive three potentiometer values. The first one controls the trigger threshold value and the second controls the rest threshold value. We can set these values with the message "set $1 10 $2 10", which will override the arguments of the object. This value can be between –1 and 1, since the oscillators of the synthesizer will control the triggering of the envelopes. The third potentiometer will control the duration of the envelope, as we've seen before (in Chapter 4).

As in Chapter 4, create the envelope you want and click the "dump" message. [ggee/envgen] will output the breakpoint values and positions out its right outlet and we'll save them in a message, which will be banged when the patch opens. This way, we can save our envelopes.

[pd envelope2] is identical to [pd envelope1]. Only the names of the [receive]s change their numbering from 1 to 2, and a [+ 1] should go below [r $0-first_inlet].

The get_inlet_switch Subpatch

There is a small difference in the [pd get_inlet_switch] subpatch of this abstraction and the [pd get_inlet_outlet_switch] of the other abstractions. The subpatch name says it, we don't need to know the matrix outlet number this module will take input from, so we're excluding it. The subpatch is shown in Figure 10-33.

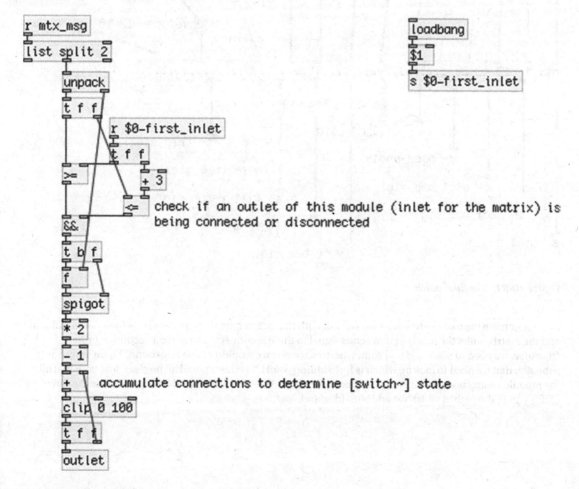

Figure 10-33. Contents of the get_inlet_switch subpatch

465

The Final Patch

Now that we've built the four module we'll use in this project it's time to build the final patch. It is shown in Figure 10-34. We have ended up with a matrix with 13 rows/outlets and 13 columns/inlets. There's not need to have the same amount of rows and columns, it just happened the way we built the interface. They are completely independent of each other, so if you want to build something different, don't worry about this relationship.

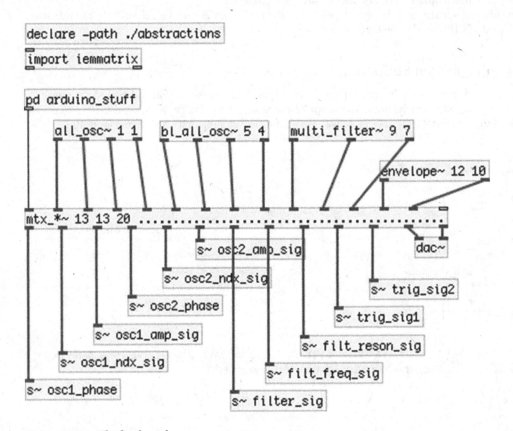

Figure 10-34. *The final patch*

We can see the four abstractions we've built with the matrix inlet they connect to as their first argument, and the matrix outlet the first [s~] that sends signal to them is connected to as their second argument. Of course, we need to place a [dac~] somewhere otherwise we wouldn't have any sound. There' only one subpatch that we need to look at, which is [pd arduino_stuff]. The rest should all be clear. Just note that all the module abstractions we've built have been placed in a directory called abstractions inside the directory of this patch, the path of which we add with [declare].

The arduino_stuff Subpatch

Figure 10-35 shows [pd arduino_stuff]. We're using the [serial_write] abstraction, but this time we provide five arguments to it. The fifth argument is the first character of the data stream from the Arduino. We choose a different one from 192 that we've been using so far, for a reason that will be explained when we go through the [pd get_switches] subpatch inside this subpatch. We're also using a different baud rate (57,600) because the 115,200 is probably too fast for the Arduino to handle. You can try for yourselves. There are three subpatches in total in [pd arduino_stuff], which we need to explain.

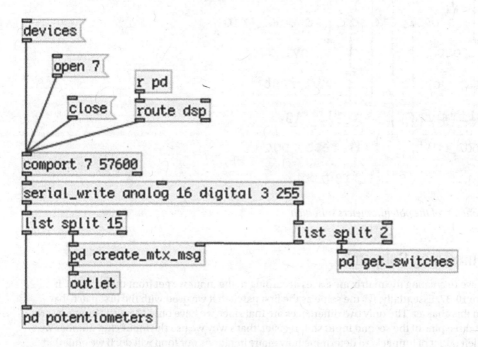

Figure 10-35. *Contents of the arduino_stuff subpatch*

The only thing we might need to explain here is that we'll be using fifteen potentiometers in total, and the connection state byte split in two. Then we'll receive the connected input chip, the connected output pin, and finally a byte that will hold the states of the three switches of our circuit (one for the scaling of each oscillator and one for turning on the DSP). This should clarify the arguments passed to [serial_write].

The potentiometers Subpatch

There's nothing special about this subpatch. It's shown in Figure 10-36. We're just diffusing of the potentiometers of our circuit to their destinations.

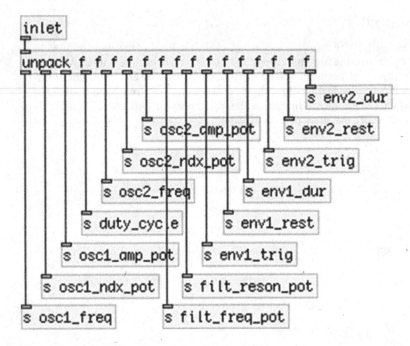

Figure 10-36. *Contents of the potentiometers subpatch*

The create_mtx_msg Subpatch

In this subpatch, we're creating the matrix messages according to the input we get from the Arduino. It is shown in Figure 10-37. Essentially, it's the same as the first patch that we used with the first patch-bay matrix we built in this chapter. The only two differences are that since we have only 13 matrix inputs, we're not using the last three pins of the second input shift register, that's why we use the input chip number we receive from the left outlet of [unpack] to determine how many iterations our loop will do. If we omit that and iterate eight times for both chips, we'll go through the three switches our circuit will have, since they will be wired to the second input shift register, and that will create unexpected behavior.

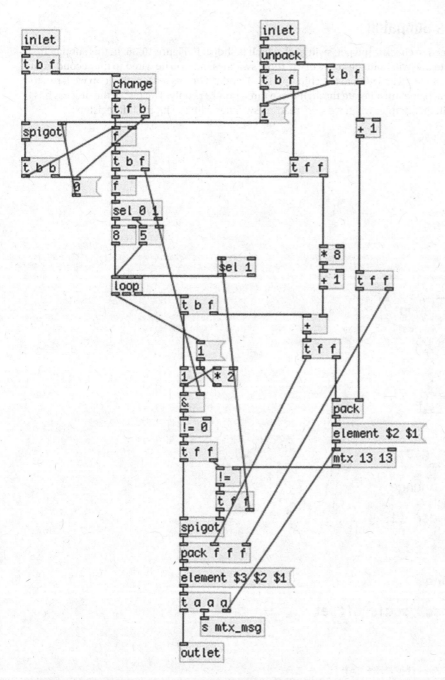

Figure 10-37. *Contents of the create_mtx_msg subpatch*

The other difference is that apart from updating [mtx] (which, by the way, now has 13 rows and 13 columns, as many as [mtx_*~] in the parent patch) and sending the matrix message to the [outlet] (instead of [print]), we also send it to [s mtx_msg], which is received in every module. So we can use the matrix messages to control the DSP of each module and its subpatches.

The get_switches Subpatch

The last subpatch we need to check is [pd get_switches], which is shown in Figure 10-38. In this subpatch, we'll receive a single byte that will hold the states of the three switches that will be wired to the second input shift register. These switches will be wired to the chip's pins 5, 6, and 7. For this reason, the loop we'll run to determine the state of each one must iterate through these three numbers only. The rest of the process has been explained earlier in this chapter when we were reading switches with the input shift registers.

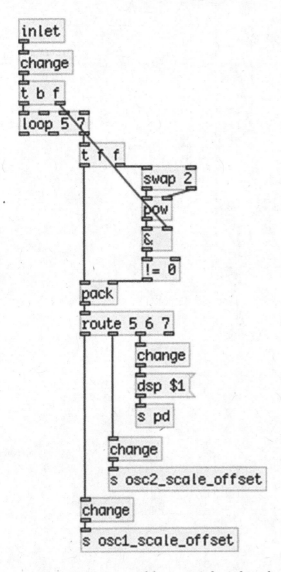

Figure 10-38. Contents of the get_switches subpatch

The first of these switches will control the scaling and offset of the first oscillator. The second switch will control the scaling and offset of the second oscillator. The last switch will control the DSP of Pd. In Figure 10-35 we can see a [r pd] connected to [route dsp], like we've used it in some other chapters already. That value will control an LED of our circuit so we can have visual feedback for when the DSP is on.

Let's come back to the [pd arduino_stuff] subpatch shown in Figure 10-35. Since we're receiving the states of all three switches in one single byte, and because this byte will have its most significant bit (the rightmost bit) set to 1, when the third switch is in the on position (these switches are wired to the last three pins of the second input shift register), this byte will potentially get the same value as our start character. To be precise, we need to test which decimal values the eight possible combinations of the three switches can get, and determine whether it is indeed possible that one of these combinations will be 192. Instead of going through this process, we can just change the start character simply by providing a fifth argument to the [serial_write] abstraction, and by setting the same value in the definition of the data transfer array in the Arduino code. We'll be masking this byte with all other bits set to 0 in the Arduino code, so there's no way it will ever reach 255. We'll see it in detail when we go through the Arduino sketch.

This concludes the Pd patch, so now we can move on to the final version of the Arduino code.

The Final Arduino Code

Listing 10-8 shows the final code that works with the final Pd patch. We've already seen most of the stuff, but there are a few additions that are project-specific, such as the switches and LEDs attached to the shift registers and that we can read them independently from the patch-bay matrix inputs and outputs.

Listing 10-8. Final Arduino Code

```
1.   #include <SPI.h>
2.   // define a log base 2 function
3.   #define LOG2(x) (log(x) / log(2))
4.
5.   // multiplexer variables
6.
7.   // change this number if using more chips
8.   const int num_of_mux_chips = 1;
9.   const int num_of_pots[num_of_mux_chips] = { 15 };
10.  const int num_of_mux_channels = 16;
11.  const int num_ctl_pins = LOG2(num_of_mux_channels);
12.  int control_pins[num_ctl_pins];
13.  // create an array to hold the masking value to create binary numbers
14.  // these values will be set in the setup function
15.  int control_values[num_ctl_pins];
16.
17.  // shift register variables
18.
19.  const int input_latch_pin = 9;
20.  const int output_latch_pin = 10;
21.  const int num_of_input_chips = 2;
22.  const int num_of_output_chips = 2;
23.  byte input_bytes[num_of_input_chips] = { 0 };
24.  byte output_bytes[num_of_output_chips] = { 0 };
25.  int num_of_output_pins[num_of_output_chips] = { 8, 5 };
```

```
26.  // the 2D array must have as many rows as the total output pins
27.  byte connection_matrix[13][num_of_input_chips] = { 0 };
28.  bool connection_detected = false;
29.  int connected_chip, connected_pin;
30.  // set the pin number of the first switch on the last input chip, starting from 0
31.  int first_switch = 5;
32.  // set position of the DSP LED
33.  int dsp_led = 7;
34.  // set the pin number of the first LED on the last output chip, starting from 0
35.  int first_led = 5;
36.  // set the number of switch that control LEDs without receiving data from the serial port
37.  int num_of_ctl_switches = 2;
38.  // set a mask for the last input byte to properly detect connections
39.  byte switch_mask = B11100000;
40.
41.  // set the total number of bytes to be transferred over serial
42.  // num_of_pots * 2 to get 10bit resolution
43.  // two bytes for the connection byte
44.  // one for the input chip that is connected
45.  // one for the output pin that is connected
46.  // one for the three switches
47.  // and one for the start character
48.  const int num_of_data = 36;
49.  byte transfer_array[num_of_data] = { 255 };
50.
51.
52.  void get_input(){
53.    digitalWrite(input_latch_pin, LOW);
54.    digitalWrite(input_latch_pin, HIGH);
55.    for(int i = 0; i < num_of_input_chips; i++)
56.      input_bytes[i] = SPI.transfer(0);
57.  }
58.
59.  void set_output(){
60.    digitalWrite(output_latch_pin, LOW);
61.    for(int i = num_of_output_chips - 1; i >= 0; i--)
62.      SPI.transfer(output_bytes[i]);
63.    digitalWrite(output_latch_pin, HIGH);
64.  }
65.
66.  byte check_connections(int pin){
67.    byte detected_connection;
68.    for(int i = 0; i < num_of_input_chips; i++){
69.      // copy input byte because we need to mask the last one here
70.      // but we also need it unmasked in the main loop function
71.      byte copy_byte = input_bytes[i];
72.      // mask the last byte to exclude switches
73.      if(i == num_of_input_chips - 1) copy_byte &= (~switch_mask);
74.      if(copy_byte != connection_matrix[pin][i]){
75.        detected_connection = copy_byte;
76.        connected_chip = i;
```

```
77.          connection_detected = true;
78.          connection_matrix[pin][i] = copy_byte;
79.          break;
80.        }
81.      }
82.    return detected_connection;
83. }
84.
85. void setup() {
86.   SPI.begin();
87.
88.   pinMode(input_latch_pin, OUTPUT);
89.   pinMode(output_latch_pin, OUTPUT);
90.
91.   set_output();
92.
93.   // set the value masks, the control pins and their modes
94.   for(int i = 0; i < num_ctl_pins; i ++){
95.     control_values[i] = pow(2,(i+1)) - 1;
96.     control_pins[i] = (num_ctl_pins - i) + 1;
97.     pinMode(control_pins[i], OUTPUT);
98.   }
99.
100.  Serial.begin(57600);
101. }
102.
103. void loop() {
104.   int index = 1;
105.   byte detected_connection;
106.
107.   if(Serial.available()){
108.     byte in_byte = Serial.read();
109.     bitWrite(output_bytes[num_of_output_chips - 1], dsp_led, in_byte);
110.   }
111.
112.   // first go through the potentiometers
113.   for(int i = 0; i < num_of_mux_chips; i++){
114.     // this loop creates a 4bit binary number that goes through the multiplexer control
pins
115.       for (int j = 0; j < num_of_pots[i]; j++){
116.       for(int k = 0; k < num_ctl_pins; k++){
117.         // this line translates the decimal j to binary
118.         // and sets the corresponding control pin
119.         digitalWrite(control_pins[k], (j & control_values[k]) >> k);
120.       }
121.       int val = analogRead(i);
122.         transfer_array[index++] = val & 0x007f;
123.       transfer_array[index++] = val >> 7;
124.     }
125.   }
126.
```

```
127.   // then go through the shift registers
128.   for(int i = 0; i < num_of_output_chips; i++){
129.     for(int j = 0; j < num_of_output_pins[i]; j++){
130.       int pin = j + (i * 8);
131.       bitSet(output_bytes[i], j);
132.       set_output();
133.       delayMicroseconds(1);
134.       get_input();
135.       detected_connection = check_connections(pin);
136.       bitClear(output_bytes[i], j);
137.       if(connection_detected){
138.         connected_pin = pin;
139.         break;
140.       }
141.     }
142.     if(connection_detected) break;
143.   }
144.
145.   if(connection_detected){
146.     transfer_array[index++] = detected_connection & 0x7f;
147.     transfer_array[index++] = detected_connection >> 7;
148.     transfer_array[index++] = connected_chip;
149.     transfer_array[index++] = connected_pin;
150.     connection_detected = false;
151.   }
152.
153.   // write the masked byte with the three switches
154.   transfer_array[num_of_data - 1] = input_bytes[num_of_input_chips - 1] & switch_mask;
155.
156.   // control the two LEDs according the two switches
157.   // copy the pin variables so we can post-increment them
158.   int first_switch_pin = first_switch;
159.   int first_led_pin = first_led;
160.   for(int i = 0; i < num_of_ctl_switches; i++){
161.     int switch_val = bitRead(input_bytes[num_of_input_chips - 1], first_switch_pin++);
162.     bitWrite(output_bytes[num_of_output_chips - 1], first_led_pin++, switch_val);
163.   }
164.
165.   Serial.write(transfer_array, num_of_data);
166. }
```

In line 9, we define an array called num_of_pots. We'll use that in the main loop function to iterate only through the fifteen pins of the multiplexer. Even though this array has only one element, it's good to define this variable as an array in case we want to use more multiplexer chips. In line 10, we still define the num_of_mux_channels so that we can use the preprocessor directive LOG2 to get the num_ctl_pins in line 11.

Line 31 defines a variable that holds the pin number of the last input shift register where the first switch of our circuit is attached to (this switch will control the scaling and offset of the first oscillator). Then in line 33, we define the pin number of the last output shift register where the DSP LED will be attached to (it's the last pin of the chip). In line 35, we define the first pin the first LED of our circuit will be attached to on the last output shift register. In line 37, we define how many switches in our circuit will control LEDs without having data from the serial line intervening. In line 49, we define the transfer_array with its first byte initialized to 255, like we did in the Pd patch.

Using Binary Numbers for Masking

We jumped line 39 deliberately, because we need to explain it separately. In this line we define a binary number, hence the capital B in the beginning of it. This way, we can define binary numbers in the Arduino language. This number represents the pins of the last input shift register where the three switches are attached to. These are the last three pins, that's why the first three digits of this number are 1s and the rest are 0s. Shift registers map their bytes this way, the Least Significant Bit (referred to as LSB, it's the rightmost digit) represents the state of the first pin, and the Most Significant Bit (referred to as MSB, it's the leftmost digit) represents the state of the last pin.

We'll need this value to mask the last byte of the input_bytes array. If we don't mask it, and we change the position of a switch but we don't make a new connection, this byte will change and the test we make in the check_connections function to see whether we have a change in the input_bytes array will succeed, even if no new connection has occurred. This will result in storing this byte to the transfer_array. Consequently, it will trigger the loop that detects connections in Pd, as shown in Figure 10-33. Anyway, in Pd, we only go through the connection pins, so the patch wouldn't try to make a new connection, but the loop would be triggered. Masking this byte will prevent this loop from being triggered, because we use [change], and the positions of the switches won't affect this last byte at all.

Using a binary value instead of a decimal is very convenient in such cases because we don't need to really think arithmetically. With such a number, we can clearly represent our circuit, since we can just use 1s for the chip pins we're interested and 0s for the pins we don't need.

The check_connections Function

We've introduced a small modification in the check_connections function. What we do is copy the byte of the input_bytes array because if we're iterating through its last byte, we need to mask it with the binary reversed value the switch_mask variable hold. We reverse this value by using the tilde character (~). This is another bitwise operation that swaps the bits of a value, replacing all 1s with 0s and vice versa. The switch_mask variable has 1s in the bits of the switches, but we need to mask the last byte of the input_bytes array the other way round, having 1s in the bits of the connection pins.

If we don't copy this byte, we'll modify it and we won't be able to detect the switch positions anymore, as the bits of the switches will have 0s no matter what we do with them. We could have stored a binary value with the switch pins set to 0, but I found it a bit more intuitive to set these pins to 1, as it makes it clearer that we need to isolate them.

The Main loop Function

In the main loop function, the first thing we do is check if we've received data from the serial line. This will happen only when we turn the DSP on in Pd. If we do receive data, we'll set the state of the DSP LED using the bitWrite function. This function is very similar to the bitSet and bitClear functions we've already used, but it takes three arguments, the byte to write to one of its bits to, the bit of the byte, and the value to write to that bit. The DSP LED is wired to the last output shift register, so we use the last byte of the output_bytes array. The bit of that byte is set by the dsp_led variable, and the value of that bit is set by the byte we receive from the serial line.

After that, we go on and read the fifteen potentiometers of our circuit, and then we check if we have a new connection in the patch-bay matrix. In line 154, we store the last byte of the input_bytes array to the transfer_array. We'll use that byte to retrieve the states of the three switches of our circuit. For this reason we mask it with the switch_mask variable, this time without inverting it (so defining it this way and not inverted was not without a reason after all). Again, this is not really necessary, but it will prevent the loop that detects the switch states from being triggered, saving some little CPU in Pd. We've already seen how we detect the switch states when I explained the [pd get_switches] subpatch inside the [pd arduino_stuff] subpatch of the final Pd patch.

Controlling the LEDs with the Switches of the Shift Registers

In lines 158 and 159, we make copies of the first_switch and first_led variables, because we want to post-increment them. If we used these variables immediately, we would change their values and after the first iteration of the loop function, they would already be beyond the bounds of the shift register pins. In line 160, we run a loop for the two switches that control their LEDs without serial data intervening. This time we use a combination of the bitRead and the bitWrite function to read each switch value and write it to its LED. The former is similar to the latter, only it takes two arguments instead of three: the byte to read a bit from and the bit of the byte to read. It shouldn't be too hard to understand how this loop works. Finally, we write the transfer_array to serial line. This concludes the Arduino code for this project.

The Final Circuit

You'll probably know how to build the circuit already, but it is shown here anyway. Figure 10-39 illustrates it with breakouts for the multiplexer and the output shift registers (refer to previous sections of this chapter for wiring the ICs). The potentiometers are not included in the circuit because it would become rather messy. You should know how to wire them by now. The three switches wired to the last three pins of the second input shift register and the three LEDs wired to the last three pins of the second output shift register are included. Take care to connect one pin of the switch to the voltage (if it's an ON/ON switch, this should be the middle pin). You can now test your circuit with the Pd patch, you've made a patch-bay matrix synthesizer!

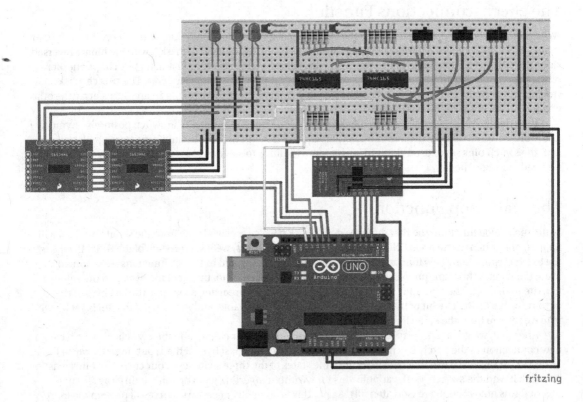

Figure 10-39. *The final circuit*

Making an Enclosure

We've already made a few enclosures in this book, but with this one, there are a few tips and tricks that you might find helpful. First, since this project is highly customizable, you could argue if the setup of the hardware interface should be project specific, or if it should be more generic so that it can host other implementations of the software. This means that you should decide where to place all the components on the enclosure. Having all the female banana terminals placed according to the connections of the software they represent, and the potentiometers grouped according to the modules of the software they control is very helpful for the specific project, but limits the hardware interface to this project's software. On the other hand, making a generic layout can be a bit confusing, as you should remember what each component is for, and the layout won't help in telling that. Figure 10-40 shows my approach to the hardware layout of the patch-bay matrix and the other components. This figure includes some connections between banana terminals to show how they are to be connected. As you can see, I chose to have the patch-bay in the middle so I can use the same enclosure for another type of synthesizer. This way, I'm not limited to the software built in this chapter. I've also used 16 potentiometers instead of 15 to have as many as the multiplexer provides in case I want to use this enclosure for another project. It's up to you which route you'll take in designing this.

Figure 10-40. *A patch-bay matrix synthesizer layout*

Another thing you might want to have is access to the Ethernet, in case you embed a computer in the enclosure. A good idea is to use an Ethernet gender changer (this is actually a female-to-female Ethernet port) and a flexible Ethernet cable, so you can connect your embedded computer's Ethernet to the gender changer, which on its other side is a female Ethernet, thus giving you access to the port of the embedded computer.

Lastly, concerning the banana terminals, it is a good idea to use different colors for different purposes. In Figure 10-40, I've used black bananas for input (matrix and shift register output) and red for output (matrix and shift register input). You might want to use a third color for control signal inlets (the phase inlet of the oscillators, for example), especially if you make your layout based on the specific project. All the female banana terminals are wired to the free pins of the input and output shift registers, which are pins

A to H on the breakouts and D0 to D7 on the input IC and QA to QH on the output IC. By free I mean all these pins that are not wired to switches or LEDs (refer to Figure 10-39). Using cables with male banana terminals enables us to make connections between input and output pins, much like we did with the jumper wires on the breadboard version of this project.

You also might want to include a diode in each patch cord. Banana terminals can be stacked so that you can connect more than one patch cord to a female banana. Using diodes in the patch cords will prevent the current flowing from one patch cord to another in case of stacked bananas. In Figure 10-40, if no diodes were included in the patch cords, output pin 9, which is connected to input pin 7, would seem like it's also connected to input pin 1. That's because while the output pin 9 would be HIGH, its current would flow through the cable of the patch cord to input pin 7. But that input pin has two banana terminals stacked. This would result in the current flowing further to output pin 7 and from there to input pin 1. Try to follow the connections in Figure 10-40 and you'll understand. The white patches in the cables include diodes, which prevent the current flowing from an input to an output. The ring of the diode is oriented toward the black banana terminal, which connects to the output shift registers.

Shutting down the Embedded Computer

As you can see in Figure 10-40, I've included an additional push button (on the right side of the three switches). I'm using this to shut down the embedded computer, like we've done in previous chapters. This push button is connected to one of the free digital pins of the Arduino, using the internal pull-up resistor. (This means that in the circuit, its one leg should connect to the Arduino digital pin, and the other to ground.) I suggest that you read this switch at the end of all data (this means that you must modify the array index the three switch byte is stored at), using the exclamation mark to reverse its value (remember, the pull-up resistor reads HIGH when the button is not pressed).

As always, you need to write a script to kill Pd and shut the embedded computer down, which you can call with [shell]. I'm using the DSP switch to control whether this button's values go through or not, so that I can't accidentally shut down the computer while the DSP is running. Figure 10-41 shows the addition to the [pd get_switches] subpatch. The rest are left for you to implement yourself, as you should be pretty familiar with the process by now. Note that you'll have to change the fourth argument to [serial_write], since it will now receive 4 bytes as "digital" values.

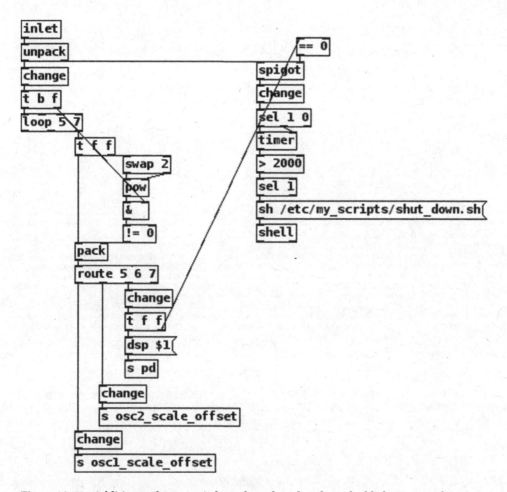

Figure 10-41. *Addition to the get_switches subpatch to shut the embedded computer down*

Conclusion

We have built a rather complex instrument that comprises many levels, in both the software and the hardware. You have been introduced to techniques and devices that expand the capabilities of the Arduino. You've seen that you can use very simple principles, like setting a digital pin HIGH or LOW, read these voltages from another digital pin, and create a much more complex interface.

This last project was probably the most difficult of all the projects in this book. Still, the programming level increased chapter by chapter, so by reaching this chapter, you should now have the understanding and tools to follow the process of making an instrument. You have seen a lot of different techniques and approaches to making musical interfaces. Now you should be able to create your own projects, solve problems, and come up with original ideas in both the technical and the creative processes.

Index

■ V

■ W

■ X, Y, Z

Get the eBook for only $5!

Why limit yourself?

Now you can take the weightless companion with you wherever you go and access your content on your PC, phone, tablet, or reader.

Since you've purchased this print book, we're happy to offer you the eBook in all 3 formats for just $5.

Convenient and fully searchable, the PDF version enables you to easily find and copy code—or perform examples by quickly toggling between instructions and applications. The MOBI format is ideal for your Kindle, while the ePUB can be utilized on a variety of mobile devices.

To learn more, go to www.apress.com/companion or contact support@apress.com.

Printed in the United States
By Bookmasters